14.50

K

THE QUATERNARY IN BRITAIN

Essays, reviews and original work
on the Quaternary published in honour of
LEWIS PENNY
on his retirement

Related Pergamon Titles of Interest

Books

ANDERSON
The Structure of Western Europe

ANDERSON & OWEN
The Structure of the British Isles, 2nd edition.

BOWEN
Quaternary Geology

GRAY & LOWE (*Editors*)
Studies in the Scottish Lateglacial Environment

LOWE, GRAY & ROBINSON (*Editors*)
Studies in the Lateglacial of North-West Europe

OWEN
The Geological Evolution of the British Isles

Journals

Computers & Geosciences

Geoforum

Quaternary Science Reviews

Full details of all books and journals and a free specimen copy of any Pergamon journal available on request from your nearest Pergamon office.

Photo courtesy of Hull Daily Mail

DR L.F. PENNY

THE QUATERNARY IN BRITAIN

Essays, reviews and original work
on the Quaternary published in honour of
LEWIS PENNY
on his retirement

Edited by

JOHN NEALE
Department of Geology
University of Hull

and

JOHN FLENLEY
Department of Geography
University of Hull

PERGAMON PRESS
OXFORD · NEW YORK · TORONTO · SYDNEY · PARIS · FRANKFURT

U.K.	Pergamon Press Ltd., Headington Hill Hall, Oxford OX3 0BW, England
U.S.A.	Pergamon Press Inc., Maxwell House, Fairview Park, Elmsford, New York 10523, U.S.A.
CANADA	Pergamon Press Canada Ltd., Suite 104, 150 Consumers Rd., Willowdale, Ontario M2J 1P9, Canada
AUSTRALIA	Pergamon Press (Aust.) Pty. Ltd., P.O. Box 544, Potts Point, N.S.W. 2011, Australia
FRANCE	Pergamon Press SARL, 24 rue des Ecoles, 75240 Paris, Cedex 05, France
FEDERAL REPUBLIC OF GERMANY	Pergamon Press GmbH, 6242 Kronberg-Taunus, Hammerweg 6, Federal Republic of Germany

---

First edition 1981

British Library Cataloguing in Publication Data

The Quaternary in Britain.
1. Geology, Stratigraphic - Quaternary
2. Geology, Great Britain
I. Penny, Lewis II. Neale, John
III. Flenley, John
551.7'9'0941 QE696

ISBN 0 08 026254 6

In order to make this volume available as economically and as rapidly as possible the authors' typescripts have been reproduced in their original forms. This method unfortunately has its typographical limitations but it is hoped that they in no way distract the reader.

Printed in Great Britain by A. Wheaton & Co. Ltd., Exeter

Preface

In September 1980, Dr. L.F. Penny retired after 31 years of service dedicated to the University of Hull. He took over the Department of Geology after the first year of its existence, when he inherited one English and two African students. By the time he handed over to the first Professor in 1962, he had built it up to be the equivalent of any other English Provincial University Department. An excellent teacher, he was much loved by students, not only for his teaching but for his helpful advice coupled with sound common sense and a fine sense of humour. His research work was mainly concerned with the British Quaternary; often carried out on foot and bicycle in the early days. His keen mind, coupled with painstaking work and accurate observation gave him a justified reputation for soundness and balanced judgement among his colleagues and fellow research workers. This was reflected in the award of a medal for Pleistocene Research by the University of Helsinki, and in 1979 by the award of the Sorby Medal by the Yorkshire Geological Society. His services to local geology had previously been recognised by his honorary life membership of the Hull Geological Society. This book arose from a meeting of the editors on the campus late in 1979. The response from colleagues, fellow research workers and friends to an invitation to contribute to a volume in his honour was most gratifying and bears eloquent testimony to the affection and respect in which he is held. We hope that this volume will prove useful to those interested in the Quaternary and prove a fitting monument to a valued colleague and friend.

John Neale   John R. Flenley

Contents

Introduction

J. R. Flenley

The primary purpose of this volume is to honour Lewis Penny, whose contribution to Quaternary Studies has been marked in several ways, including his election as President of the Quaternary Research Association in 1971. The book does, however, have a further purpose: the addition of another rock to the pyramid of Quaternary knowledge. Since the individual papers inside these covers may appear somewhat more like a series of pebbles than a consolidated rock, a few words of introduction may seem appropriate.

I propose to attempt this not so much by pointing out the links between the papers as by considering the nature of the subject to which they contribute. In the last twenty years or so we have seen the gradual emergence of Quaternary Studies as a *subject*. Before about 1960 there had been an enormous amount of Quaternary research done, but it was done largely by geologists, botanists and geographers working independently of each other. There were, of course, outstanding exceptions to this, notably the Sub-department of Quaternary Research at Cambridge, although even that had a strong botanical bias.

The Quaternary Research Association grew from the Quaternary Research Group, an informal organization of those who felt the need to break down barriers between traditional subjects. From its start in the nineteen-sixties, the Q.R.A. had the same ideal. It was achieved particularly by the annual field meetings in which biologists, geologists, geographers, archaeologists and others could hold on-the-spot discussions of key deposits.

What is the nature of the resultant subject which we call Quaternary Studies or Quaternary Science? First, it *is* a science. This is so not so much because it uses contributions from other sciences, but because it applies objective logic in the development of its conclusions. Usually it is not an experimental science, but relies on the intelligence - or intuition - of the observer to select those points in space and time where the most revealing natural experiments are likely to have occurred. So far it has not been a particularly theoretical science. On the other hand, the data collected empirically will sooner or later have to be assembled into some theoretical framework, and this is now being attempted by the climate modellers. It is already clear that some kind of astronomical forcing is likely to be responsible for major climatic cycles (e.g. Hays, Imbrie and Shackleton 1976). The problem here is one of causation. A statistical correlation between two phenomena may be causative, or random, or, perhaps most commonly, result from the control of both phenomena by a more fundamental cause. Quaternary Science still awaits its Newton or Darwin to find that cause.

Second, the subject owes a great deal, and probably too much, to one particular assumption. This is the famous 'dictum' of Hutton and Lyell, the 'doctrine of uniformity' "The present is the key to the past" (Lyell 1850). Quaternarists are in good company here, for the whole of geology is built on the same foundation, and a somewhat sandy one it is. In its extreme form - that the range of environmental variation in the past has never exceeded that now observable - the idea is clearly unreasonable. Instead we use what Rymer (1978) calls the Methodological Principle of Uniformitarianism, which implies only that we expect the same general laws of physics, chemistry, biology and other sciences to have applied in the past as they do today. This is reasonable, but is still an assumption, and we do well to be aware of the fact.

Third, it seems to me that our subject has distinct relationships, at least by analogy, with history. History, too, is a science (Collingwood 1946), using assemblages of fragmentary factual evidence to reconstruct past events. The only difference, apart from the time-scale, is that history ultimately tries to reconstruct the subjective thoughts of people, while Quaternary research tries to reconstruct more objective environmental events. Even here, however, there is some similarity, for historians do use, perhaps without admitting it, a kind of doctrine of uniformity. They assume that mankind has had, in the past, the same kind of drives as he has today: the biological requirements of survival and reproduction, and the social impulses such as power, altruism, greed and self-sacrifice. Good historians have always been ready to admit that the age in which they live will colour their opinions as to the relative importance of these drives.

Perhaps Quaternary research can learn something from this analogy. Take, for example, the conflict of opinion which occurred concerning the climate of the Late Devensian around 13,000 B.P. (Coope 1970). The palynological evidence suggested a cold climate, the coleopteran evidence a temperate one. This conflict was chiefly resolved by, broadly speaking, the realization that plants migrate more slowly than beetles. Yet it is possible, with hindsight, to see that if rates of migration had been considered earlier in relation to the palynological evidence the conflict might not have arisen. Rates of migration are difficult to measure, so we tend to ignore them. Ridley (1930) showed, however, that aquatic plants are, in general, rapid migrators, perhaps because their propagules adhere to the feet of birds. The aquatics present around 13,000 B.P. include some of quite temperate tolerances, including *Myriophyllum verticillatum* and *Typha latifolia* which scarcely occur today in N.W. Europe, further north than central Norway (Godwin 1975). If our opinions had not been coloured by the low value placed on estimates of migration rates, the significance of this might have been realized earlier.

Fourth, Quaternary Science covers a specially significant time range. The Quaternarist tends to think in units of about a millenium, with a range from 10^1 to 10^6 years. The geologist tends to think in tens of millions of years - say 10^6 to 10^9 years. The historian thinks in decades or centuries, say 10^0 to 10^3 years. Quaternary Science is not unique: prehistory covers much the same time-scale. These scales are important because natural changes tend to occur with characteristic rates. Significant changes whose rates fall wholly or partially within the Quaternary range include variations in the Earth's orbit, biological evolution, weathering and soil formation, magnetic reversal and dispersal of organisms.

Having outlined the nature of Quaternary Science, we may consider briefly a few examples of its relevance to other subjects. Perhaps its greatest significance is to Geomorphology, for the whole surface of the Earth has been greatly modified during the last million years or so. Quaternary Science enables us to quantify rates of landform development with greater accuracy than before. Biogeography is equally indebted, for modern floras and faunas are only really explicable in historical terms. Quaternary research has contributed to the rethinking of classical ecological theory. Climax theory is called into question by pollen diagrams from interglacials, and the classical hydrosere is not supported by evidence from lake and bog stratigraphy (Walker 1970). Geology has perhaps contributed most to Quaternary Science and gained least from it. Yet Geology has gained valuable reminders of the speed of environmental fluctuation, and it may yet turn out that variable climates have been quite common in geological time. Certainly previous ice-ages have been recognized from the similarity of their 'tillites' to Quaternary tills.

The papers in this volume are arranged in the approximate chronological order of their subject matter. After Boylan's historical paper there are four which deal with pre-Devensian times. Papers 6 to 16 are chiefly concerned with events during the time of the last glaciation. Papers 17 to 21 are about the Late Devensian and Flandrian, and we conclude with Francis's paper on classification of glacial sediments.

It is inevitable in a work dedicated to Lewis Penny that there should be a geological bias to the papers. Yet, in accordance with the idea that Quaternary Science is becoming an holistic subject, many of the papers cross the borders of conventional disciplines. Shotton's paper (No. 13), for example, reviews geological, geomorphological and biological evidence from north-east England. There is a similar breadth in Gaunt's Quaternary history of the southern part of the Vale of York (No. 9). The paper by Blackham, Davies and Flenley (No. 17) attempts to use biological evidence to date a geomorphological feature, and there are several papers linking geology and geomorphology (e.g. Nos. 1, 6, 7, 8, 10). Papers which limit themselves to evidence of one kind will have significance for other branches of the subject, even though this was not the main point of the paper.

It is only too clear that there are innumerable topics omitted and many gaps in our knowledge of those aspects of the Quaternary which have been touched upon here. The editors are keenly aware that, because of this, and of their own amateurism, there will be many deficiencies in this book. If, however, it represents some small step forward, it will have achieved its principal aim, that of honouring Dr. Lewis Penny.

REFERENCES

Collingwood, R.G. (1946). *The idea of history.* Oxford, Clarendon Press, 339 pp.

Coope, G.R. (1970). Climatic interpretations of Late Weichselian Coleoptera from the British Isles. *Rev. Geogr. phys. Geol. dyn.* (2) 12, 2, 149-155.

Godwin, H. (1975). *History of the British Flora.* 2nd Edition. Cambridge, C.U.P., 541 pp.

Hays, J.D., J. Imbrie, and N.J. Shackleton (1976). Variations in the Earth's Orbit: Pacemakers of the Ice Ages. *Science 194*, 1121-1132.

Lyell, C. (1850). *Principles of Geology*, 8th Edition. London, Murray, 811 pp.

Ridley, H.N. (1930). *Dispersal of plants throughout the world.* Ashford, Reeve.

Rymer, L. (1978). The use of uniformitarianism and analogy in palaeo-ecology, particularly pollen analysis. In D. Walker and J.C. Guppy (Eds), *Biology and Quaternary Environments.* Australian Academy of Science. pp. 245-257.

Walker, D. (1970). Direction and rate in some British post-glacial hydroseres. In D. Walker and R.G. West (Eds), *Studies in the Vegetational History of the British Isles*, Cambridge, C.U.P., pp. 117-139.

1. The Role of William Buckland (1784-1856) in the Recognition of Glaciation in Great Britain*

Patrick J. Boylan

Although a number of naturalists and geologists of the eighteenth and early nineteenth centuries had speculated about the possibility of the glacial origin of some landforms and deposits, it was notuntil the 1830s that the glacial theory was put on to a sound footing by Jean de Charpentier and Louis Agassiz. As Gordon Davies (1968) has pointed out, Robert Jamieson of Edinburgh was an early supporter of the theory and key papers on the glacial hypothesis were published by Jamieson in the *Edinburgh New Philosophical Journal* - which he edited - between 1836 and 1839 including, in April 1838, an English translation of the seminal work of Agassiz, the July 1837 "*Discours de Neuchâtel*", which had previously only seen limited circulation in the *Actes* of the Société Helvetique of Neuchâtel.

The announcement of the recognition of glaciation in the British Isles in the autumn of 1840 was of particular significance for two reasons: first, because, intellectually and conceptually, there is a substantial difference between postulating that the existing complex of glaciers in the Alps had formerly extended over a larger area than they do today, as Charpentier and Agassiz proposed, and the claim made in 1840 that widespread evidence of a former glaciation could be identified in both highland and lowland Britain, where no glacier occurs today; second, because of the international standing of British geology and, particularly, the Geological Society of London at that time.

William Buckland was born in 1784 in Devon, England, and, after studying at Oxford University, remained there as a tutor, a lecturer and, from 1819, as the first Professor of Geology and Mineralogy in the University. Schooled in the tradition of close observation and recording so actively promulgated by the Geological Society of London in its early years, Buckland was a prodigious field worker and geological traveller, both in Britain and on the continent of Europe. According to legend, his horses soon learned to stop automatically at every natural rock outcrop or quarry that he passed on his travels, and certainly he was one of the largest contributors to the Geological Society's first geological map of Great Britain. Philosophically, Buckland is frequently characterised as a classic "Catastrophist" because of his powerful advocacy - particularly between 1819 and the late 1820s - of the Diluvial Theory. In fact, however, it would probably be much more accurate to describe Buckland as a classic "Actualist": in broad terms at least, he accepted the uniformity of geological processes, but could not accept a concept of strict uniformity in the rate and scale at which these natural geological processes took place.

---

*Based on a contribution to the 8th Symposium of the International Committee on the History of Geological Sciences (INHIGEO) on the theme "Regional influences on the origin and development of geological theories", held in Münster and Bonn, West Germany, 12 - 24 September 1978.

One of Buckland's primary interests was in what we today categorise as the Quaternary and, in Britain, he found abundant evidence which he could not explain in terms of currently observable geological processes, and hence turned to a Diluvialist position - postulating catastrophic fluviatile processes to explain the observed phenomena.

Within the Quaternary at least, Buckland's search for a catastrophic cause for many puzzling phenomena was, in fact, a reasonable one, although, as he himself eventually recognised, fluviatile processes, even on a catastrophic scale, did not offer a satisfactory explanation for the abundant field evidence with which he was familiar.

In the autumn of 1838 - perhaps prompted by the *Discours de Neuchâtel* - Buckland went to Switzerland to meet Agassiz and to examine for himself some of the evidence that was being put forward for the glacial theory. Although sceptical at first, Buckland soon found the evidence overwhelming and became an enthusiastic convert to the glacial theory, at least so far as the former extension of the Alpine glaciers was concerned. Moreover, Buckland began to recognise direct parallels between the glacial phenomena of northern Switzerland and features of the superficial geology in Britain which he had, so far, found hard to explain.

There is little doubt that, on discussing this evidence with Agassiz, Buckland became convinced that at least parts of Britain had been relatively recently glaciated. He did not, however, react publicly immediately, even though the glacial theory offered at least an observable mechanism for catastrophic - in the most literal sense - geological change. Buckland must have suspected that, despite the papers published by Jamieson in the *New Philosophical Journal*, the geological establishment in Britain, particularly the Geological Society itself, was not likely to be very sympathetic to what might well be interpreted as old-fashioned Catastrophism in a slightly modified guise - with imaginary glaciers taking the place of Biblical or Wernerian Floods.

Consequently, Buckland appears to have said little or nothing about his views even in private, until Agassiz (who was very highly thought of in Britain because of his palaeontological work, especially on fossil fishes) was able to visit Britain again, this time to see Buckland's glacial evidence.

The opportunity finally came in 1840, by which time Buckland was serving his second term as President of the Geological Society. Agassiz sent in advance a paper "On the polished and striated surfaces of the rocks which form the beds of glaciers in the Alps" which was read to the Geological Society on 10th June 1840 (Agassiz, 1840). In September, Agassiz finally arrived in Britain and made his way more or less directly to Glasgow to meet up with Buckland who was attending the meeting of the British Association for the Advancement of Science. Agassiz gave four papers to the Association's meeting, including one entitled "On Glaciers and Boulders in Switzerland", (Agassiz, 1841), which discussed in particular the action of glaciers in striating (scratching) and polishing rocks. Agassiz also found ample evidence of glaciation in the area around Glasgow. Immediately after the British Association meeting ended on 23 September, Agassiz, accompanied by Buckland for most of the time, set out on an extended tour of the Scottish highlands looking mainly for evidence of glaciation, but also visiting important fossil fish localities and collections. Agassiz recognised abundant evidence of glaciation in the highlands. The route of this famous geological journey has been reconstructed by Gordon Davies (1968) and George White (1970), and by 4th October, as George White has pointed out, Buckland was so confident that he wrote to Professor Fleming in Aberdeen ... "we have found abundant Traces of Glaciers round Ben Nevis ...", and, three days later, the leading Edinburgh newspaper, *The Scotsman*, published a report of the discoveries in the Ben Nevis area in the form of a letter written by Agassiz to Jamieson.

Buckland had, however, been carrying out fieldwork on his own account, and continued to do so after he and Agassiz parted company. Before returning south, Buckland visited Charles Lyell at his Scottish home, Kinnordy House in Forfarshire, and quickly converted Lyell to the glacial theory by confidently demonstrating the glacial origin of many features immediately around Lyell's own house.

Agassiz, Buckland and Lyell all returned to London, and it seems clear that Buckland made careful plans for presenting what was likely to be a highly controversial interpretation to the Geological Society. What finally emerged was an arrangement under which, over three fornightly meetings of the Geological Society, three substantial papers would be given. Agassiz opened the proceedings on the opening night of the Geological Society's year with the whole of a paper entitled "On glaciers, and the evidence of their having once existed in Scotland, Ireland and England" (Agassiz, 1840 - 41), which ranged widely over his evidence in support of the glacial theory, comparing Alpine evidence with the observations that he had recently made in the Scottish highlands, England and Ireland. Buckland's contribution was titled "On the evidences of glaciers in Scotland and the North of England" (Buckland, 1840 - 41), and was so long that it was divided into two parts and read over the three meetings of the Society, on the 4th and 18th November and 2nd December 1840. The third contribution, given in the middle of Buckland's, was a paper by Lyell entitled "On the geological evidence of the former existence of glaciers in Forfarshire" (Lyell, 1840 - 41), given on 18th November and 2nd December. In a way, the inclusion of a contribution by Lyell seems superfluous, particularly since it was Buckland who first recognised extensive evidence of glaciation in Forfarshire, convincing Lyell, and described some of the key sites in his own memoir. Possibly, Lyell himself was anxious to make such a contribution, but it seems more likely that, at least in part, Lyell's public statement on his conversion to the glacial theory was urged on him by Buckland in order to strengthen the case being put forward. In fact, Lyell's conversion to the glacial theory was very shortlived indeed, and within less than a year he had almost completely reverted to his previous theory that the glacial deposits were, in fact, of marine origin, with erratic boulders being interpreted as dropstones from icebergs.

Buckland's two-part Memoir to the Geological Society gives a very clear picture of the large scale of his fieldwork in connection with the glacial theory, and also gives an excellent insight into his inductive method His paper refers to more than eighty localities at which he had recognised various glacial phenomena, not counting the sites that he had also visited with Agassiz and Lyell, and which were described in their respective papers.

In the course of a major review of Buckland's scientific work, all of the glacial localities known to have been visited by Buckland have been identified, visited and re-examined by the author, with a view to the eventual production of a field guide to localities of significance in the history of the glacial theory in Britain. It is not possible, in a contribution of this length, to do more than refer to a representative selection of the sites on which Buckland based his case for the adoption of the glacial hypothesis, and which illustrate both the range and accuracy (in terms of present-day interpretations) of Bucklands's observations.

Buckland began his Geological Society paper by giving a brief history of his involvement with the glacial theory, and referred to a number of puzzling features that he had seen up to thirty years earlier and which, he believed, could, at last, be explained in terms of the glacial theory.

The first site that he described was particularly important in a number of respects. This was a small-scale ridge running from National Grid Ref. NX 923948 to NX 920955, about 900 m long and never more than 5 m high, which crosses the valley at an altitude of approximately 150 m, just above Crichhope (or Crickhope) Linn near Thornhill, Dumfriesshire. This is not a very mountainous part of Scotland - indeed, the higher mountains of the Scottish Highlands are almost 200 km to the north. Buckland discovered this feature while travelling north from Dumfries to Glasgow for the British Association meeting, and by analogy with similar features that he had seen in Switzerland, he interpreted the ridge as a small terminal moraine. This southerly locality, visited independently by Buckland a few days before the opening of the British Association meeting on 18th September 1840, therefore probably ranks as the first glacial locality to be fully recognised and understood in Britain, pre-dating by several days the discoveries of Agassiz in central Glasgow. The Crichhope Linn locality is still remarkably little changed from Buckland's description: the moraine is intersected by a small road, and the unstratified gravelly till core of the moraine can still be seen in natural sections. Considering the historic significance of the locality, there has been remarkably little interest in it in the 140 years since Buckland's original publication of the site. The area was mapped on the 6" scale by H.M. Skae of the Geological Survey of Scotland in 1868, and the original Field Slip is deposited in the I.G.S Library, Murchison House, Edinburgh (Dumfries 31 NE). Skae mapped the moraine as a kame over a patch of gravel, and a similar explanation was given by J. Horne in the Sheet Memoir (Geological Survey of Scotland, 1877). Only Charlesworth (1926) appears to have recognised the historical significance of the locality and provided an acceptable 20th century interpretation - that the feature is, indeed, a moraine associated with a lobe of a major Nithsdale Valley glacier.

An important group of lowland sites that also illustrate Buckland's work was just north of Kirremuir in Forfarshire in the low-lying valley of the Tay, but on the Highland margin. Around Cortachy and Pearsie, and within 2 km of Lyell's house at Kinnordy are a series of sand and gravel kames, which Buckland interpreted as morainic features. Good examples can be observed from a view point on the minor road from Kinnordy to Glen Prosen at NO 365575, especially on the south side of the Carity Burn between Newmill (NO 364577) and Chapeltown (NO 378584), and a section in a sandy ridge can be examined by the roadside at NO 379588. One of the most remarkable glacial features of the area is seen at NO 383586, where the minor road from Prosen Bridge to Pearsie runs along the top of a steep-sided gravel ridge, approximately 40 m high and only 15 m across at the narrowest point, lying between the Carity Burn to the south and the Prosen Water to the north. Buckland interpreted these features as "... produced by glaciers, and modified in part subsequent by water" (Buckland, 1840 - 41, p. 333). In this, Buckland appears to have been closer to a full recognition of their fluvioglacial origin than was Lyell (1840 - 41, p. 339), who appears to have regarded them as lateral moraines of glaciers, although the published summary of Lyell's contribution is often unclear about individual localities (unlike those of Agassiz and Buckland).

Buckland also attributed a glacial origin to the massive sand and gravel "ridges" between the Highland margin and the River Ericht around Blairgowrie (e.g. in National Grid squares NO 1945, NO 2045 and NO 2145). Similarly, he argued that the "... transverse barriers forming a succession of small lakes in the valley of the Lunanburn ... to be moraines" (Buckland, 1840 - 41, p. 334). These natural dams can be seen particularly well from NO 129436 (the barrier between Loch of Drumellie and Loch of Clunie), from NO 050450 (the barrier between Loch of Butterstone and Loch of Lowes) and from NO 045446 (Loch of Lowes and Loch of Craiglush).

The abundant evidence of glaciation in the Scottish Highlands seen by Agassiz and Buckland on their tour following the British Association meeting in Glasgow was mainly summarized in the paper of Agassiz (1840 - 41), but Buckland added many other localities in the Highlands to which he returned independently after he and Agassiz parted company in Aberdeen (almost certainly on 9th October 1840 - see White, 1970), and following his subsequent visit to Lyell at Kinnordy.

Returning from Forfarshire to the high road from Perth to Inverness at Dunkeld,

Buckland identified as "moraines" the kames and dissected fluvioglacial deposits around Murthly Castle on the south bank of the Tay about 6 km S.E. of Dunkeld (in Grid squares NO 0639 and NO 0640). He also argued that "The vast congeries of gravel and boulders on the shoulder of the mountain, exactly opposite the gorge of the Tumel [sic] ... was lodged there by glaciers which descended the lateral valley of the Tumel from the north side of Schiehallion and the adjacent mountains ..." (Buckland, 1840 - 41, p. 334). Sections in these deposits of stony till can again be seen as a result of recent road widening on the A9 through the Pass of Killiecrankie (e.g. at NN 916614).

In the same area, Buckland identified "mammillated, polished and striated slate rocks" in several localities, e.g. immediately above the Linn of Tummel falls (NN 905600) and on the north side of Loch Tummel at Bohally (NN 784591). (The Bohally locality has been flooded as a result of the Loch Tummel hydroelectric scheme, but a closely comparable "mammillated and striated" roche moutonnée can be studied in the same area at NN 680586).

Buckland also studied in some detail the fresh glacial topography around Schiehallion, citing in particular clear evidence of glacial scouring, including striations, on porphyry dykes in two places by the side of the Military Road near Braes of Foss (at NN 725567 and NN 745560).

Continuing to the south, he recognised as "moraines or the detritus of moraines" the dissected fluvioglacial deposits and kames at, for example, Taymouth Castle (occupying most of Grid squares NN 7745 and NN 7846), and at Fortingall (Grid squares NN 7346 and NN 7446). Continuing along the Military Road south of Aberfeldy Buckland recorded "A remarkable group of moraines ... [of] thirty or forty round-topped moraines, from thirty to sixty feet high, are crowded together like sepulchral tumuli ... composed of unstratified gravel and boulders", (Buckland 1840-41, p. 334 - 5). These are clearly identifiable within the hummocky moraine topography of the upper part of Glen Cochill (spelt "Glen Coefield" in Buckland's text), e.g. around NN 886452, NN 887445 and NN 893440.

Other areas of Scotland cited by Buckland as demonstrating widespread glacial action ranged from Ben Wyvis, Easter Ross, in the north, to the English border near Berwick on Tweed in the south, including localities in the Trossachs, Stirling, and the Edinburgh area, the latter including both Calton Hill (NT 263742) and the north-west face of Castle Rock, Edinburgh (NT 252736), as well as what is now known as "Agassiz's Rock" on the south side of Blackford Hill (BT 259703).

Buckland's observations of glacial phenomena in northern England were equally wide-ranging, and, again, only a few examples can be given here. He recognised, for example, the moraine complex damming the southern margin of Lake Windermere (SD 376865 and SD 370860), the drumlin fields at Edenhall, east of Penrith (e.g. Grid square NY 5632) and near Barrow in Furness (Grid squares SD 2270 to SD 2769), as well as the glacial origin of tills in many localities, for example between Lancaster and Kendal, and in Northumberland, both on the coast and inland: "Immediately below the vomitories of the eastern valleys of the Cheviots, enormous moraines ... cover a tract four miles from north to south, and two miles from west to east and the high road ... [winds] among cultivated mounds of them from near Wooler [Grid square NT 9926], through North and South Middleton [NU 0025, NT 9923] and by West and East Lilburn to Rosedean and Wooperton [NT 0320]". (Buckland, 1840-41, p. 346).

However, perhaps the most interesting site and description in Northumberland is of a natural section cut by the College Burn through a substantial morainic feature near Kirknewton (NT 905294). Here Buckland recognised substantial contortions within the structure of the deposit which he interpreted as the result of the morainic material being "... severed from their original position, moved forward, and contorted by the pressure of a glacier, which descended the deep trough of the College Burn from the northern summit of the Cheviots" (Buckland, 1840 - 41, p. 346).

Buckland also recognised evidence of glacial erosion in northern England in the form of polished and striated surfaces, e.g. at Fox Howe near Ambleside (NY 363051), near Rydal (e.g. NY 366057 and NY 348066) and in the Pass of Wythburn (e.g between NY 337085 and NY 325127). He also attributed to glacial action dispersal of erratics, such as the occurrence of Lake District rocks on Walney Island, Barrow in Furness (seen today on both the foreshore at SD 174681, and displayed in the entrance to a small public park at SD 187687), and the widespread dispersal of Shap granite, both within north-west England and over Stainmore into County Durham.

Overall, Buckland's contribution to the three sessions at the Geological Society was a very powerful one, and one that was very much in accordance with the tradition of the Society - i.e. close observation, careful, neutral description of those observations, and interpretation drawn from them. Because of the wealth of detailed observations covering Scotland, the English Lake District, Northern England, and even into Lancashire, Buckland's contribution must have been the most telling of the three contributions in terms of the Geological Society's traditional expectations.

However, Buckland went further than this. As early as 1821 in his classic work on the fossil bone deposit at Kirkdale Cave, Yorkshire, Buckland had produced a brilliant inductive interpretation of the origin of the broken bones in the Cave, together with predictions about the likely behaviour patterns of hyaenas which were not fully vindicated until the 1960s (Boylan, 1967, 1972). In his Geological Society paper on the glacial theory, Buckland carried out a similar inductive test. Before going to the area around Comrie near Crieff, Scotland, he had "... tested the value of the glacial theory by marking in anticipation on a map the localities where there ought to be evidences [sic] of glaciers having existed, if the theory were founded on correct principles. The results coincided with the anticipations". (Buckland, 1840 - 41, p. 335). In this area alone, Buckland cited thirteen localities at which he had identified glacial activity of all kinds, entirely in accordance with his expectations. These included rounded and striated rock surfaces and roches moutonées near Lawers Farm, Comrie (NO 785224), Funtullich (NO 749221) in Glen Lednock, and near the Falls of Turret (NO 837244); and 'moraines' (including kames, kame terraces and hummocky moraine) in Glen Turret around NO 825258 and NO 855252, and in Glen Lednock at, for example, Invergeldie (NO 742277), Glenmaik (NO 726281), and around Kingarth (spelt "Kenagart" in Buckland's text), e.g. at NO 756250 and at NO 765244.

Even more remarkably, using a large-scale map, he presented the sceptics in the Geological Society with a series of predictions about the pattern of glaciation in parts of the English Lake District which he had <u>not</u> checked at all himself. The sceptic or unbeliever could go and test Buckland's predictions for himself.

Despite the powerful arguments put forward by Agassiz, Buckland and Lyell, and Buckland's careful preparation of the case, the response to the papers from leading members of the Society, led by Murchison, Greenough and Whewell was almost universally hostile, as contemporary accounts show. Much the most complete and valuable record is the detailed notes of the palaeontologist, S.P. Woodward (Sub-Curator of the Geological Society at the time), of the meeting of 18th November 1840, when Buckland completed the reading of the first part of his memoir. These notes were edited and published much later by his son, H.B. Woodward (1883)

In his response to the discussion, Buckland insisted that he had "... been a "sturdy" opponent of Professor Agassiz when he first broached the glacial theory, and having set out from Neuchâtel with the determination of confounding and ridiculing the

Professor. But he went and saw all these things and returned converted". (Woodward, 1883). Buckland went on to "... condemn the tone in which Mr Murchison had spoken of the 'beautiful' terms employed by the Professor to designate the glacial phenomena. That highly expressive phrase 'roches moutonees', which he had done well to revive, and that other 'beautiful designation', the glacier remanie! remanie! remanie! continued the Doctor most impressively, amidst the cheers of the delighted assembly, who were, by this time, elevated by the hopes of soon getting some tea (it was a quarter to twelve P.M.) ..." (Woodward, 1883). Moreover, Buckland flouted the Presidential conventions of the Society in closing the meeting and " ... concluded - not as we expected, by lowering his voice to a well-bred whisper, 'Now to', etc., - but with a look and tone of triumph he pronounced upon his opponents who dared to question the orthodoxy of the scratches, and grooves, and polished surfaces of the glacial mountains (when they should come to be d————d) the pains of *eternal itch,* without the privilege of scratching!" (Woodward, 1883).

The argument was not, of course, over, and most of the leading members of the Society remained sceptical about, or positively hostile to, the glacial theory for many years. Murchison, who succeeded Buckland as President of the Geological Society, devoted no less than seventeen pages of his first presidential Anniversary Address to an attack on the glacial theory and its proponents, especially Agassiz and Buckland (Murchison, 1842, pp. 671 - 87). Moreover, as is well-known, Lyell abandoned the theory almost immediately and remained decidedly unenthusiastic until the end of the 1850s. Nevertheless, the reasoned arguments and evidence put forward had a considerable effect on some of the younger members of the Society present at the meetings of November and December 1840, and who were ultimately to push forward the understanding of glaciation in subsequent decades.

Despite the hostility and overt ridicule, in the long term the episode, carefully organised by Buckland, had a significance far outside Great Britain. In particular, Buckland's demonstration of the existence of widespread evidence of glaciation as far south as the lowland areas of northern England was powerful evidence in support of Agassiz's concept of the glaciation being the result of a continental-scale ice cap (analogous with that of Greenland, in the view of Agassiz) rather than a mere extension of existing relatively small-scale mountain glaciers. So far as Buckland was concerned he had at last found the explanation for the "Catastrophic" Quaternary features with which he had been familiar for a quarter of a century or more.

Two puzzles remain. First, even though Buckland was, at the time, the President of the Geological Society as well as one of its most distinguished members, and one of the leading academics in the whole field of geology, none of the three Geological Society papers were, in fact, published by the Society, except in the form of the formal abstract in the *Proceedings*. Admittedly, the *Transactions* were already heavily committed and somewhat in arrears at that time, but it does seem surprising that such a revolutionary paper put forward by such a distinguished serving President was not published in full by the Society.

Second, there is an interesting mystery about the failure of either Buckland or Lyell to refer to a clear and very interesting glacial geology feature which both of them had obviously seen - the Loch of Kinnordy (NO 360543), within the grounds of Lyell's family home. This is a very clear kettle-hole lying in a hollow in the surface of the glacial deposits, 'perched' above the local water-table. Possibly, Buckland felt that Lyell ought to reveal the fact that his own home stood on 'moraines' by this glacially-dammed lake, or perhaps both felt that, tactically, such a revelation would either weaken the combined case, or embarrass Lyell, or both.

Perhaps the 'tactical' explanation for this strange omission is most likely, because it seems probable, from the surviving evidence, that Buckland's handling of the scientific politics of the announcement of the glacial theory in Britain, particularly in relation to the Geological Society, was almost as fascinating to observe - and certainly as thoroughly prepared - as the overwhelming scientific case for the adoption of the glacial hypothesis put forward in 1840 by Buckland, the doyen of British Quaternary Studies, with the most able support of Agassiz, the leading continental worker in the field, and Lyell, the most brilliant and influential of the younger generation of British geologists.

REFERENCES

Agassiz, L. (1840). On the polished and striated surfaces of the rocks which form the beds of glaciers in the Alps. *Proc. geol. Soc. Lond.*, 3 (71), 321 - 322.

Agassiz, L. (1840 - 41). On glaciers, and the evidence of their having once existed in Scotland, Ireland and England. *Proc. geol. Soc. Lond.*, 3 (72), 327 - 332.

Agassiz, L. (1841). On glaciers and boulders in Switzerland. *Rep. Br. Ass. Advmt Sci., Glasgow, 1840*, 113 - 114.

Boylan, P.J. (1967). Dean William Buckland, 1784 - 1856: A pioneer in cave science. *Stud. Speleol.*, 1, 237 - 253.

Boylan, P.J. (1972). The scientific significance of the Kirkdale Cave hyaenas. *Yorks. phil. Soc., Ann. Rep. for 1971*, 38 - 47.

Buckland, W. (1840 - 41). On the evidences of glaciers in Scotland and Northern England. *Proc. geol. Soc. Lond.*, 3 (72), 332 - 337; 345 - 348.

Charlesworth, J.K. (1926). The glacial geology of the Southern Uplands, west of Annandale and Upper Clydesdale. *Trans. R. Soc. Einb.*, 55, 1 - 23.

Davies, G.L. (1968). The tour of the British Isles made by Louis Agassiz in 1840. *Ann. Sci.*, 24, 131 - 146.

Geological Survey of Scotland. (1877). *Explanation of Sheet 9, Kirkudbright (North Eastern Part) and Dumfries-shire (South-Western Part)*. H.M.S.O., Edinburgh, 53 pp.

Lyell, C. (1840 - 41). On the Geological Evidence of the former existence of Glaciers in Forfarshire. *Proc. geol. Soc. Lond.* 3 (72), 337 - 345.

Murchison, R.I. (1842). Anniversary Address of the President. *Proc. geol. Soc. Lond.*, 3 (86), 637 - 687.

White, G.W. (1970). Early Discoverers XXVII. Announcement of glaciation in Scotland. William Buckland (1784 - 1856). *J. Glaciol.*, 9, 143 - 145.

Woodward, H.B. (1833). Dr Buckland and the Glacial Theory. *Midl. Nat.*, 6, 225 - 229.

Patrick J. Boylan, Esq.,
Director of Museums and Art Galleries,
96 New Walk,
Leicester,
LE1 6TD.

2. British Pre-Devensian Glaciations

J. A. Catt

INTRODUCTION

Inspired by the evidence of multiple Quaternary glaciation in the Alps and other areas abroad, Quaternary geologists have now struggled for almost a century to subdivide Britain's glacial succession on the basis of age. Simple stratigraphic procedures, which had proved successful in subdividing and correlating deposits of older geological periods, showed even in late Victorian times that parts of England, such as Lincolnshire (Jukes-Browne, 1885), were invaded more than once by large ice sheets. Later, Quaternary studies were broadened by a range of relative dating techniques (geomorphological, palaeontological, archaeological and sedimentological), which also suggested repeated glaciation of many areas, but these often gave conflicting and misleading results, especially when used in isolation and without careful stratigraphic control.

Since 1950, palaeobotanical studies in particular have achieved considerable advances in the subdivision and correlation of Quaternary deposits, and radiocarbon dating has helped clarify the chronology of glacial episodes during the last 30,000 years or so. However, the exact ages and extents of earlier British glaciations remain obscure, partly because there are few methods of absolute dating suited to a wide range of older deposits, and partly because much of the sedimentary evidence has been destroyed by subsequent erosion.

Space forbids a comprehensive review of all the evidence for earlier glaciations. It is possible only to recall some of the more important problems involved in recognising, dating and correlating them, and to look for ways in which solutions to these problems may be found.

HOW SHOULD WE SUBDIVIDE THE BRITISH QUATERNARY?

Because the Quaternary was too short for major evolutionary changes in most plant and animal groups, Mitchell and others (1973) based their subdivision on lithological features of sediments and fossil assemblages, which reflect climatic and related environmental changes. However, this is not a completely satisfactory way of distinguishing the various stages they recognised, as similar fossil assemblages or lithological features may be of different ages; also, dissimilar assemblages at widely separated sites may be synchronous, because some animal and plant groups migrate more rapidly than others in response to changing climate. In addition, many terrestrial Quaternary deposits, such as those of glacial and fluvial origin, are often stratigraphically complex, and correlation difficulties have frequently been underestimated; it is consequently possible that some stages may be misplaced in the sequence, and others have yet to be recognised. However, despite these possible failings, the 14 stages given by Mitchell and others provide at least a useful temporary framework within which to discuss the British Quaternary.

A more reliable table of pre-Devensian stages will probably emerge only when many absolute dates are available for deposits beyond the present range of radiocarbon dating. Several promising new techniques are currently being developed (e.g. measurement of thermoluminescence in aeolian deposits, of amino-acid diagenesis in bones, and of the short-lived transient isotopes formed during breakdown of ^{238}U and ^{235}U to Pb in carbonate rocks), but few results are yet available for British deposits.

The deep ocean sediments now provide a useful "yardstick" of dated worldwide climatic fluctuations, against which to judge the often patchy record of terrestrial Quaternary successions. They have certainly released us from the straight-jacket of classic Quaternary subdivisions, such as the four-fold Alpine sequence, but correlation between oceanic and terrestrial successions is difficult, and we are still a long way from matching the remarkable correlation achieved, for example, with palaeosols in the loess of central Europe (Kukla, 1977). Even a few absolute dates from older parts of the British Quaternary could greatly improve the situation, providing the stratigraphic relationships between the dated deposits and others are soundly established. In his work on Quaternary deposits, Lewis Penny always emphasised care in measuring and recording successions, and in tracing and correctly interpreting stratigraphic relationships. The value of absolute dates or any other laboratory analysis of deposits is always limited by the quality of the field interpretation, and neglect of this aspect only leads to confusion. Rather than wait for new dating techniques to solve our problems for us, we must therefore continue struggling to solve them by careful field observation, as this will always be an indispensable basis for Quaternary subdivision and correlation.

One aspect of field observation, which has been neglected in Britain but is crucial to correct stratigraphic interpretation, is the recognition and interpretation of subaerial weathering profiles. The value of buried soils in subdividing deposits that are superficially similar, in correlating them from site to site, and in verifying the height and configuration of old land surfaces, is admirably exemplified by the work of Rose and Allen (1977) in southern parts of East Anglia. However, there is also some merit in the careful examination of unburied soils on Quaternary deposits, as certain macro- and micro-morphological features are regarded as interglacial in origin (Catt, 1980), and the relationships of these to periglacial soil structures or Flandrian soil features can often be established. This is especially useful if profiles with Devensian or other features can be traced laterally into buried soils within a succession of Quaternary deposits. One difficulty in recognising buried soils is that some processes of sedimentation and diagenesis can locally produce successions resembling subaerial weathering profiles. However, Valentine and Dalrymple (1976) point out that the two can often be distinguished if they are traced laterally, as true subaerial soils vary according to characteristic (catenary) relationships with the landscape in which they formed.

HOW SHOULD WE RECOGNISE OLDER GLACIATIONS?

The sedimentary and geomorphological features, which are universally accepted as evidence for quite recent glaciations (e.g. during the Late Devensian stage in Britain), are all ultimately prone to modification or destruction by various denudation processes, especially in upland areas and during periods of cold climate. The evidence for progressively older glaciations is therefore increasingly patchy and doubtful, and confined to lowland areas which were probably affected only during the more extensive invasions.

In such lowland areas glaciers typically deposit large amounts of outwash sand and gravel, and tills that are usually finer in texture and lithologically less variable than in upland areas. The composition of these deposits, for example in terms of stone content or sand mineralogy, often contrasts with that of the underlying bedrock and other (non-glacial) Quaternary formations, and scattered remnants may be identified and correlated by such means. However, confusion can arise through non-glacial reworking of earlier glacial sediments. Because of these and other difficulties, we shall probably never be certain exactly how many times Britain was glaciated during the Quaternary, nor exactly how far the glaciers extended on many occasions.

The assemblage of features that seems the most persistent and reliable indicator of older glaciations in lowland Britain is the occurrence of a widespread diamicton (till) containing far-travelled stones, the finer and softer examples of which are often striated, and the frequent association of this diamicton with gravels containing a related suite of stones. Individual features of this assemblage are not of themselves reliable indicators of glaciation; for example, some diamictons of restricted lateral occurrence can form by solifluction or as temperate mudflows, and striation of individual stones can also result from solifluction. Other characters of the diamicton, such as lamination, clast orientation, microfabric and relationships with other sediments, may clarify the exact mode of deposition, but allowances must also be made for possible post-depositional disturbance and pedogenic alteration. Other individual features often given as evidence for older glaciations, such as river diversions and scattered erratics, are clearly much less reliable, but in some areas they are the only remaining evidence, and cannot be completely ignored.

HOW MANY QUATERNARY GLACIATIONS AFFECTED BRITAIN?

The precise age of the Tertiary-Quaternary boundary has still to be decided, but if we accept Bowen's suggestion (1978) of 1,610,000 years B.P. (the Olduvai-Matuyama palaeomagnetic reversal), then the oceanic cores indicate we should think in terms of about seventeen major Quaternary cold periods. By no means all of these necessarily caused glacial conditions in Britain, but we should clearly be wrong to restrict our attention to anything like the last three or four, as did earlier generations. Compared with the oceanic sequence, the succession of only seven cold stages (with seven intervening warm interglacial stages) for the whole of the British Quaternary (Mitchell and others, 1973) seems at first sight to be a distinctly patchy record. However, some of the additional subsidiary (interstadial) warm periods recognised in the terrestrial record may not be distinguishable in the oceanic sediments from interglacials, so the British sequence may be fuller than it seems; but how full it is and where the gaps in it occur are both still uncertain.

The distribution and approximate age of deposits resulting from the last major glaciation of Britain, during the Late Devensian of Mitchell and others (1973) or stage 2 of oceanic cores such as V28-238 (Shackleton and Opdyke, 1973), is now quite well known, thanks in no small measure to the work of Lewis Penny himself (Penny, 1959, 1964, 1974; Penny, Coope and Catt, 1969). The ice reached its maximum (Fig. 1a) about 18,000 B.P., at least in eastern England, and did not disappear completely until about 14,000 B.P. The Middle and Early Devensian (approximating to stages 3 and 4 respectively of the oceanic cores) were generally quite cold, with evidence of periglacial features on several occasions, but there is no incontrovertible evidence for glaciation during these periods anywhere in Britain. The Early Devensian glaciations claimed for parts of West Yorkshire (Edwards, Mitchell and Whitehead, 1950) and Lincolnshire (Straw, 1958, 1961a, 1979a) have been disputed by Penny (1974) and Madgett and Catt (1978) respectively, and clear stratigraphic relationships to deposits of the preceding warm Ipswichian stage (stage 5 of oceanic cores) have never been demonstrated. The Lincolnshire deposits are probably part of the Late Devensian succession, and those near Leeds are perhaps pre-Devensian in age.

Glacial deposits of older Quaternary cold stages are best known in areas south of the Devensian limit, such as the Midlands, East Anglia, south Wales, south-west England and southern Ireland. North of the Devensian limit they are largely restricted to hollows, valleys and lowland areas, because the upland areas were probably denuded of earlier Quaternary sediments and also Tertiary or interglacial soils by repeated glacial and periglacial scouring. Where they emerge from beneath the Devensian cover, the pre-Devensian glacial deposits occur characteristically as hilltop and plateau remnants (e.g. the "older drift" of the north Midlands and south Yorkshire) left after a period of extensive dissection preceding the Late Devensian glaciation. Further south, progressively older glacial deposits occur at the surface in a crude off-lapping relationship, which suggests either that some of the earlier ice sheets invaded only parts of the south, or that northern areas were repeatedly subject to more extensive erosion.

Ocean core V28-238 suggests that before the Ipswichian, which probably began 128,000 B.P., there was a long cold period (stage 6) lasting for about 67,000 years. It seems reasonable to equate this with the Wolstonian Stage of Mitchell and others (1973), especially as Szabo and Collins (1975) obtained isotopic ages for bones from Ipswichian deposits at Stutton (Suffolk) and the preceding warm (Hoxnian) interglacial stage at Clacton (Essex), which agree well with the age ranges of stages 5 and 7 respectively in core V28-238. However, the correlation and dating of stages preceding the Hoxnian is at present more speculative.

The Wolstonian type sequence, which is provided by deposits capping interfluves south-east of Coventry (Shotton, 1953), near Leicester (Rice, 1968), and in temporary sections along the intervening M69 motorway (Shotton, 1976), shows that two ice advances affected the Midlands. These were separated by a retreat phase of about 10,000 years, during which proglacial Lake Harrison was impounded against the Jurassic scarp by ice lobes blocking the Soar, Tame and Avon valleys. The earlier advance deposited a reddish (Triassic-derived) till as far south as Warwick, whereas the later came from the north-east and deposited the grey, chalky Oadby Till as far as Moreton-in-the Marsh (Bishop, 1958). This tripartite succession, which has also been traced in Charnwood Forest (Bridger, 1975) and around Market Bosworth in western Leicestershire (Douglas, 1980), is unfortunately nowhere related stratigraphically to either younger or older interglacial deposits, so its exact age remains in doubt, even though it was chosen as the type Wolstonian, and contains a cold vertebrate fauna thought to confirm a Wolstonian age (Shotton, 1976). Better evidence for a Wolstonian glaciation is available further west, where a Bunter-rich till overlies Hoxnian deposits at Quinton (Horton, 1974), several km east of the Devensian limit near Birmingham. There is also good evidence in eastern England, where the Basement Till underlies the Ipswichian raised beach at Sewerby (Catt and Penny, 1966), but overlies flint gravels containing Acheulian artefacts and Hoxnian mammal remains, both probably derived, at Welton-le-Wold near Louth (Alabaster and Straw, 1976).

Tills and gravels, which underlie Hoxnian lake sediments at numerous sites and overlie freshwater deposits and soils attributed to the next older (Cromerian) interglacial, and are therefore Anglian in age, are widespread in East Anglia and extend as far south as Essex and Hertfordshire. At Corton cliff (Suffolk), the type site for the Anglian Stage, two lithologically distinct tills occur, separated by the shelly Corton Sands. The sands were originally thought to be interstadial, but they probably represent only a local, brief ice-free interval, because most of the shells are derived, and there is petrographic evidence for mixing of the lower (Cromer) and upper (Lowestoft) tills in north-west Norfolk (Banham, Davies and Perrin, 1975), and also stratigraphic evidence in the Norwich area that the Cromer Till ice lobe prevented the eastward spread of the Lowestoft Till ice (Cox and Nickless, 1972). Gibbard (1977) showed that two ice advances separated by a short retreat phase also occurred near the Anglian limit in Hertfordshire, but here the two tills are lithologically indistinguishable. Till and gravels with Welsh erratics underlie the Hoxnian interglacial deposits at Nechells (Kelly, 1964) and Quinton (Horton, 1974) near Birmingham. These are probably Anglian in age, but could be older as there are no fossiliferous deposits beneath them.

There is no direct evidence (e.g. till) for a pre-Cromerian glaciation in Norfolk and northern Suffolk, though a North Sea glacier during the Baventian Stage has been postulated on the evidence of far-travelled pebbles, such as *Rhaxella* chert from eastern Yorkshire (Hey, 1976), a typical glacial heavy mineral assemblage (Funnell and West, 1962), and some aspects of the fossil flora (Turner, 1975) in the coastal Baventian deposits. However, in south-east Suffolk, Essex and Hertfordshire, there is slightly firmer evidence. Large erratics of Welsh volcanic rocks (Hey and Brenchley, 1977) are present in the Kesgrave Sands and Gravels, which underlie the Cromerian soil (Rose and Allen, 1977); these are very abundant and must have been brought part of the way by a glacier, even if final deposition was by a periglacial equivalent of the proto-Thames flowing through the Vale of St. Albans. Western erratics have also been reported from weathered till-like deposits along the proto-Thames valley in western Essex and south-east Hertfordshire (the Pebbly Clay Drift of Thomasson, 1961), on the Chiltern dipslope south-west of St. Albans (the Chiltern Drift of Wooldridge and Linton, 1955) and in Oxfordshire (the Northern Drift of Arkell, 1947). The last of these also contains ice-scratched stones and erratics from Norway and northern England, and is older than the Cromerian deposit at Sugworth (Briggs and others, 1975).

Among Thames gravels older than the Anglian Winter Hill Terrace, Green and McGregor (1978) identified three separate periods when ice seems to have introduced far-travelled stones from north-western areas into Hertfordshire; these are the Westland Green, Higher Gravel Train and Lower Gravel Train stages. Hey (1965) correlated the Westland Green stage with gravels in the upper Thames occurring at levels above the Northern Drift, and Evans (1971) suggested that the Northern Drift is Baventian and equivalent to the Lower Gravel Train. However, it is perhaps more likely that the Lower Gravel Train was deposited during the Beestonian Stage of Mitchell and others (1973), which Evans dismissed as equivalent to a minor cold period indicated around 350,000 B.P. in some of the early analysed oceanic cores. The Northern Drift glaciation might then be Beestonian, and the Higher Gravel Train and Westland Green Gravel, Baventian or older.

The Pebbly Clay Drift, Chiltern Drift and Northern Drift are all thin, weathered throughout, and have undergone a complex history of cryoturbation and pedogenesis (Bullock and Murphy, 1979) similar to the Cromerian Valley Farm palaeosol in Suffolk. This alteration would have destroyed many critical till characteristics, and it is even possible that the stony clays originated partly by illuvial clay enrichment of gravels. The Beestonian Northern Drift glaciation is therefore slightly less definite than the later Anglian and Wolstonian advances, but is in turn rather more definite than the Baventian and the other possible early glaciation affecting the Thames catchment. Including the Late Devensian, southern Britain therefore seems to have experienced six separate glacial episodes, at least two of which comprised two separate advances in certain areas. The occasional erratics found in older marine deposits, such as the East Anglian Crags, are insufficient evidence for yet earlier glaciations in Britain, though they may well indicate cold conditions. The largest are often Scandinavian and could have been carried by icebergs calved from early glaciers there, and the small examples could be beach pebbles transported by drifting floes of sea-ice.

WHERE WERE THE ICE MARGINS?

Because of the subsequent erosion, ice marginal features (e.g. end-moraines, kames, meltwater channels) of pre-Devensian glaciations are rarely well preserved, and one must rely almost entirely on the stratigraphic and sedimentological evidence for the extents of earlier ice sheets. In practice, this usually means drawing ice margins to enclose all till or gravel patches thought to be of a certain age from their lithology or stratigraphic position. However, because of the unknown extent of denudation, this is often little better than guesswork. Even less reliable are the limits of the earliest phases (Figs. 1d-1f), for which there is little or no evidence of direct glacial deposition.

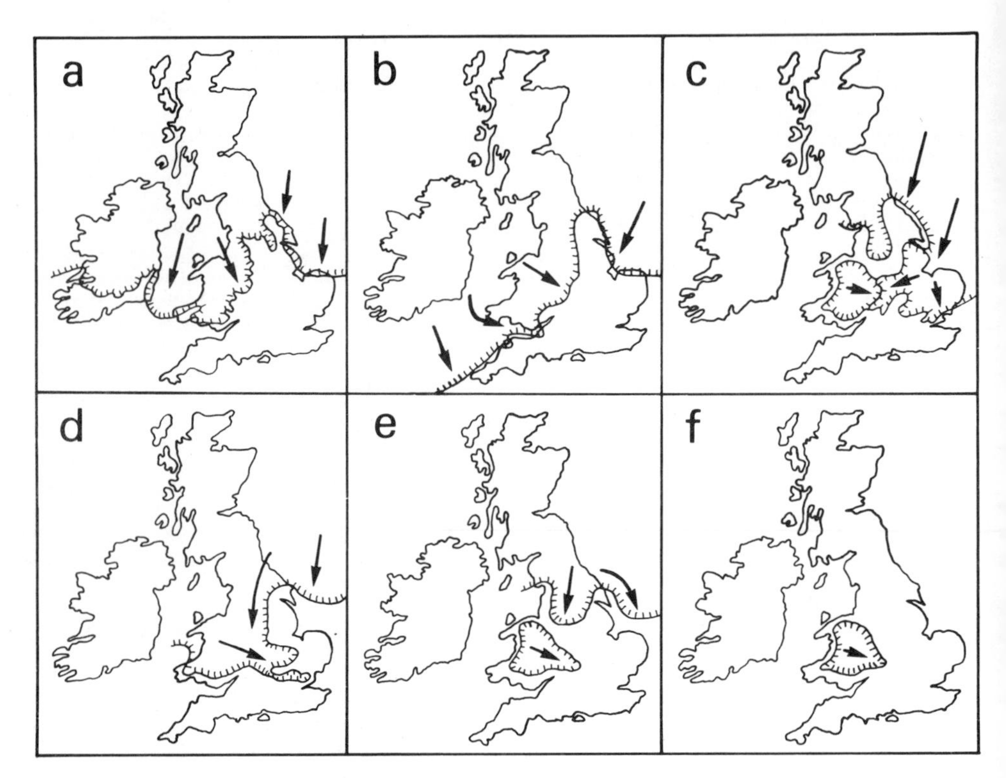

Fig. 1. Possible ice margins and principal ice movement directions during the Devensian (a), Wolstonian (b), Anglian (c), Beestonian (d), Baventian (e) and another early Quaternary stage (f).

Conventional glacier dynamics suggest that much of northern Britain would have been under a considerable thickness of ice when glaciers reached as far south as the Thames catchment. However, in Yorkshire and areas to the south, it is likely that the pattern of ice movement near the glacier margins was often controlled by other factors, such as topography, ice surges, and in eastern England the westward pressure of Scandinavian ice. Consequently we cannot assume that during any particular glaciation ice covered the whole land area north of the southernmost occurrence of glacial deposits. During the Late Devensian glaciation, higher ground, such as the southern Pennines, North York Moors and Yorkshire Wolds, was not invaded, and Boulton and others (1977) suggest that ice only reached Lincolnshire and north Norfolk because of a surge southwards from the Tees estuary. Similarly localised glacial deposition is likely to have occurred also in earlier periods; a surge could well explain emplacement of the Wolstonian Basement Till on the east coast, and Perrin, Rose and Davies (1979) suggested that the Anglian glacier invaded eastern England largely via a gap in the Chalk escarpment now occupied by The Wash. The latter suggestion contradicts earlier views (Straw, 1979b) that all the pre-Devensian tills of Lincolnshire and the east Midlands were deposited by Wolstonian ice sweeping southwards from Yorkshire. However, it does explain the northward diversion of the R. Trent from an original eastward course via the Lincoln or Ancaster gaps, and the absence of similar tills (or indeed of any proven Wolstonian glacial deposits other than the coastal Basement Till) north of the Humber and west of the Trent. It also suggests that much of the extremely patchy, undated "older drift" of Derbyshire and eastern Yorkshire could be Pre-Anglian, or at least somewhat older than the Wolstonian age assumed by Mitchell and others (1973, Tables 4 and 5) and most other workers.

A Wolstonian age has been similarly assumed for the pre-Devensian glacial deposits left by Irish Sea ice on the Scillies (Mitchell and Orme, 1967), Lundy (Mitchell, 1968) and various coastal areas bordering the Bristol Channel (Stephens, 1966; Edmonds, 1972; Bowen, 1977; Gilbertson and Hawkins, 1978), though all could be older. The oldest recognisable terrace of the Severn (the Woolridge Terrace) contains abundant flint (Hey, 1961), and probably originated as outwash of the ice that deposited chalky till from Leicestershire to Moreton-in-Marsh. The western margin of this ice stream may have extended to Tewkesbury (Beckinsale and Richardson, 1964), where it probably encountered ice from Wales and the Welsh borderland. Kellaway and others (1975) suggested that the English Channel and much of southern England were also glaciated in the Wolstonian. However, their evidence relies too heavily on scattered occurrences of foreign rocks, which could have been introduced by human or other non-glacial action, and on improbable reinterpretations of Quaternary features long accepted as non-glacial (e.g. the Goodwood-Slindon raised beach in Sussex). It is unlikely that the Irish Sea ice invaded south-west England further than the northern coastal strip of Cornwall, Devon and Somerset.

A Munsterian (= Wolstonian) Irish Sea glacier was also regarded as responsible for deposition of the shelly Ballycroneen Till, which overlies the 8 m raised beach at several sites on the south-east coast of Ireland. The beach is correlated with the Gortian Interglacial, which has usually been equated with the Hoxnian in England (Mitchell and others, 1973, Table II), though Warren (1979) suggested it is Ipswichian, in which case the Ballycroneen Till could be Devensian. Other tills south of the Late Devensian glacial limit (approximating to the South Ireland End-Moraine of Charlesworth, 1928), and containing erratics mainly from inland parts of Ireland, also overlie the 8 m beach, but Synge (1977) concluded that they were often soliflucted or slumped over the beach; the glaciation/s during which they were originally deposited could therefore be older than the beach. In view of the current doubts about the exact age of the Gortian interglacial deposits and their stratigraphic relationships with the various tills, it is quite impossible to state with certainty when any part of Ireland was glaciated before the Late Devensian, though southern areas must have undergone at least one earlier glaciation.

Kellaway and others (1975) reinterpreted the Clay-with-flints of southern England as the weathered remains of Anglian or older tills. However, they ignored the many features of composition and distribution which suggest that this deposit formed by *in situ* weathering, slumping and cryoturbation of a mixture of basal Tertiary sediment (usually Reading Beds) and Upper Chalk (Catt and Hodgson 1976). The quartzite erratics quoted as evidence for the glacial origin of Clay-with-flints in the Chilterns in fact occur in the Chiltern Drift of Wooldridge and Linton (1955), which has been misleadingly included in the Clay-with-flints by some authors, probably because it shows similar weathering features.

The Anglian ice margin in Essex and Hertfordshire is probably quite precisely defined by the southern limit of the chalky Lowestoft Till, which underlies Hoxnian lake deposits at numerous sites (Sparks and others, 1969 ; Turner, 1970; Gibbard, 1977). However, the similar till in Bedfordshire, north Buckinghamshire and Northamptonshire, probably deposited by a large ice lobe occupying the broad clay vale between the Chalk and Middle Lias scarps, is not dated by any interglacials, and is apparently continuous with the chalky Oadby Till in Leicestershire, which is part of the Wolstonian type sequence. This suggests either that similar chalky tills were deposited during both the Wolstonian and Anglian glaciations, or that the Wolstonian type sequence is actually Anglian in age. Believing the former to be true, Straw (1979b) drew the Wolstonian limit along the Chiltern scarp and through Norwich to Cromer, but Hoxnian deposits at Barford (Cox and Nickless, 1972) and Narborough (Stevens, 1959) suggest that this is too far east, at least in Norfolk. In contrast, Perrin, Rose and Davies (1979) showed that the composition of all the chalky tills in East Anglia, the east Midlands and Lincolnshire can reasonably be explained by the single Anglian glacial advance via The Wash, the ice then fanning out northwards, westwards and southwards. Transferring the pre-Devensian Wragby, Calcethorpe, Heath and Belmont Tills of Lincolnshire from the Wolstonian (Straw, 1979b) to the Anglian helps explain why the sequence at Kirmington contains a pre-Hoxnian "lead-coloured" till (Stather, 1905) resembling the Wragby Till, but no Wolstonian glacial deposits. The transfer is not invalidated by the superposition of Calcethorpe Till on the Wolstonian Basement Till at Welton-le Wold (Alabaster and Straw, 1976), as earlier exposures (Straw, 1961b) showed the Calcethorpe Till also overlying Devensian till, indicating that it had been at least partly soliflucted there from higher surfaces on the Lincolnshire Wolds. However, some of the erratics in the Wragby and Heath Tills are from north Lincolnshire, and suggest a southward rather than the northward or westward ice movement postulated for these tills by Perrin, Rose and Davies (1979).

It is consequently impossible at present to draw definite Anglian and Wolstonian ice margins in the east Midlands. Those shown in Fig. 1b and c are tentative suggestions, based partly on the known distribution of Hoxnian deposits over chalky till, and partly on the petrographic correlation of the chalky ("Wolstonian") till of the east Midlands with the Anglian Lowestoft Till in East Anglia and the Wragby and other tills in Lincolnshire (Perrin, Rose and Davies, 1979). If the chalky till of the east Midlands is Anglian, then a grey chalk- and flint-free Lower Boulder Clay occurring beneath it near Kettering (Hollingworth and Taylor, 1946) and Buckingham (Horton, 1970) may be an unweathered equivalent of the Beestonian Northern Drift of the Oxford area; if so, the Beestonian glacier would have invaded parts of the east Midlands as well as the proto-Thames valley (Fig. 1d). However, if the chalky till of this area is Wolstonian, the Lower Boulder Clay could be correlated with the Anglian Lowestoft Till, and the absence of chalk and flint would be explained by the NW-SE ice movement often postulated for the Lowestoft advance. Detailed petrographic studies of the Lower Boulder Clay and Northern Drift might help clarify which of these two possible correlations is correct.

Comparison of soil profiles developed on the chalky till in various areas might also help resolve the present confusion over whether it resulted from one or two glaciations. Recent Soil Survey mapping has shown that soils with well-developed interglacial features (the palaeoargillic horizons of Avery, 1973) occur on the Lowestoft Till in Essex (Sturdy and others, 1979) and parts of Suffolk and west Norfolk, and on the Calcethorpe Till in Lincolnshire (Heaven, 1978, 73-81). In Essex and Lincolnshire there is evidence for a period of soil disturbance, probably by cryoturbation and solifluction, between two phases of interglacial pedogenesis, which suggests that the tills are Anglian. Such complex profiles have yet to be found on the chalky tills further west, though at present it would be premature to conclude that the tills of the Midlands are therefore younger, as many areas have yet to be surveyed in detail, and the extent of periglacial erosion of earlier interglacial soils is difficult to assess.

CONCLUSIONS

Considerable advances have been made in the last 2-3 decades in our understanding of Quaternary stratigraphy in relation to climatic change, but we are still a long way from knowing where and when glaciers existed in Britain before the Late Devensian. Glaciations before the Beestonian are likely to remain elusive and indeterminate, but careful fieldwork and laboratory analysis of representative samples of glacial sediment could help clarify the extents of the English Wolstonian, Anglian and Northern Drift (Beestonian) glaciations. In particular, more could be done to identify and evaluate weathering profiles, and to distinguish diamictons deposited and redeposited in various ways. In the past, failures in both of these aspects have resulted in major misinterpretations of stratigraphic relationships within the glacial sequence. Petrographic studies of till sheets are also helpful in correlation, but closely-spaced samples and careful stratigraphic controls are needed to overcome the difficulties resulting from lateral variation and the possibility of similar deposits being left by successive advances over the same ground.

REFERENCES

Alabaster, C. and A. Straw (1976). The Pleistocene context of faunal remains and artefacts discovered at Welton-le-Wold, Lincolnshire. *Proc Yorks. geol. Soc.*, 41, 75-94.

Arkell, W.J. (1947). *The Geology of Oxford*. Oxford Univ. Press. viii+267 pp.

Avery, B.W. (1973). Soil classification in the Soil Survey of England and Wales. *J. Soil Sci.*, 24, 324-338.

Banham, P.H., H. Davies and R.M.S. Perrin (1975). Short field meeting in north Norfolk. *Proc. Geol. Ass.*, 86, 251-258.

Beckinsale, R.P. and L. Richardson, (1964). Recent findings on the physical development of the Lower Severn valley. *Geogrl. J.*, 130, 87-105.

Bishop, W.W. (1958). The Pleistocene geology and geomorphology of three gaps in the Midland Jurassic escarpment. *Phil. Trans. R. Soc.*, B 241, 255-306.

Boulton, G.S., A.S. Jones, K.M. Clayton and M.J. Kenning (1977). A British ice-sheet model and patterns of glacial erosion and deposition in Britain. In F.W.Shotton (Ed.), *British Quaternary Studies : Recent Advances*. Oxford Univ. Press. Chap. 17, pp. 231-246.

Bowen, D.Q. (1977). The coast of Wales. In C. Kidson and M.J. Tooley (Eds), *The Quaternary History of the Irish Sea*. Geol. J. Spec. Issue, 7, pp. 223-256.

Bowen, D.Q. (1978). *Quaternary Geology. A Stratigraphic Framework for Multidisciplinary Work*. Pergamon Press, Oxford. xi + 221 pp.

Bridger, J.F.D. (1975). The Pleistocene succession in the southern part of Charnwood Forest, Leicestershire. *Mercian Geol.*, 5, 189-203.

Briggs, D.J., D.D. Gilbertson, A.S. Goudie, P.J. Osborne, H.A. Osmaston, M.E. Pettitt, F.W. Shotton, and A.J. Stuart (1975). A new interglacial site at Sugworth. *Nature, Lond.*, 257, 477-479.

Bullock, P. and C.P. Murphy (1979). Evolution of a paleo-argillic brown earth (Paleudalf) from Oxfordshire, England. *Geoderma*, 22, 225-252.

Catt, J.A. (1980). Soils and Quaternary Geology in Britain. *J. Soil. Sci.*, 30, 607-642.

Catt, J.A. and J.M. Hodgson (1976). Soils and geomorphology of the Chalk in south-east England. *Earth Surf. Processes*, 1, 181-193.

Catt, J.A. and L.F. Penny (1966). The Pleistocene Deposits of Holderness, East Yorkshire. *Proc. Yorks. geol. Soc.*, 35, 375-420.

Charlesworth, J.K. (1928). The glacial retreat from central and southern Ireland. *Q. Jl geol. Soc. Lond.*, 84, 293-344.

Cox, F.C. and E.F.P. Nickless (1972). Some aspects of the glacial history of central Norfolk. *Bull. geol. Surv. Gt Br.*, 42, 79-98.

Douglas, T.D. (1980). The Quaternary Deposits of western Leicestershire. *Phil. Trans. R. Soc.*, B 288, 259-286.

Edmonds, E.A. (1972). *The Pleistocene history of the Barnstaple area.* Inst. geol. Sci. Rept, 72/2, iv + 12 pp.

Edwards, W., G.H. Mitchell and T.H. Whitehead (1950). *Geology of the district north and east of Leeds.* Mem. geol. Surv. U.K., vi + 93 pp.

Evans, P. (1971). Towards a Pleistocene time-scale. Part 2 of *The Phanerozoic Time Scale - A Supplement.* Geol. Soc. London. Spec. Publ., 5, pp. 123-356.

Funnell, B.M. and R.G. West (1962). The Early Pleistocene of Easton Bavents, Suffolk. *Q. Jl geol. Soc.* 113, 125-141.

Gibbard, P.L. (1977). Pleistocene history of the Vale of St. Albans. *Phil. Trans. R. Soc.* B 280, 445-483.

Gilbertson, D.D. and A.B. Hawkins (1978). The Pleistocene succession at Kenn, Somerset. *Bull. geol. Surv. Gt Br.*, 66, iv + 41 pp.

Green, C.P. and D.F.M. McGregor (1978). Pleistocene gravel trains of the River Thames. *Proc. Geol. Ass.*, 89, 143-156.

Heaven, F. (1978). Soils in Lincolnshire III. Sheet TF 28 (Donington-on-Bain). *Soil Surv. Record*, 55, viii + 152 pp.

Hey, R.W. (1961). The Pleistocene history of the Malvern Hills and adjacent areas. *Proc. Cotteswold Nat. Fld Club*, 33, 185-191.

Hey, R.W. (1965). Highly quartzose Pebble Gravels in the London Basin. *Proc. Geol. Ass.*, 76, 403-420.

Hey, R.W. (1976). Provenance of far-travelled pebbles in the pre-Anglian Pleistocene of East Anglia. *Proc. Geol. Ass.*, 87, 69-82.

Hey, R.W. & P.J. Brenchley (1977). Volcanic pebbles from Pleistocene gravels in Norfolk and Essex. *Geol. Mag.*, 114, 219-225.

Hollingworth, S.E. and J.H. Taylor (1946). An outline of the geology of the Kettering district. *Proc. Geol. Ass.*, 57, 204-233.

Horton, A. (1970). *The drift sequence and subglacial topography in parts of the Ouse and Nene basin.* Inst. geol. Sci. Rept., 70/9, iv + 30 pp.

Horton, A. (1974). *The sequence of Pleistocene deposits proved during the construction of the Birmingham motorways.* Inst. geol. Sci. Rept., 74/11, iv + 21 pp.

Jukes-Browne, A.J. (1885). The boulder clays of Lincolnshire. *Q. Jl geol. Soc. Lond.*, 41, 114-132.

Kellaway, G.A., J.H. Redding, E.R. Shephard-Thorn, and J.P. Destombes (1975). The Quaternary history of the English Channel. *Phil Trans. R. Soc.* A 279, 189-218.

Kelly, M.R. (1964). The middle Pleistocene of north Birmingham. *Phil. Trans. R. Soc.*, B 247, 533-592.

Kukla, G.J. (1977). Pleistocene Land-Sea correlations. I. Europe. *Earth Sci. Rev.*, 13, 307-374.

Madgett, P.A. and J.A. Catt (1978). Petrography, stratigraphy and weathering of Late Pleistocene tills in East Yorkshire, Lincolnshire and north Norfolk. *Proc. Yorks. geol. Soc.*, 42, 55-108.

Mitchell, G.F. (1968). Glacial gravel on Lundy Island. *Proc. R. geol. Soc. Cornwall*, 20, 65-69.

Mitchell, G.F. and A.R. Orme (1967). The Pleistocene deposits of the Isles of Scilly. *Q. Jl geol. Soc. Lond.*, 123, 59-92.

Mitchell, G.F., L.F. Penny, F.W. Shotton and R.G. West (1973). A correlation of Quaternary deposits in the British Isles. *Geol. Soc. Lond. Spec. Rept.*, 4, 99 pp.

Penny, L.F. (1959). The Last Glaciation in East Yorkshire. *Trans. Leeds geol. Ass.*, 7, 65-77.

Penny, L.F. (1964). A review of the Last Glaciation in Great Britain. *Proc. Yorks. geol. Soc.*, 34, 387-411.

Penny, L.F. (1974). Quaternary. In D.H. Rayner and J.E. Hemingway (Eds), *The Geology and Mineral Resources of Yorkshire.* Yorkshire Geological Society. pp. 245-264.

Penny, L.F., G.R. Coope and J.A. Catt (1969). Age and insect fauna of the Dimlington Silts, East Yorkshire. *Nature, Lond.*, 224, 65-67.

Perrin, R.M.S., J. Rose and H. Davies (1979). The distribution, variation and origins of pre-Devensian tills in eastern England. *Phil. Trans. R. Soc.*, B 287, 535 -570.

Rice, R.J. (1968). The Quaternary deposits of central Leicestershire. *Phil. Trans. R. Soc.*, A 262, 459 - 509.

Rose, J. and P. Allen (1977). Middle Pleistocene stratigraphy in south-east Suffolk. *J. geol. Soc.*, 133, 83 - 102.

Shackleton, N.J. and N.D. Opdyke (1973). Oxygen isotope and palaeomagnetic stratigraphy of Equatorial Pacific core V28 - 238: oxygen isotope temperatures and ice volumes on a 10^5 year and 10^6 year scale. *Quat. Res.*, 3, 39 - 55.

Shotton, F.W. (1953). Pleistocene deposits of the area between Coventry, Rugby and Leamington and their bearing upon the topographic development of the Midlands. *Phil. Trans. R. Soc.*, B 237, 209 - 260.

Shotton, F.W. (1976). Amplification of the Wolstonian Stage of the British Pleistocene. *Geol. Mag.*, 113, 241 - 250.

Sparks, B.W., R.G. West, R.B.G. Williams and M. Ransom (1969). Hoxnian interglacial deposits near Hatfield, Herts. *Proc. Geol. Ass.*, 80, 243 - 267.

Stather, J.W. (1905). Investigation of the fossiliferous drift deposits at Kirmington, Lincolnshire, and at various localities in the East Riding of Yorkshire. *Rep. Br. Ass. Advmt Sci.*, 1904, 272 - 274.

Stephens, N. (1966). Some Pleistocene deposits in north Devon. *Biul. Peryglac.*, 15, 103 - 114.

Stevens, L.A. (1959). The interglacial of the Nar valley, Norfolk. *Q. Jl geol. Soc.*, 115, 291 - 315.

Straw, A. (1958). The glacial sequence in Lincolnshire. *East Midl. Geogr.*, 1 (9), 29 - 40.

Straw, A. (1961a). Drifts, meltwater channels and ice margins in the Lincolnshire Wolds. *Trans. Inst. Br. Geogr.*, 29, 115 - 128.

Straw, A. (1961b). Notes on certain sections in glacial drift in Lincolnshire. *Trans. Lincs. Nat. Un.*, 15, 106 - 109.

Straw, A. (1979a). An early Devensian glaciation in eastern England? *Quaternary Newsl.*, 28, 18 - 24.

Straw, A. (1979b). The geomorphological significance of the Wolstonian glaciation of eastern England. *Trans. Inst. Br. Geogr.*, 4 (New Series), 540 - 549.

Sturdy, R.G., R.H. Allen, P. Bullock, J.A. Catt and S. Greenfield (1979). Paleosols developed on Chalky Boulder Clay in Essex. *J. Soil Sci.*, 30, 117 - 137.

Synge, F.M. (1977). The coasts of Leinster. In C. Kidson and M.J. Tooley (Eds), *Quaternary History of the Irish Sea.* Geol. J. Spec. Issue, 7, pp. 199 - 220.

Szabo, B.J. and D. Collins (1975). Ages of fossil bones from British interglacial sites. *Nature, Lond.*, 254, 680 - 682.

Thomasson, A.J. (1961). Some aspects of the drift deposits and geomorphology of south-east Hertfordshire. *Proc. Geol. Ass.*, 72, 287 - 302.

Turner, C. (1970). The Middle Pleistocene deposits at Marks Tey, Essex. *Phil. Trans. R. Soc.*, B 257, 373 - 440.

Turner, C. (1975). The correlation and duration of Middle Pleistocene interglacial periods in north-west Europe. In K.W. Butzer and G.L. Isaac (Eds), *After the Australopithecines: Stratigraphy, Ecology and Culture change in the Middle Pleistocene.* Mouton, The Hague. pp. 259-308.

Valentine, K.W.G. and J.B. Dalrymple (1976). Quaternary buried paleosols: a critical review. *Quat. Res.*, 6, 209 - 222.

Warren, W. (1979). Stratigraphic position and age of the Gortian interglacial deposits. *Bull. geol. Surv. Ireland*, 2, 315 - 332.

Wooldridge, S.W. and D.L. Linton (1955). *Structure, surface and drainage in south-east England.* (2nd Edition), London, G. Philip & Sons Ltd., viii + 176 pp.

Dr. J.A. Catt,
Rothamsted Experimental Station,
Harpenden,
Herts. AL5 2JQ.

3. Pleistocene Deposits and Superficial Structures, Allington Quarry, Maidstone, Kent

B. C. Worssam

INTRODUCTION AND PREVIOUS WORK

The present account arises from a study, initiated by Dr W.A. Read as District Geologist, East Anglia and South-East England Unit, IGS, of the causes of periodic local but rapid and deep subsidences that occur in the built-up area of Maidstone. Examination of a particularly spectacular subsidence in Queen's Road, Maidstone, in 1974 led Dr Read to the conclusion that the hollow lay at the intersection of two gulls formed by cambering in the Hythe Beds formation, and that the subsidence had resulted from the sudden collapse of a metastable infill of silt. A search for good surface exposures of similar material in the vicinity led to Allington Quarry. Examination of samples by Dr D. McCann (IGS Engineering Geology Unit) and Dr J.A. Catt (Rothamsted Experimental Station) has shown that some at least of the silt infilling fissures has the physical properties of loess. It may well, therefore, constitute part of the loess deduced (Perrin, Davies and Fysh, 1974; Catt, 1977) to have formerly covered a large part of southern England. The present paper examines the mechanism of cambering in the light of exposures provided by Allington Quarry between 1976 and 1979. Mr F.G. Berry (IGS East Anglia and South-East England Unit) undertook a study of the palaeontology of the silt, based on some samples collected by Dr Read and others collected by himself. His results are given as an Appendix.

Allington Quarry lies north-west of Maidstone, west of the lane leading to Allington Church from the main Maidstone-London road. Its position is shown in Fig. 1, together with the outlines of two other quarries, Aylesford Quarry and Bensted's Quarry, as they existed in 1949. Aylesford Quarry has since been worked southward to the railway line west of Allington Quarry, and Bensted's Quarry has been filled in and built over.

The total depth of Allington Quarry is some 15 m. In 1978/79 only its south side had a sheer face of this height. The quarry was then being worked northward and eastward on three levels or 'benches'. The heights of successive faces, from the quarry floor upwards, were approximately 3.5 m, 5 m and 6.5 m (Fig. 2).

The quarry is in the Hythe Beds formation of the Lower Greensand (Lower Cretaceous, Aptian stage). Around Maidstone this comprises alternate beds of hard grey sandy limestone or 'ragstone' up to about 0.5 m thick, worked for roadstone, and slightly cemented, compact, pale grey argillaceous calcareous sand or 'hassock' up to 1 m thick. Chert lenses occur in both ragstone and hassock in the upper part of the formation. The detailed succession in Allington Quarry resembles that recorded (Worssam, 1963, fig. 10 pp. 35-37) from Bensted's Quarry, and the same quarrymen's names can be applied to many of the limestone beds (Fig. 2).

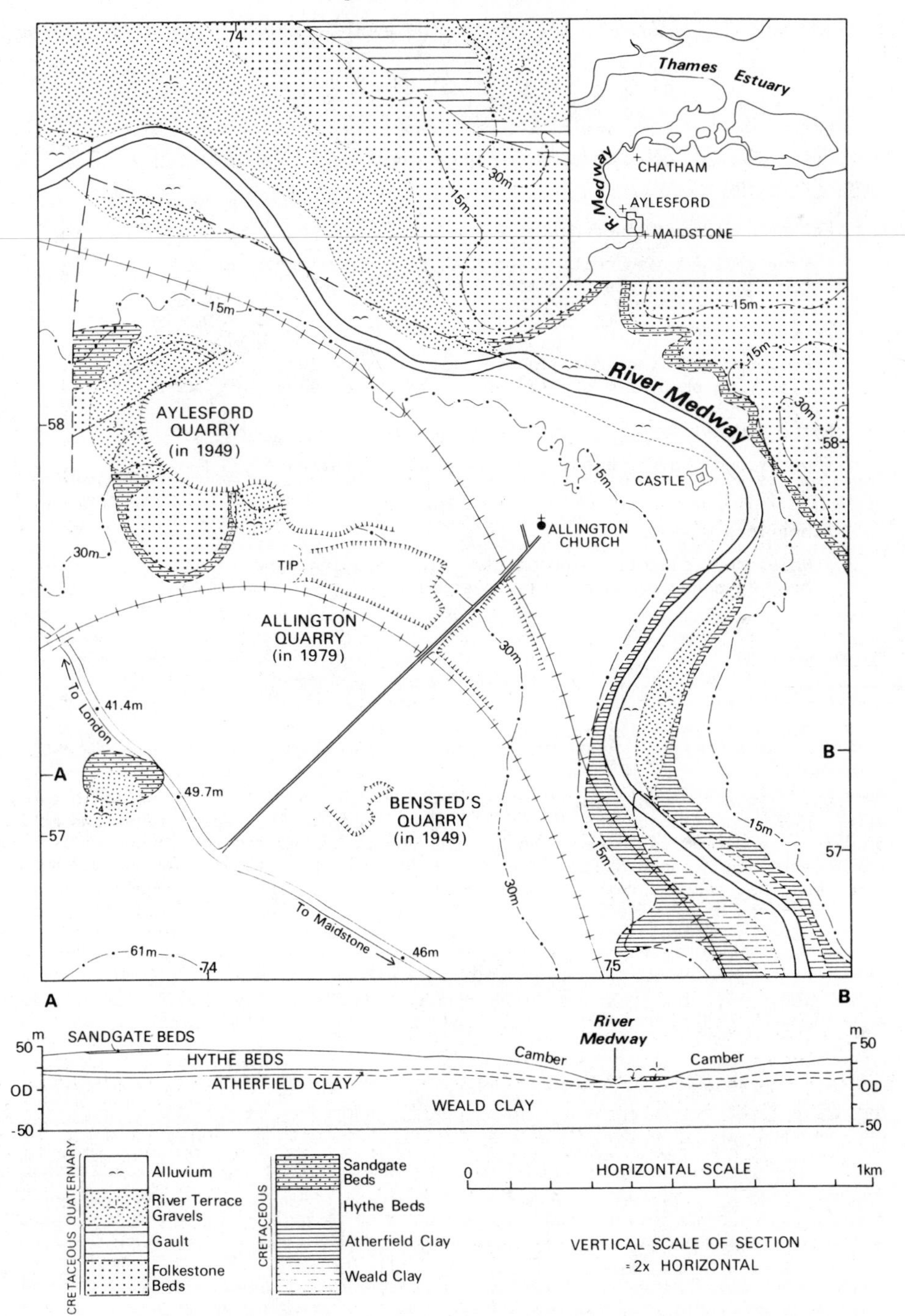

Fig. 1. Geology of the area around Allington Quarry, compiled from 1:10,560 Geological Survey maps by D.A. Gray and B.C. Worssam.

The succession at outcrop in the area to the south and west of the River Medway in Fig. 1 is:

	Thickness (m)
Lower Greensand	
Folkestone Beds (lower part):	
Fine-grained glauconitic sand with some thin layers (up to 25 mm) of tabular chert, seen for about	6
Sandgate Beds:	
Fullers earth (montmorillonite clay), 1 m or less, overlying about 2 m of glauconitic silty to sandy clay, about	3
Hythe Beds:	
Described above; basal beds not exposed, estimated total	21
Atherfield Clay:	
Pale grey, slightly glauconitic clay overlying dark grey to chocolate-brown clay, estimated	10
Wealden Series	
Weald Clay:	
Grey clay and silty clay, seen for about	5

Vertical fissures, some open but most filled with a variably stony yellow-brown silt, were a conspicuous feature of Allington Quarry. They were widest and most closely spaced towards the east end of the quarry. They tapered downwards, so that the lowest face exposed fewer and narrower fissures than the higher faces. Many of them could be traced across the benches or across the floor of the quarry, as straight-sided belts of silt, in places up to 5 m wide, with approximately NW-SE alignment. Some could be seen to die out along their length. Since the faces in Fig. 2 are oblique to the trend of the fissures this exaggerates their width, while the middle and lower faces show the fissures offset to the east from their positions on the top face; the apparent bends in some fissures on the middle face also result from them being viewed obliquely, the face being steeply sloping rather than vertical.

Silt-filled fissures are characteristic of Hythe Beds quarries around Maidstone. They were first described by the local quarry owner and amateur geologist W.H. Bensted (1862, pp. 447,450). He attributed them to erosion by a rapidly-flowing precursor of the Medway, their filling having taken place when its flow slackened.

Topley (1875, pp. 179-184) regarded the fissures as pipes; and the 'brickearth' filling them as old alluvium let down into the Hythe Beds by solution. Examples he observed were mostly in the north-eastern part of Maidstone, close to the Sandgate Beds outcrop, and were lined with clay and fuller's earth of the Sandgate Beds. He regarded a Sandgate Beds clay cover as necessary for the formation of the pipes by concentrating solution of the Hythe Beds along pre-existing fissures.

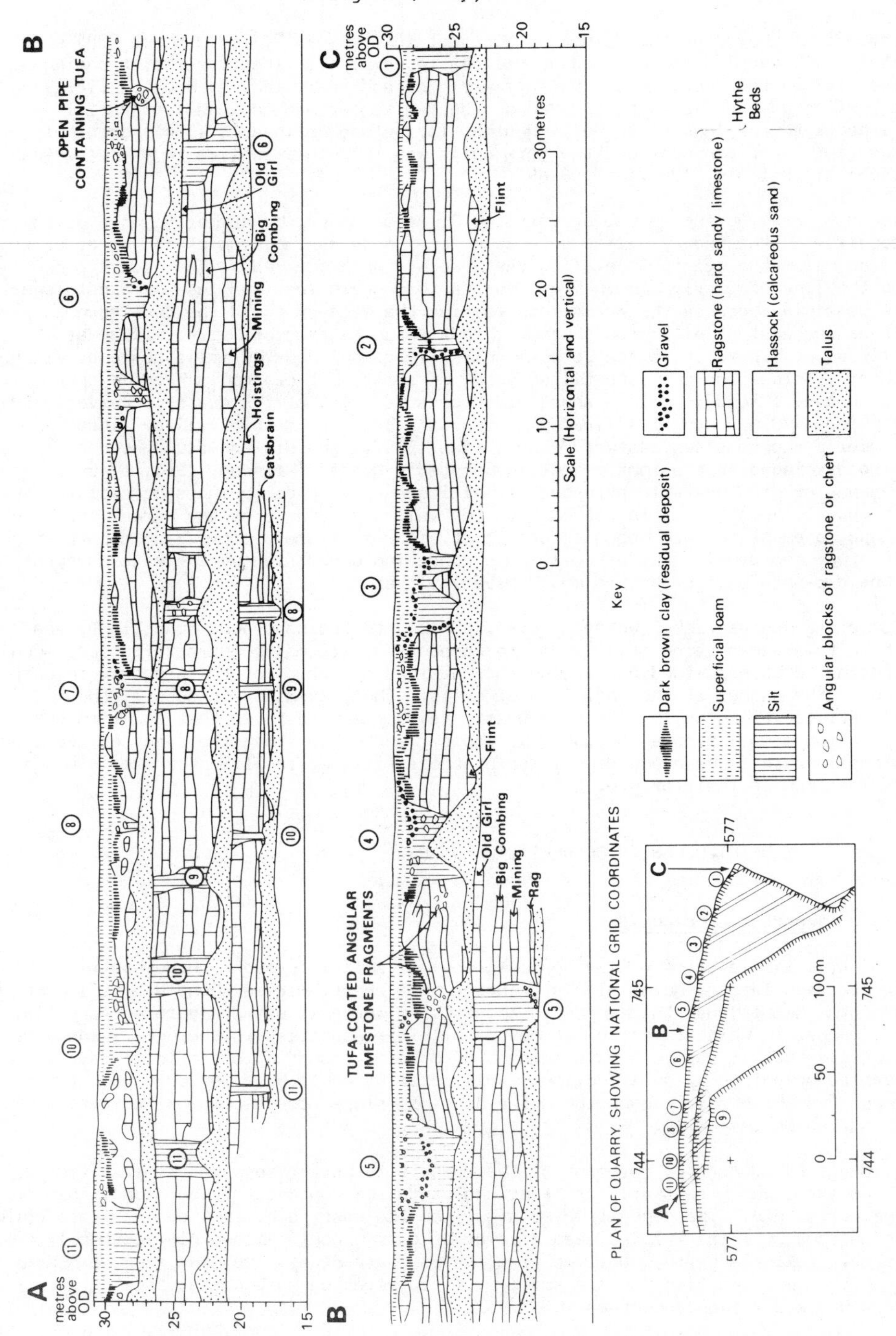

Fig. 2. North face of Allington Quarry in 1978, and a plan of the quarry to show the position of gulls.

Geological survey of the Maidstone district in 1946-1951 (1-inch Sheet 288) showed that the Hythe Beds in the Medway valley are cambered. The fissures, trending parallel to the contours of the valley slopes and following, as Bensted (1862) had described, all the bends of the Medway, seemed to correspond in all essential respects to the 'gulls' of Hollingworth, Taylor and Kellaway's (1944) original description of cambers, and the term gulls was therefore applied to them (Worssam, 1963; in Gallois, 1965, pp. 68-69).

Broadly speaking the amount of vertical lowering caused by cambering at a given locality in the Medway valley depends on the depth of incision of the river at that place below the original level of the base of the Hythe Beds. The beds are cambered on the line of Section AB in Fig.1 but farther north less and less vertical lowering is possible, because the normal dip carries the base of the Hythe Beds down to river level due east of Allington Church. Dip-and-fault structure, i.e. valleyward tilting of blocks of strata between gulls, is conspicuous by its absence at Allington Quarry, unless shown by the tilted block of strata between gulls 3 and 4 on the north face (Fig.2). Only a short distance south east of the quarry, however, old workings 500 m south of Allington Church, and between the two railway lines, formerly showed steep eastward dips (Osman, 1907), and dip-and-fault structure was also developed to a slight extent in Bensted's Quarry (Worssam, 1963, p. 36). The absence of dip-and-fault structure at Allington Quarry can perhaps be explained by the quarry being a little too high up the valley side west of the Medway for downslope movement to have begun to affect it. The exposures provide an opportunity to study the development of gulls in a condition unmodified by the mass movement that dip-and-fault structure would have involved.

Cambering was ascribed (Worssam, 1963, p. 129) to frozen-ground conditions, and while this explanation still seems to account for its major characteristics, much remains to be explained concerning the detailed operation of the process, in particular the manner of the infilling of gulls. Thus, examined in detail, the silt-filled fissures in Allington Quarry reveal many features that point to the operation of solution as favoured by Topley, and others that indicate, at least on a small scale, subaqueous deposition of their filling, recalling Bensted's suggested explanation of their origin.

DRIFT DEPOSITS OF ALLINGTON QUARRY

(i) Superficial deposits.

A superficial layer of yellowish-brown loam or silt is present throughout. In places it contains large angular blocks of Hythe Beds chert. It is about 0.5 to 1 m thick over the Hythe Beds, thickening to about 1.5 m where it extends across the gulls. The manner in which this deposit covers all irregularities in the underlying beds suggests that it may be a mere remnant of a large amount of superficial detritus carried downslope by solifluction. This detritus would have included much former loess from higher levels on the Hythe Beds dip-slope, corresponding to that now preserved in the gulls.

Evidence for downslope movement of the deposit is better seen on the east face of the quarry, which is nearly at right angles to the contours of the slope, than on the north face. The deposit over a gull on the east face, in the south-east corner of the quarry (Fig. 3a), includes a layer of dark greenish-brown clay (probably largely fuller's earth), derived from the Sandgate Beds. The clay sags downwards over the gull but also shows a small overfold directed northwards, as if its movement had a downslope component.

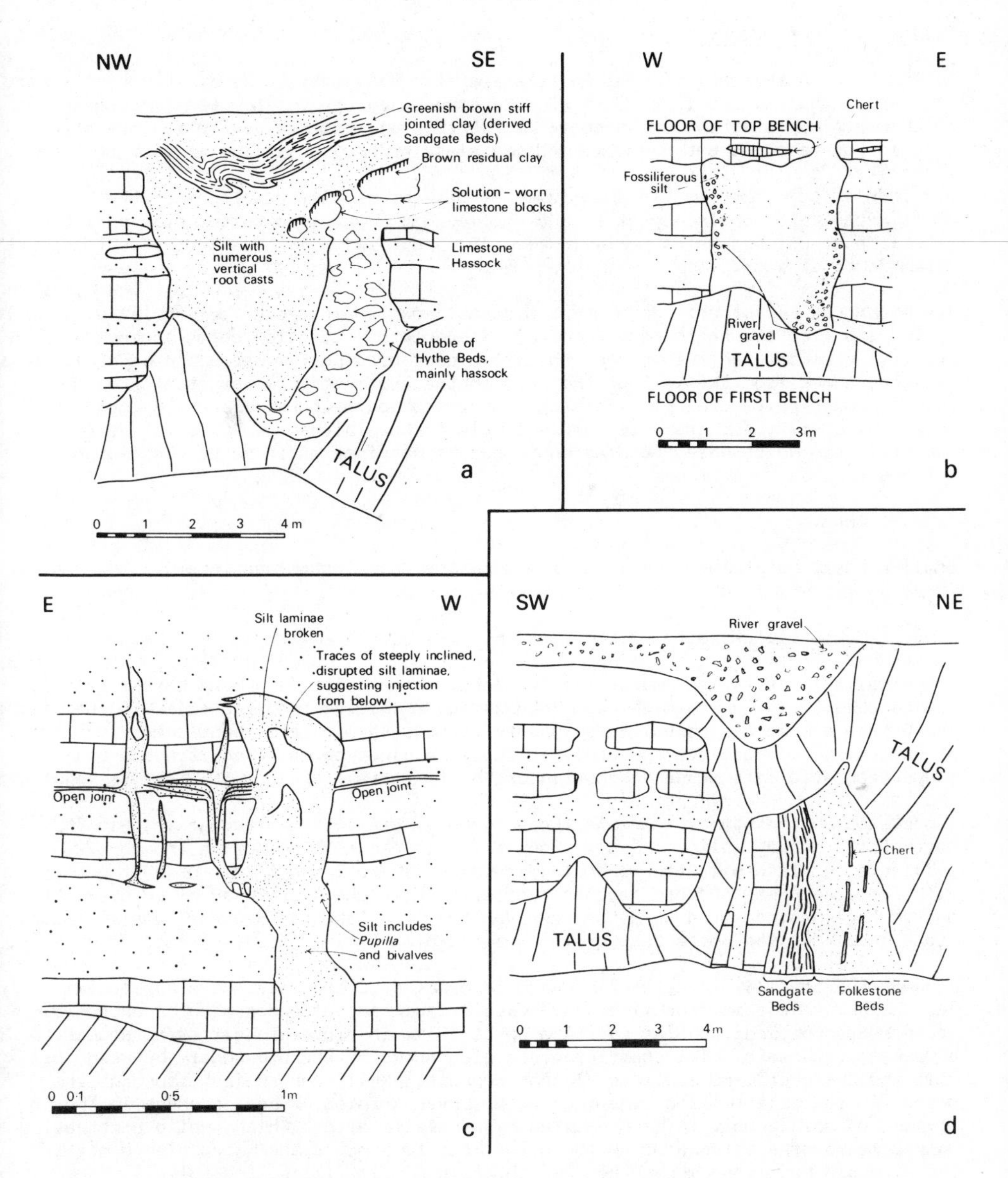

Fig. 3. Exposures at various places, in and adjacent to Allington Quarry

a. South-east corner of the quarry, looking east (in 1976).
b. Exposure (1977) in middle stage on north side of quarry, near gull No. 8 of Fig.2.
c. Silt-filled cavity near the bottom of the quarry, looking south (exposed 1976).
d. Exposure in an old quarry on the south-east side of the lane leading to Allington Church, on a face alongside the lane and backing on to the east face of Allington Quarry (7462 5754). Based on a field sketch made in 1949 by D.A.Gray.

(ii) Residual deposits.

At the base of this superficial layer where it overlies Hythe Beds, though not where it overlies gulls, occurs a layer of dark brown clay up to a few centimetres in thickness, in places including chert fragments. It seems to be the residue after solution of ragstone and hassock, and may also include some illuvial clay that has percolated downwards through the overlying loam. It is believed to be analogous in its mode of formation to the thin layer of dark brown clay developed at the base of drift deposits on Chalk (Chartres and Whalley, 1975) and also to the clay enriched B-horizon of parabraunerde soils (Kukla, 1977, p. 326). It thus indicates temperate climatic conditions.

The brown clay layer extends a short distance down the sides of some gulls e.g. gulls Nos. 2 and 3 on the north face (Fig. 2). Its date of formation seems to have been relatively recent. For instance, the brown clay draping the large blocks of limestone detached from the wall of the gull in the south-eastern corner of the quarry (Fig.3a) must have developed after the incorporation of these blocks in the surrounding silt, for they are hardly likely to have broken away without developing some tilt, in which case the clay would not be confined as it is to their upper surface.

(iii) Tufa.

Small deposits of tufa occur in cavities on the north face (see later).

(iv) Gull-fill deposits.

A typical gull is filled mainly with yellowish-brown silt, and many have a gravelly lining about 0.15 to 0.3 m thick. In gull No. 2 (Fig.2) and also in No. 5 the silt includes a downward-sagging gravelly layer some 0.15 m thick, about 2 m below the ground surface. The layer possibly indicates a minor phase of solifluxion or slope-wash at a late stage in the filling of fissures.

The clasts in the gravel resemble those commonly present in river gravels in this part of the Medway valley, mainly angular flints about 50 mm across and rounded pebbles of Wealden siltstone and clay ironstone 10 to 20 mm in diameter, together with lesser amounts of worn fragments of Hythe Beds chert 20 to 30 mm in diameter and of fine-grained iron-sandstone derived from the Sandgate Beds or basal Folkestone Beds. On the north face gravel was not present in the gulls westward of No.8.

Beneath the gravelly layer at 2 m depth in gull No. 2 the silt shows horizontal lamination. Above it, the silt is vertically jointed but appears to be unbedded; it includes vertical tubular cavities up to 10 mm in diameter that are presumed to be recent rootcasts. This particular section shows no sharp boundary between this silt and the overlying superficial loam deposit. Molluscan shells, abundant in places in the silt filling other gulls, may have originally been present in the filling of No. 2, only to be dissolved by decalcification, which, while penetrating only a metre or so downwards on the Hythe Beds to produce the brown clay described above, would have gone much deeper in the porous silt filling the gulls.

One of the gulls shown in Fig.2 that has yielded fossiliferous silt is No. 8 on the second face. This was narrow on the top face and it may be that some constriction in its width, behind the talus at the foot of the top face, had obstructed the downward progress of decalcification. An effect of this sort seems to have led to the preservation of the fossiliferous silt in a section seen in a nearby part of the quarry in June 1977 (Fig. 3b). Shells were there present in silt near the top of what must originally have been a cavity roofed over by a cherty limestone bed except for a gap about 1 m wide.

The mollusc fauna of Maidstone 'brickearths' (Worssam, 1963, pp. 106-7) is that typical of loess in general, dominated by *Pupilla muscorum* and *Succinea oblonga*, traditionally taken as indicators of cold steppe conditions (but see Appendix). While Fig. 3b may not represent a typical situation, an abundant mollusc fauna is unlikely to have lived *in-situ* in such a cavity, 5 m below the ground surface, nor are the shells likely to have been blown into it together with wind-borne silt. Instead, the sharp, irregular and unweathered junction between hassock and silt, with no parting plane, suggests that the silt had been carried into the cavity as a suspension in water, the water possibly draining off through the hassock so that silt filled the cavity to the top, leaving no air-space. Another section showing evidence of injection of fossiliferous silt into a cavity is illustrated in Fig. 3c. The exposure was on the lowest face of the quarry (at 7436 5764) and hence some 12 to 14 m below the ground surface. There were evidently two stages of filling here, an earlier one when silt-bearing groundwater had flowed gently in, giving fine lamination in a widened horizontal joint, and a later rapid injection filling a large cavity apparently from below, as shown by upward-arching bedding traces in the silt and disruption of the horizontal laminae of the silt in the smaller cavity.

CAVITIES

(i) Vertical sub-cylindrical cavities (pipes)

One open cavity was to be seen in the face illustrated in Fig. 2. This, at 3 m west of B, and exposed at 3 m below the ground surface, had the form of a pipe, about 0.7 m in diameter. It extended vertically downwards behind the face, which here had a steeply sloping profile. The walls were of roughly fractured limestone and hassock, showing no smoothing by solution. Angular blocks of limestone and hassock about 0.2 m in diameter, lightly in contact with each other, formed a loose filling to the cavity. Some had a thin coating of hard tufa, and some soft white powdery tufa also occurred between the blocks. The cavity extended downwards for at least 1.5 m, beyond which it could not be seen or probed because of the filling. At two other points on the face, 24 m and 30 m east of B, the continuity of a limestone bed was interrupted by a cluster of angular limestone fragments, some with a coating of tufa. No distinct cavities were present at these points, though they may have been quarried away.

Above each of the three points (see Fig.2) the superficial loam projects downwards, with a thickening of the brown clay at its base. The pipe-like shape of the cavity suggests it was caused by solution by downward-migrating groundwater. This solution seems unlikely to have been accomplished by water that had slowly percolated through the surface loam and its brown clay basal layer. Instead, during periods of abnormally heavy rainfall, downward flow of excess subsoil water may have been concentrated at a few widely spaced points, with the brown clay layer having served to direct the water flow to these points, a function that Topley had envisaged would be fulfilled by a cover of Sandgate Beds clay. The thickening of the clay layer may have resulted from slight flowage of the clay. The tufa coating the loose limestone blocks could have been precipitated from calcium bicarbonate-charged water that entered the cavity when slow downward percolation was resumed. Mr Berry (Appendix) reports that the soft tufa is biogenic.

(ii) Elongate cavities

Open fissures 0.15 to 0.2 m wide were numerous towards the west end of the south face of the quarry, although some loam-filled gulls occurred there also. The fissures occurred about half-way up the face, which unlike the north face was not divided into stages. They appeared to be elongated in a NW-SE direction, like the gulls. A number of them were roofed over by chert lenses in limestone and hassock beds at the horizon of the Flint (at the bottom of the top stage of the quarry's north face) and went down as far as the Rag limestone (at the bottom of the middle stage on the north face - see Fig. 2). They were thus about 5 m in height. The Rag was underlain by a conspicuous hassock bed, of a darker grey colour than those above it, as if it was more clayey and hence retained moisture. It could have acted as a minor aquiclude, leading to the flow of groundwater, and hence the formation of fissures by solution, being concentrated in the 5 m thick group of strata above it.

(iii) Open joints

A third form of discontinuity of the Hythe Beds consists of open vertical joints about 10 mm wide. They were fairly closely spaced, at approximately 0.3 m intervals, in the limestone beds but not present in the intervening hassocks. On the south face alignments between 155° and 175° were measured. The most likely mode of origin seems to be one involving mechanical disruption and connected with the formation of gulls under permafrost conditions (see below).

DATE OF FORMATION OF ALLINGTON GULLS

On geomorphological grounds the gulls at Allington must post-date deposition of river gravels of the Third Terrace of the Medway. This terrace surface on the Gault Clay outcrop on the north side of the valley 1.5 km north of the quarry is 33 m (100ft) above OD, the same height as the ground surface at the quarry. On the south side of the valley, two patches of Third Terrace gravel, partially quarried away, were mapped in 1949 1 km north-west of the present quarry (Fig. 1). The surface height of the smaller patch was 33 m O.D. The age of the terrace is not certain, but in a section in the deposit north of the river (Gray in Worssam, 1963 p. 119), fluviatile gravel was interbedded with chalky gravel containing blocks of sarsen stone up to 0.5 m across, suggesting cold-climate conditions of deposition. The terrace may date from the glacial stage preceding the Ipswichian, for at a much lower level (15 m O.D.) is the Second Terrace at Aylesford, with its warm climate fauna. This terrace is either Ipswichian or early Devensian, depending on whether its fauna is regarded as *in-situ* or derived. Both interglacial (Ipswichian) and periglacial (Devensian) climatic conditions may have played a part in the development of the gulls. The undisturbed First Terrace below the cambers (Fig.1 section AB), probably late Devensian, gives a date for the cessation of cambering.

The loess filling the gulls is most probably of Devensian age. Such a date is indicated by the fauna recently recovered (Appendix) and it would not be at variance with the recorded vertebrate fauna of Maidstone brickearths (Worssam, 1963, p. 107), which includes *Bos sp.*, *Cervus elaphus*, *C. tarandus*, *Coelodonta antiquitatis*, *Hyaena spelea* and *Mammuthus primigenius*.

MODE OF ORIGIN OF GULLS

Previous accounts, mentioned above, suggest that gulls in the Maidstone area owe their formation to the growth of ground-ice wedges in the rocks. Most ice-wedge casts reported from southern England occur in sand or gravel deposits. They appear as vertical, narrowly V-shaped clefts, mostly less than 4.5 m in depth, infilled with fine-grained material of surface origin washed in when the ice-wedge melted. On flat ground ice-wedge casts occur in polygons (patterned ground). On valley sides the patterned ground takes the form of stripes running down the slope (West, 1968, ch. V.pl.10). Of these features, only the downward-tapering form and type of filling correspond strictly to features shown by Allington Quarry gulls.

Some gulls show irregularities of form that are difficult to explain as resulting simply from the forcing apart of rocks on either side by the growth of ice-wedges. Thus one on the south face was seen to have a flat floor formed by a continuous limestone bed about 5 m below the ground surface. Others, e.g. that in Fig. 3b, are roofed over at least locally. Thirdly, not all the tapering gulls have straight walls. No. 2 on the north face shows a constriction about 2 m below ground surface where a particularly thick limestone bed projects inwards. This would indicate some process other than simple mechanical widening of a fissure, for in that case a projection on one side should be matched by a recess on the other. In these instances, therefore, some process of removal of Hythe Beds material must have operated, and can hardly be other than one of solution, presumably operating under periglacial climatic conditions if only because of the large scale of structures compared with that of those resulting from recent solution activity.

In other instances only the former presence of ground-ice seems adequate to explain the observed structure. An example is provided by the section illustrated in Fig. 3d, based on a field sketch made in 1949 by D.A. Gray. The block of strata on the right-hand side of the sketch presumably bridged over a cavity similar to that in Fig. 3b, but it is hard to envisage how it could have assumed the vertical position in its virtually intact state, unless the beds were either frozen or let down very gradually over a thawing body of ground-ice.

The association in the Medway valley in general of widened silt-filled fissures with camber slopes does suggest that ground-ice played an essential role in the cambering process. Because ice-wedge casts more than 4.5 m deep are rare, ice-wedges penetrating from the surface downwards may have been less important in the initiation of gulls than vertical segregations of ground ice formed initially in fissures, perhaps similar to those exposed in the south face, with the ice fed by groundwater held up by clayey hassocks of the basal part of the Hythe Beds. Only later may the fissures have extended up to the surface and downwards to the Atherfield Clay.

On cambered slopes, the lateral pressure exerted by the growth of ice segregations can be envisaged as the cause of downslope movement of intervening blocks of strata. At a site such as Allington Quarry, above the camber slope, this pressure may have caused up-arching of the strata in the intervening blocks, resulting in widened joints in the limestone beds. The granular texture of the hassocks may have enabled them to accommodate to the stress without breaking along joints. Upon collapse of the arch after melting of the ground-ice the joints may have been prevented from closing by friction between beds.

The mode of filling of the Maidstone gulls is far from being fully explained. On the simplest assumption, if the loess of southern England is of pleniglacial date (cf. Kerney, 1963; Catt, 1977), then the ground was frozen when it was accumulating; around Maidstone it would have formed a blanket over a developing camber surface that concealed large ground-ice segregations. Only later, as the ground-ice melted, would those parts of the loess that overlay gulls have subsided, the rest being swept down-slope as a hillwash, contributing in passing to the 'superficial loam' of Fig. 2, and ultimately being carried out to sea, like the greater part of the former southern England loess cover (Catt, 1977). Some of the ground-ice segregations may have extended underneath outliers of Sandgate and Folkestone Beds; this would explain the foundering (Worssam, 1963, p. 131) of an outlier immediately south-east of the quarry (cf. Fig. 3d). The gravelly lining of gulls could be interpreted as a solifluction deposit derived from river terraces at higher levels, and which covered the Hythe Beds dip-slope before the arrival of the loess.

The continuing presence of deep-seated cavities and their infilling by the washing-in of unconsolidated material is indicated by the injection structures of Figs. 3b and 3c, while the subsidences reported from time to time around Maidstone suggest this to be a process still in operation.

ACKNOWLEDGEMENTS

I am indebted to my co-workers mentioned in the Introduction, in particular to Dr W.A. Read, on whose initiative this study was undertaken. Thanks are due to Messrs. Amey Roadstone Corporation Limited for allowing access to the quarry on a number of occasions. The paper is published by permission of the Director, Institute of Geological Sciences.

APPENDIX: BIOTA OF ALLINGTON QUARRY SILTS

F.G. Berry

Ten samples of silt have been washed, and the specimens registered in the I.G.S. collections; the following three samples yielded the most extensive range of biota:

1. 'Gull-fill I' (Gull No. 8 on middle stage of north face of quarry as shown in Fig.2). Collected 22nd September 1978. Sample weight 20 kg.

   Succinea oblonga Draparnaud (558 apices); *Pupilla muscorum* (Linné) (194 apices); *Pisidium obtusale lapponicum* Clessin (6 valves); *P. casertanum* (Poli) (one with additional cardinal tooth 2 x C_3); *Trichia sp. indet.*; slug granules (very numerous); Limacid shells (20); *Candona neglecta* Sars (ostracods, 16 valves); *Gasterosteus* (stickleback) spine (1).

2. 'Gull-fill A' (exposure near floor of quarry, illustrated in Fig.3c). Collected 24th June 1976. Sample weight 250 gm.

   Succinea oblonga; Pupilla muscorum; Pisidium casertanum (1 valve); *Trichia sp. indet.;* slug granules; *Candona neglecta.*

3. 'Gull-fill G', sample C, c. 3½ m below ground level (TQ. 7452 5764). Collected 16 August 1978. Sample weight not recorded:

   Succinea oblonga; Pupilla muscorum; Trichia hispida (Linné); *Microtus* cf. *gregalis* (narrow-necked vole, Lower M1 molar).

Most of the samples yielded abundant rhizocretes and slug granules. Of special interest are *Pisidium obtusale lapponicum*, an arctic-alpine form now extinct in Britain (Kerney 1977, p.34) and the vole compared to *M. gregalis*, another indicator of cold conditions. The variety of *Pupilla muscorum* is the longer and broader form described by Kerney (1963).

The ostracods were identified by Miss Diane Gregory, of the I.G.S. Palaeontology Unit, who reported: "The ostracods submitted are identified as *Candona neglecta* Sars. This is a species which inhabits shallow fresh to oligohaline waters. It can withstand dessication and is fairly eurythermal. Although widespread in Europe, Central Asia, and North Africa, it is not found in Norway. For this reason I would consider full-glacial Devensian temperatures may have been too low for this species to survive and therefore tentatively suggest an interstadial or Holocene origin".

Although loess is commonly regarded as indicating a dry steppe environment the shell fauna from Allington suggests damp conditions. *Succinea oblonga* and *Trichia hispida* are hygrophile to mesophile, while the large variety of *Pupilla muscorum* may have colonised wetter habitats than smaller varieties of this species. In all, the biota give the impression of pools of varied depth in the fissures, plus intermittent running water at times of melting snow etc.; the abundance of rhizocretes, which are not likely to be derived, points to an ephemeral vegetation.

Tufa from the open cavity on the north face at 3 m west of B (Fig. 2) was a 'root-tufa' of biogenic origin. It yielded: *Cecilioides acicula* (Müller) (burrowing snail, probably Holocene); *Pupilla muscorum* and *Succinea oblonga* (possibly derived); beetle fragments; woodlouse fragments. This list tends to confirm the Holocene origin suggested by the field relations of the tufa.

REFERENCES

Bensted, W.H. (1862). Notes on the geology of Maidstone. *Geologist*, 5, 294-301, 334-341, 378-382, 447-450.

Catt, J.A. (1977). Loess and coversands. In F.W. Shotton (Ed.), *British Quaternary Studies: Recent Advances*. Oxford Univ. Press. pp. 221-229.

Chartres, C.J., and W.B. Whalley (1975). Evidence for Late Quaternary solution of Chalk at Basingstoke, Hampshire. *Proc. Geol. Ass.*, 86, 365-372.

Gallois, R.W. (1965). *The Wealden District*. 4th. Edit., British Regional Geology, HMSO, London. 101 pp.

Hollingworth, S.E., J.H. Taylor and G.A. Kellaway (1944). Large scale superficial structures in the Northampton Ironstone Field. *Q. Jl geol. Soc. Lond.*, 100, 1-35.

Kerney, M.P. (1963). Late-glacial deposits on the Chalk of south-east England. *Phil. Trans. R. Soc.*, B 246, 203-254.

Kerney, M.P. (1977). British Quaternary non-marine Mollusca: a brief review. In F.W. Shotton (Ed.), *British Quaternary Studies: Recent Advances*. Oxford Univ. Press. pp. 31-42.

Kukla, G.J. (1977). Pleistocene Land-sea correlations. I. Europe. *Earth Sci. Rev.*, 13, 307-374.

Osman, C.W. (1907). Excursion to Aylesford and Allington. *Proc. Geol. Ass.*, 20, 104-114.

Perrin, R.M.S., H. Davies and M.D. Fysh (1974). Distribution of late Pleistocene aeolian deposits in eastern and southern England. *Nature, Lond.*, 248, 320-324.

Topley, W. (1875). *The geology of the Weald*. Mem. geol. Surv. Gt Br., xiv + 503pp.

West, R.G. (1968). *Pleistocene Geology and Biology*. Longman, London. vi + 379 pp.

Worssam, B.C. (1963). *Geology of the country around Maidstone*. Mem. geol. Surv. Gt Br., viii + 152 pp.

B.C. Worssam Esq.,
Institute of Geological Sciences,
Keyworth,
Nottingham. NG12 5GG.

4. The 'Gipping Till'* Revisited

F. C. Cox+

In an account of the Middle and Late Quaternary history of East Anglia, Bristow and Cox (1973) put forward an interpretation of the evidence for several glacial episodes that had implications for the glacial history of the British Isles as a whole. The Geological Society's 'Quaternary Correlation: British Isles' (Mitchell and others, 1973) included these somewhat unorthodox views as a dissenting minority

Mitchell and others (1973)		Bristow and Cox (1973)		Straw (1979)	
Stage	Sediments deposited	Stage	Sediments	Stage	Sediments
FLANDRIAN	Peats, muds, alluvial deposits	FLANDRIAN	Channel fill deposits	FLANDRIAN	Devensian deglaciation deposits
DEVENSIAN	Brown boulder clay Hunstanton boulder clay, Terrace gravels	DEVENSIAN	Hunstanton Till, Terrace gravels Head	DEVENSIAN	Tills of North Lincs. Hunstanton Till
IPSWICHIAN	Terrace gravels, → organic silts associated with river valleys	IPSWICHIAN + HOXNIAN	Organic silts and clays ←	IPSWICHIAN	Organic clays
WOLSTONIAN	Late glacial silts, and clays Terrace gravels Chalky boulder clays of the Midlands			WOLSTONIAN	Chalky Boulder Clay of East Anglia Chalky-Jurassic Boulder Clays of the Midlands
HOXNIAN	Organic silts and clays in enclosed hollows ↗			HOXNIAN ↖	Organic clays at Hoxne
ANGLIAN	Chalky boulder clay, Lowestoft Till, North Sea Drift, associated glacial gravels	ANGLIAN	Chalky Boulder Clay, Norwich Brickearth, and associated outwash deposits	ANGLIAN	Jurassic Boulder Clay of East Anglia

Table 1: The Glacial Stratigraphy of East Anglia and adjacent areas

* Bristow and Cox (1973) 'The Gipping Till': a reappraisal of East Anglian glacial stratigraphy.

+ Published by permission of the Director, Institute of Geological Sciences.

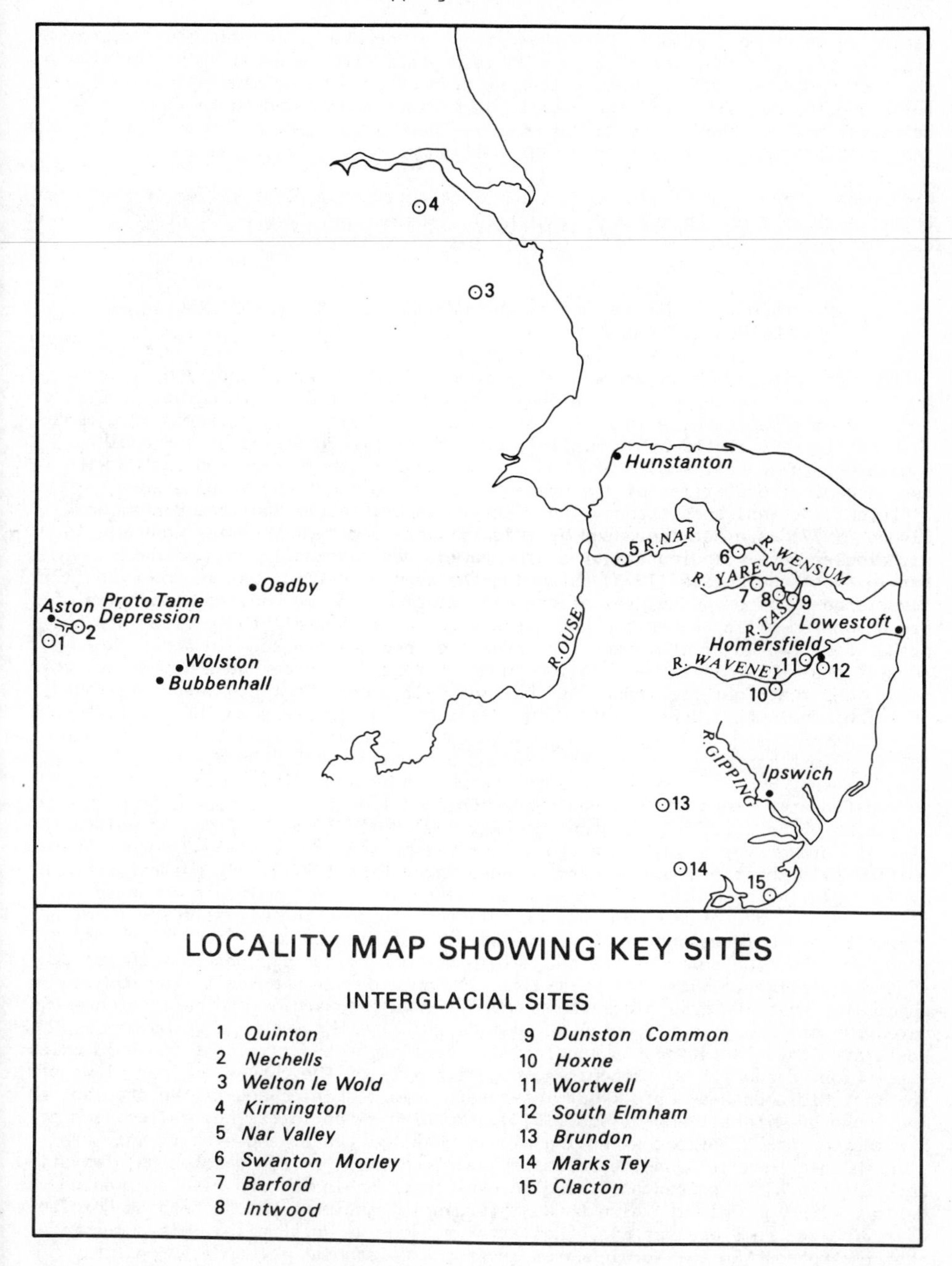

report in the section on Eastern England. Although there was considerable discussion of the views put forward and little evidence was produced to contradict them, at present only one of the ideas contained in the original account has gained general acceptance. This is that there is only one chalky boulder clay in East Anglia. The suggested correlation of this deposit with the chalky boulder clay of the East Midlands, and by implication the absence of a major glacial period following the deposition of the Hoxnian lake clays and preceding the Ipswichian deposits, has not been generally accepted. These ideas are in direct conflict with the stratigraphical correlation of Mitchell and others (1973) (Table 1).

With these general points in mind, I have attempted a critical review of the literature that has been published during the last decade on the glacial stratigraphy of East Anglia and nearby regions (Fig. 1).

THE UNIFIED NATURE OF THE CHALKY BOULDER CLAY OF EAST ANGLIA AND THE EAST MIDLANDS

In the same year as the Bristow and Cox paper, Perrin, Davies and Fysh (1973) published their research into the physical, chemical, and mineralogical properties of tills from East Anglia and the East Midlands. Their analyses revealed a remarkable consistency in the heavy mineral and mechanical properties of the chalky boulder clays of East Anglia and the East Midlands; these were contrasted with the mechanical properties of the Cromer Till, which was found to possess a different internal consistency. In justifiable criticism, Shotton, Banham and Bishop (1977) were not impressed by this evidence and took the view that similar lithologies merely indicate like source rocks, not necessarily synchroneity. Perrin, Rose and Davies (1979) extending the work of Perrin, Davies and Fysh (1973) give large numbers of analyses of the mineralogical and mechanical properties of pre-Devensian tills in Eastern England and conclude that all these tills are the products of two major ice sheets which were penecontemporaneous in age. They suggest that these tills are Anglian in age although they do not substantiate this view. One must therefore conclude that in their sense 'Anglian' means pre- the 'Hoxnian' deposits. Bowen (1978, Fig. 2-13, p. 37) figures a map of limits to the different British glaciations. He points out that there is doubt as to the age of the ice limit mapped in south-west England and states that "the Wolstonian ice margin is not known". No such doubts are shared by West (1977, Fig. 12.3, p. 280), who confidently links Bowen's south-west England line to one crossing to East Anglia which he entitles 'Wolstonian' Fig. 2). However, much of West's Wolstonian line is drawn over ground that the writer has mapped in Norfolk and across the unified till sheet figured by Perrin, Rose and Davies (1979). It passes across a boulder clay plateau which borehole evidence shows to be a multiple sequence of tills and sand and gravel commonly associated with deep tunnel valleys of over 60 metres in depth (Woodland, 1970). Nowhere in this region do we see an outwash sandur of the type described at the margin of the Anglian Boulder Clay in Norfolk (Cox and Nickless, 1972; Straw, 1973). In fact, one is tempted to the obvious conclusion that if three glaciations and only two ice margins can be located - the Devensian has been clearly defined (Edmonds, 1977) - then perhaps we have postulated one glaciation too many! In an account of the limits of the Wolstonian glaciation, Straw (1979) recognises the uniformity of the chalky boulder clays of the East Midlands and Norfolk but interprets this glacial event as 'Wolstonian' in the sense of Mitchell and others (1973). Hoxnian sites in the Nar Valley, and at Kirmington, are dismissed as having highly uncertain ages, and even if they are Hoxnian then they are considered to represent enclaves in the Wolstonian glaciation. Phillips (1976) is quoted as giving an example of an Ipswichian site at Swanton Morley, although Phillips also shows interglacial sediments at Barford as Hoxnian overlying the same boulder clay that Straw regards as Wolstonian. His suggestion that the outwash in the Norwich area is from a Wolstonian ice sheet completely ignores the detailed stratigraphy of this area based upon purpose-drilled boreholes and a detailed six-inch survey described by Cox and Nickless (1972) (see also Cox and Nickless, 1974, a comment on Straw, 1973). In a similarly detailed study of local stratigraphy, Gallois (1978) confirms the uniformity of the Chalky-Jurassic

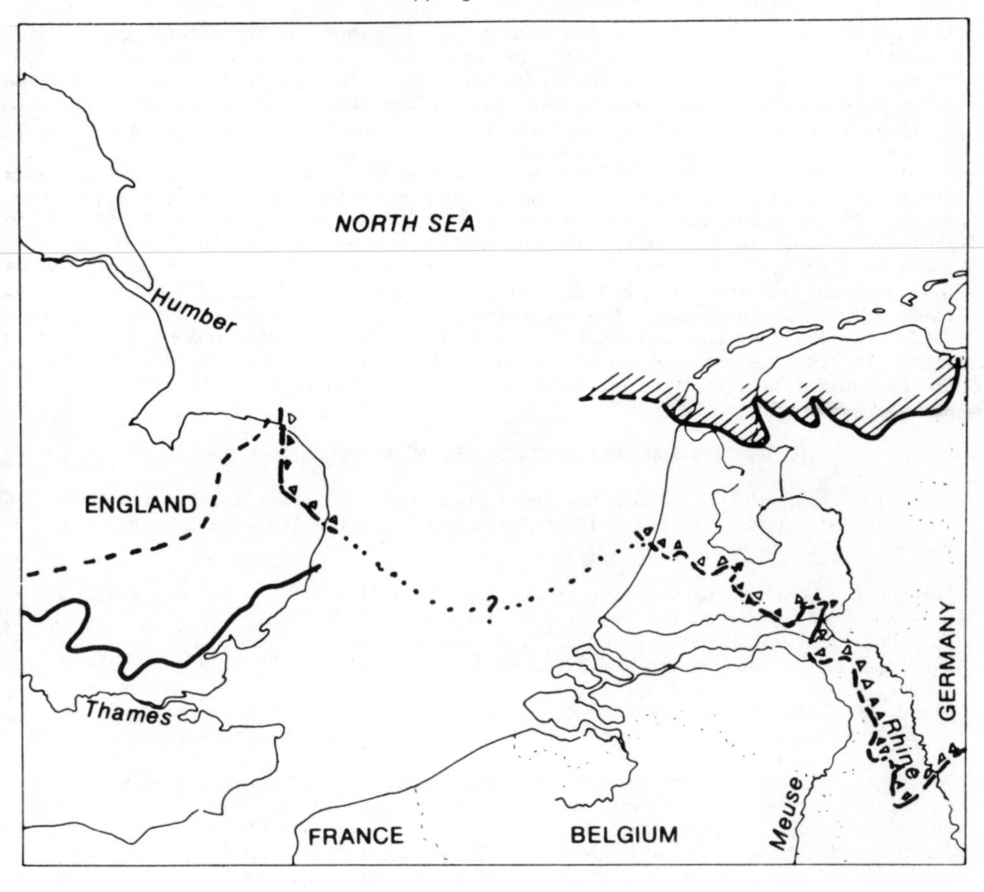

Fig. 2

The limit of the Chalky Boulder Clay in Britain (Bristow and Cox, 1973)

The limit of the Potklei in the Netherlands (Zagwijn, 1979)

The limit of the Wolstonian glaciation (West, 1977)

The limit of Scandinavian ice in Britain and the Netherlands (Bristow and Cox, 1973 and Oele and Schüttenhelm, 1979)

boulder clay of west Norfolk, equates it with similar deposits in Lincolnshire, and in particular demonstrates the field relationships of the Nar Valley Clay which with marine clays and the Nar Valley Freshwater Beds overlie the Chalky-Jurassic till complex. Straw's conclusions simply ignore the published evidence.

It is salutary to quote Shotton, Banham and Bishop (1977) who state: "Nowhere outside a stratigraphical table is an Ipswichian sequence known to occur vertically above a Hoxnian sequence, however. Further, although cold indicating fossils and cryoturbated sediments not uncommonly overlie and underlie Hoxnian and Ipswichian rocks, respectively, no Hoxnian section is capped by independently dated pre-Devensian beds, and no Ipswichian sequence rests upon undoubted post-Anglian deposits".

Despite this statement Shotton and others (1977) favour the interpretation of Mitchell and others (1973), in which the East Midlands Chalky Boulder Clay is younger than the East Anglian Chalky Boulder Clay. In support of this thesis they cite the Bubbenhall Clay which is said to be older than the Oadby Chalky Till (which in their view is of Wolstonian age) and therefore the Bubbenhall Clay is a deposit of the Anglian glaciation. They do admit, however, that this deposit has not been related to the Hoxnian and may therefore be an early glacial phase of the Wolstonian. The Paxford Gravel and the Stretton Sand which underlies the 'Wolstonian' till at Moreton in the Marsh has yielded a mammalian fauna which Shotton (1973) has ascribed to the Hoxnian. In view of the current controversy as to the nature of the Ipswichian interglacial (see below) the evidence of vertebrate faunas may be unreliable as a stratigraphical indicator and for this reason such an interpretation is contentious. In conclusion, the suggested correlation by Shotton, Banham and Bishop (1977 Fig. 19.3) of the Lowestoft Chalky Boulder Clay with the Bubbenhall Clay contrasts with the correlation of the Lowestoft Chalky Boulder Clay with the Oadby Chalky Boulder Clay of the East Midlands by Perrin, Davies and Fysh (1973).

THE HOXNIAN AND IPSWICHIAN INTERGLACIAL DEPOSITS

Since 1973 several papers have been published that deal specifically with these interglacial deposits in the British Isles and in particular with the problems raised by Bristow and Cox (1973).

Perhaps the most useful starting point is at Hoxne itself. In 1971 excavations were commenced there at the Oakley Park Pit under the direction of J.J. Wymer in collaboration with R. Singer and B. Gladfelter. Details of this work have been published in various papers and field guides (Gladfelter, 1975; Wymer, 1974; Singer, Wymer and Gladfelter, 1973). It was contended by Bristow and Cox that the clays and silts at this site (in common with all the other Hoxnian sites in East Anglia) were deposited as valley lake deposits developed in association with former subglacial tunnel valleys and that these valleys were the precursors of the modern drainage. It has not been possible to demonstrate a clear stratigraphical distinction between the 'Ipswichian' and Hoxnian Lake Clays in East Anglia. Wymer (1974) goes further and suggests that the doubt expressed by Evans (1971) on the validity of the correlation based on pollen deposits at Hoxne and those at Clacton may indicate the presence of an un-named warm period following the 'Anglian'. However, the upward passage of the Hoxnian Lake Clays at Hoxne from the Lowestoft Till with no marked stratigraphical break would seem to rule out this suggestion, but the archaeological evidence from Clacton does seem to indicate an earlier industry than that suggested by the tools at Hoxne. Just to confuse things further Wymer states that "if it were not for the pollen profile, the molluscan evidence would put the estuarine beds of Clacton into the Ipswichian"! So here we have a clear example of the Pleistocene dilemma. All the evidence appears to conflict.

Gladfelter (1975) has looked closely at the stratigraphy of the Hoxne site and has made additional studies of the sediments at a number of other 'Hoxnian' localities in East Anglia. At Hoxne he goes to great lengths to identify 'Wolstonian' deposits in the sediments that post-date the lake clays. He quotes Funnell (1955) as suggesting that the main terrace deposits of the Waveney Valley are Wolstonian based upon their "morphology, elevation and texture". By far the strongest reason for this conclusion was that they post-date the Hoxne Lake Clays! The terrace deposits in the Waveney Valley are said to pre-date the Ipswichian deposits (Gladfelter, 1975, p. 237) but no evidence is given for this conclusion. This suggestion contrasts strongly with the evidence from Swanton Morley in the Wensum Valley where Ipswichian clays are dredged up from excavations within a first terrace (Phillips, 1976). The 'Ipswichian' deposits in the Waveney Valley have been described by Sparks and West (1968). Samples were taken from a sewer trench.

A figure (p. 472) showing a long profile of the surface topography at the Wortwell site gives no indication of the stratigraphical relationships of the deposits at this point. It is suggested by the authors that the grab samples were: "derived from a low spread of gravel, possibly a terrace, at 48-49 feet (14.5 - 14.8 m) A.O.D., and post-date the higher terrace, the Homersfield Terrace, of this part of the Waveney Valley". From the section and profile given in this paper, beds 2 to 4 could easily be the feather edge of the terrace deposits which overlie Bed 1 (the source of the Ipswichian material). Further investigations are clearly needed in the Waveney Valley before the problem of the relationships of the interglacial deposits and the terrace sequence can be solved. Perhaps the most significant aspect of the arguments that Gladfelter presents is the tendency to validate the more generally accepted glacial stratigraphy by reference to that same stratigraphy. For example: "The surface sand and gravel overlying the Hoxne Lake Clays must be Wolstonian in age because they pre-date the Ipswichian Lake Clays" It is just this problem of a mental constraint imposed by a pre-conceived pattern of events that prevents an objective assessment of the evidence. What the evidence may indicate is that the 'Hoxnian' Lake Clays pre-date the 'Ipswichian' Clays in the Waveney Valley. It does not provide evidence of a Wolstonian glaciation in East Anglia.

Within the Wensum-Yare Valley system to the north of Hoxne four interglacial sites have been recorded. Swanton Morley, Barford, Dunston Common (Phillips, 1976) and Intwood Hall (Cox and Nickless, 1972). All sites are within the present valleys. The Swanton Morley and Dunston Common deposits are either within or beneath a first terrace. The former has been identified as Ipswichian, the latter Hoxnian. The lake clays at Barford are regarded by Phillips as Hoxnian and occur beneath a first terrace (Cox and Nickless, 1972), whilst the deposits at Intwood Hall may be Ipswichian or Flandrian in age (Turner in Cox, Poole and Wood, *in press*). This latter site is overlain by alluvium. This situation is similar to the site at Marks Tey where Turner (1970) has described a 'complete' sequence of Hoxnian Lake Clays. At both localities a hollow in the boulder clay is overlain by recent alluvium. The recognition of the Hoxnian and Ipswichian deposits in East Anglia is therefore based solely on the pollen profiles and is unrelated to either topography or stratigraphy. For a consideration of the reliance that may be placed upon these profiles we would do well to examine recent papers on the vertebrate fauna of the Ipswichian deposits.

Sutcliffe (1975) questioned the tendency of the many workers on Quaternary problems rigidly to follow the sequence recommended by the Geological Society of London (Mitchell and others, 1973). He made the point that by slavishly following this table, correlations are made with other sites and that any alternative sequence of events is not considered. He suggested that there may be more than one warm phase in the 'Ipswichian' and that similarly there may be more than one cold episode in the period designated 'Wolstonian'. Taking up Evans' point in discussion of Bristow and Cox (1973), that the Hoxnian deposits may include sediments equivalent to Evans' warm cycle 5W (Evans, 1971), Sutcliffe indicates that the mammalian evidence supports a more complex sequence of events than the 'Geological Society's' Stages, and that the palaeobotanical approach would not be able to distinguish between two closely spaced warm periods separated by a cold period. It is this 'hazard' that has led to the grouping of Thames terrace deposits of differing ages. West (1976) in reply to Sutcliffe, admits that the sequence of Mitchell and others may be an over simplification but comments that it was based upon the only firm evidence available at the time of publication. He agrees that too rigid an approach is undesirable, but it is clear from some of the papers I have considered here that it is precisely such an approach which forms a basis of much of the recent research. Sutcliffe (1976) gave a full account of the mammalian evidence for a subdivision of the Ipswichian deposits and concluded that there is evidence for at least two Ipswichian warm periods and therefore at least two sets of pollen assemblages that are apparently indistinguishable. Mayhew (1976) enters this discussion on the side of the 'orthodox' stratigraphy, suggesting that the differences in fauna described between various sites by Sutcliffe are indications of faunal change through a single Ipswichian Interglacial. In this he is supported by Stuart (1974) who believes that faunal changes during the Hoxnian and Ipswichian Interglacials were a response to vegetational changes which in turn reflected a rapidly changing climate.

This however is only a partial answer to Sutcliffe, for the problems under discussion do also arise from controversy over the stratigraphical interpretations. Singer, Wymer and Gladfelter (1973) submitted a number of samples from Hoxne for dating. Their results were not entirely conclusive but three samples suggested ages greater than 40,000 years.

The problem of the age of the interglacials was taken further by an application of uranium series dating using bone fragments (Szabo and Collins, 1975). Of their samples only one from Clacton gave what is described as a satisfactory closed system finite date, of 245,000 yr BP. Even this result must be treated with great caution as the evidence from their other samples illustrates the mobility of the elements measured. However, it is interesting in the light of the recent controversy on the Ipswichian Interglacial (see above) that their sample from Brundon appears to correlate with a warm period between the Hoxnian and Ipswichian.

In all the literature on the glacial stratigraphy of the British Isles there is no site described where an undoubted boulder clay can be seen overlying undoubted Hoxnian deposits and overlain by Ipswichian deposits. There are however four localities where evidence for a glacial deposit in this position might be sought.

NECHELLS AND QUINTON (Table 2)

These two locations in the Birmingham area are well documented (Kelly, 1964; Horton, 1974). Horton, from an examination of motorway boreholes, produces evidence that throws doubt on much of the earlier correlations for the glacial sequence of the Birmingham area. In particular he suggests that the boulder clay underlying the Quinton Interglacial deposits (Hoxnian) is the Upper Boulder Clay of Pickering (1957). Its relationship to the 'Hoxnian' type lake clays of Quinton cannot be reconciled with the interpretation of the Nechells (Hoxnian) interglacial deposit which Kelly (1964) thought by correlation of its overlying late glacial lake series to underlie the Aston ('Wolstonian') Till.

In a discussion of the Quinton evidence Horton (p. 18) says that "The borehole sequences provide no evidence of a pause in deposition during the infill of the proto-Tame depression at Aston. If this is proven then, in contradiction of the established sequence, the Aston Till may be older than the Nechells interglacial series, and hence the problem of correlating the Quinton sequence with the Aston till would no longer exist".

The question remains however, what is the correlative of the overlying boulder clay and glacial sand at Quinton? It could be a remnant of a 'Wolstonian' glaciation or even of the Devensian but without direct reference to the Ipswichian this problem cannot be resolved.

From the standpoint of Bristow and Cox (1973) the singular importance of Horton's work is that it gives clear evidence on palynological grounds for correlating the main glaciation of the Birmingham area with that of East Anglia.

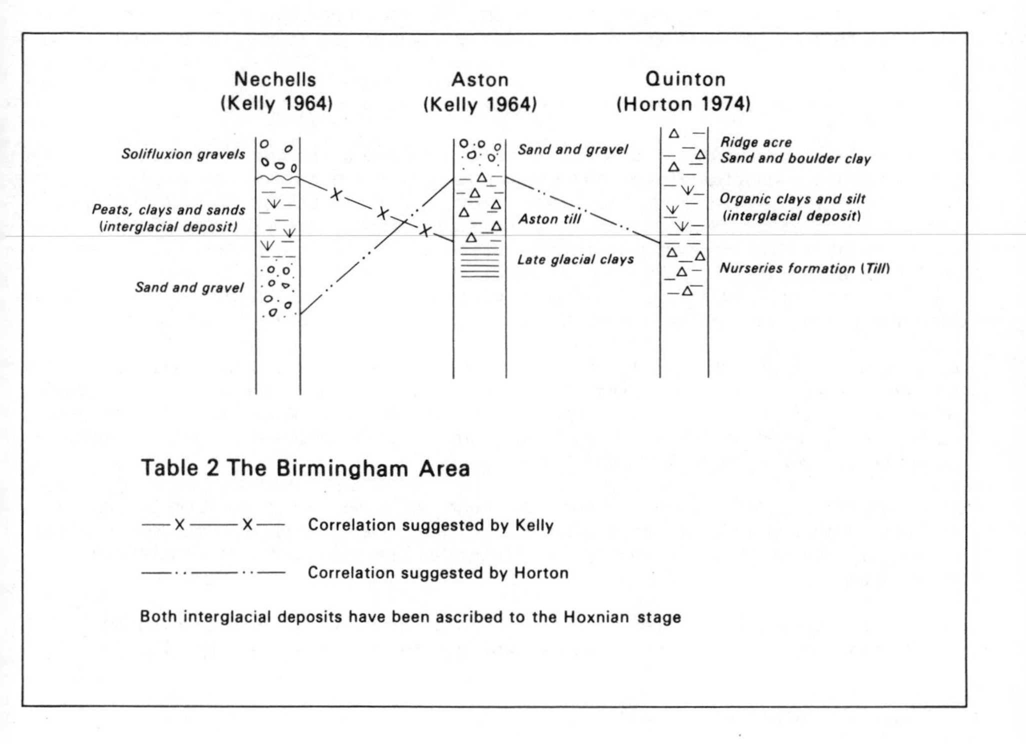

Table 2 The Birmingham Area

WELTON-LE-WOLD

Excavations at this site are described in detail by Alabaster and Straw (1976). Two till lithologies are recognised which are thought to represent two separate glacial events. Underlying the tills are glacial gravels which contain hand-axes similar in type to those found at Hoxne. If Straw's suggestion that one of the overlying tills can be correlated with the Basement Till of Holderness (Catt and Penny, 1966) is correct, then by inference this till is pre-Ipswichian but its relationship to the Hoxnian would remain unproven. However, as at Quinton there is no clear evidence relating the tills to palynologically dated Ipswichian deposits; the evidence for a 'Wolstonian' age for one of the Welton tills is solely based on the recognition by Straw of a period of valley development which separates the tills and is thought to have taken place during the Ipswichian Interglacial.

WEST NORFOLK

Gallois (1978) describes a glacial event which pre-dates the Hunstanton raised beach and post-dates the Nar Valley clays. Since this beach is thought by Gallois to be equivalent to the Sewerby raised beach of Holderness which has been described on a basis of mammalian fauna as Ipswichian (Boylan, 1967), this glacial event represented by the Lower Tottenhill Gravels may indicate a glaciation between Ipswichian and Hoxnian deposits. However, Gallois suggests that the gravels may be an early event in the Hunstanton (Devensian) glaciation.

In the light of the Birmingham evidence and the indirect nature of the Welton and Norfolk evidence the case for a Wolstonian glaciation is as yet not proven.

CORRELATION WITH THE CONTINENT

Zagwijn (1979) accepts the correlation between the Hoxnian and the Holsteinian as 'indisputable', quoting Turner and Francis in Bristow and Cox (1973) in support of this conclusion. This leads him to correlate the limit of the "Potklei" of the northern part of the Netherlands with the limit of the Chalky Boulder Clay of East Anglia. The former are varved clays associated with tunnel valleys whereas the latter is a true till clearly deposited directly from ice. In the same volume Oele and Schüttenhelm (1979) figure a map showing the maximum extension of Scandinavian land ice which they describe as Saalian in age. However they do not show the limit of Scandinavian ice as it has been mapped in East Anglia. If these two margins are compared (Fig. 2) the temptation to link them is overwhelming. Such a line would join like lithologies and it seems unlikely that the ice that reached so far south in the Netherlands left no deposits in East Anglia whereas the Elsterian ice came as far south as London but is only represented by some varved clays in the northern part of the Netherlands.

Zagwijn (1979, p. 38) states - "only one conclusion may be drawn from the above mentioned observations of Bristow and Cox (1973), namely that the maximum glacial advance in eastern Britain was during Elsterian times in terms of continental stratigraphy".

It is here suggested that there is indeed an alternative to this conclusion i.e. the Hoxnian deposits are not of equivalent age to the Holstein deposits!

SUMMING UP

I commenced this survey with a partisan interest but I have tried to weigh the evidence as objectively as possible. In vain have I searched for the line separating the 'Wolstonian' and 'Anglian' glaciations. Straw's work (1979) is perhaps the only evidence for a till that separates Hoxnian from Ipswichian deposits, but the evidence for this is tenuous and indirect. The evidence of Horton (1974) and Perrin, Davies and Fysh (1973) provides clear support for contention that the 'Wolstonian' boulder clays of the Midlands are equivalent to the 'Anglian' boulder clays of East Anglia. The detailed work at Hoxne and the other East Anglian Interglacial sites has produced no evidence to refute the interpretation of the East Anglian glacial stratigraphy outlined by Bristow and myself in 1973.

However, my own view is that the evidence suggests a long period of warm climate broken by short cold periods, with only two glacial events since the beginning of the Middle Pleistocene - one which preceded the interglacial deposits and one that follows them (i.e. pre-Hoxnian = 'Anglian' and post-Ipswichian = Devensian). It is perhaps unreasonable to ignore the problems that this interpretation introduces with respect to the Continental sequences but we must first establish a firm basis for our own stratigraphy before such a correlation can have any validity. Indeed, it is precisely the use of a rigid framework imposed from outside that prevents a truly scientific assessment of the local evidence. In conclusion I can do no better than quote Gallois (1978) who, in explanation of his deliberate avoidance of the British Standard Stages proposed by the Geological Society Quaternary Era Subcommittee (Mitchell and others 1973), said .... "the establishment of a standard sequence by adding together type sections which are geographically widely separated defies all the rules of stratigraphical nomenclature".

REFERENCES

Alabaster, C. and A. Straw (1976). The Pleistocene context of faunal remains and artefacts discovered at Welton-le-Wold, Lincolnshire. *Proc. Yorks. geol. Soc.*, 41, 75-94.

Bowen, D.Q. (1978). *Quaternary Geology. A stratigraphic framework for multidisciplinary work.* Pergamon Press, Oxford. xi + 221 pp.

Boylan, P.J. (1967). The Pleistocene mammalia of the Sewerby-Hessle buried cliff, East Yorkshire. *Proc. Yorks. geol. Soc.*, 35, 375-420.

Bristow, C.R. and F.C. Cox (1973). The Gipping Till: a reappraisal of East Anglian glacial stratigraphy. *J. geol. Soc. Lond.*, 129, 1-37.

Catt, J.A. and L.F. Penny (1966). The Pleistocene deposits of Holderness, East Yorkshire. *Proc. Yorks. geol. Soc.*, 35, 375-420.

Cox, F.C. and E.F.P. Nickless (1972). Some aspects of the glacial history of central Norfolk. *Bull. geol. Surv. Gt Br.*, 42, 78-98.

Cox, F.C. and E.F.P. Nickless (1974). The Glacial Geomorphology of Central and North Norfolk. *East Midl. Geogr.*, 6, 42.

Cox, F.C., E.G. Poole, and C.J. Wood (in press). *The Geology of the country around Norwich.* Mem. geol. Surv. Gt Br.

Edmonds, E.A. (1977). Quaternary Map of the United Kingdom, Scale 1:625,000. *Inst. geol. Sci.*

Evans, P. (1971). Towards a Pleistocene time-scale. Part 2 of *The Phanerozoic Time-Scale. - A Supplement.* Geol. Soc. Lond. Spec. Publ., 5, 123-356.

Funnell, B.M. (1955). An account of the geology of the Bungay district. *Trans. Suffolk Nat. Soc.*, 9, 115-126.

Gallois, R.W. (1978). The Pleistocene History of West Norfolk. *Bull. geol. Soc. Norfolk.* 30, 3-38.

Gladfelter, B.G. (1975). Middle Pleistocene Sedimentary Sequences in East Anglia (United Kingdom). In K.W. Butzer and G.L. Isaac (Eds), *After the Australopithecines: Stratigraphy, Ecology and Culture change in the Middle Pleistocene.* Mouton, The Hague.

Horton, A. (1974). *The sequence of Pleistocene deposits proved during the construction of the Birmingham motorways.* Inst. geol. Sci. Rept, 74/11, iv + 21 pp.

Kelly, M.R. (1964). The Middle Pleistocene of North Birmingham. *Phil. Trans. R. Soc.*, B. 247, 533-592.

Mayhew, D.F. (1976). Comments on the British Glacial-Interglacial sequence. *Quaternary Newsl.*, 19, 8-9.

Mitchell, G.F., L.F. Penny, F.W. Shotton and R.G. West (1973). A correlation of Quaternary deposits in the British Isles. *Geol. Soc. Lond. Spec. Rept*, 4, 99 pp.

Oele, E. and R.T.E. Schüttenhelm (1979). Development of the North Sea after the Saalian glaciation. In E. Oele, R.T.E., Schüttenhelm and A.J. Wiggers. *The Quaternary History of the North Sea.* Acta Univ. Ups., Symp. Ups. Annum. Quingentesimum Celbrantis, 2, ISBN 91-554-0495-2.

Perrin, R.M.S., H. Davies and M.D. Fysh (1973). Lithology of the Chalky Boulder Clay. *Nature Phys. Sci.*, 245, 101-104.

Perrin, R.M.S., J. Rose and H.D. Davies (1979). The Distribution, Variation and Origins of pre-Devensian Tills in Eastern England. *Phil. Trans. R. Soc.*, B 1024, 535-570.

Phillips, L.M. (1976). Pleistocene vegetational History and Geology in Norfolk. *Phil. Trans. R. Soc.*, B 275, 215-286.

Pickering, R. (1957). The Pleistocene geology of the South Birmingham area. *Q. Jl geol. Soc. Lond.*, 113, 223-237.

Shotton, F.W. (1973). A mammalian fauna from the Stretton Sand at Stretton-on-Fosse, South Warwickshire. *Geol. Mag.*, 109, 473-476.

Shotton, F.W., P.H. Banham and W.W. Bishop (1977). Glacial-interglacial stratigraphy of the Quaternary in Midland and eastern England. In F.W. Shotton (Ed.), *British Quaternary Studies: Recent Advances.* Oxford Univ. Press. pp. 267-282.

Singer, R., J. Wymer and B.G. Gladfelter (1973). Radiocarbon dates from Hoxne, Suffolk. *Geol. J.*, 81, 508-509.

Sparks, B.W. and R.G. West (1968). Interglacial deposits at Wortwell, Norfolk. *Geol. Mag.*, 105, 471-481.

Straw, A. (1973). The Glacial Geomorphology of Central and North Norfolk. *East Midl. Geogr.*, 5, Pt 7, 39, 333-353.

Straw, A. (1979). The geomorphological significance of the Wolstonian glaciation of Eastern England. *Trans. Inst. Br. Geogr.*, 4, 540-549.

Stuart, A.J. (1974). Pleistocene history of vertebrate fauna. *Biol. Rev.*, 49, 225-266.

Sutcliffe, A.J. (1975). A Hazard in the Interpretation of Glacial-Interglacial Sequences. *Quaternary Newsl.*, 17, 1-3.

Sutcliffe, A.J. (1976). The British Glacial-Interglacial Sequence. *Quaternary Newsl.*, 18, 1-7.

Szabo, B.J. and D. Collins (1975). Ages of fossil bones from British Interglacial Sites. *Nature, Lond.*, 254, 680-681.

Turner, C. (1970). The Middle Pleistocene deposits at Marks Tey, Essex. *Phil. Trans. R. Soc.*, B 257, 373-440.

West, R.G. (1976). The British Glacial-Interglacial Sequence. *Quaternary Newsl.*, 18, 1.

West, R.G. (1977). *Pleistocene Geology and Biology with especial reference to the British Isles.* Longman, London. 440 pp.

Woodland, A.W. (1970). The buried tunnel-valleys of East Anglia. *Proc. Yorks. geol. Soc.*, 37, 521-578.

Wymer, J.J. (1974). Clactonian and Acheulian industries in Britain - their chronology and significance. *Proc. Geol. Ass.*, 85, 391-422.

Zagwijn, W.H. (1979). Early and Middle Pleistocene coastlines in the southern North Sea Basin in E. Oele, R.T.E. Schüttenhelm and A.J. Wiggers, *The Quaternary History of the North Sea.* Acta Univ. Ups. Symp. Univ. Ups. Annum Quingentesimum Celbrantis, 2, ISBN 91-554-0495-2.

Dr F.C. Cox,
Institute of Geological Sciences,
Ring Road, Halton,
Leeds. LS15 8TQ.

5. A Contribution to the Pleistocene of Suffolk: An Interglacial Site at Sicklesmere, Near Bury St. Edmunds

R. G. West

INTRODUCTION

At some Pleistocene sites a few pollen analyses of organic sediments will give clear evidence of vegetational history and age. At other sites, pollen analyses, though they may afford evidence of vegetational history, will provide no evidence of age (e.g. Pleistocene sediments at Speeton, Yorks., (Penny & Rawson, 1969)). An example of the former occurred at a site at Sicklesmere, 3 km south-east of Bury St. Edmunds (G.R. TL 874609).

Like many other sites in East Anglia studied palaeobotanically in the last thirty years, the site at Sicklesmere (Fig. 1) was recorded by the officers of the Geological Survey in the last century. Bennett and Blake (1886) described the site at Oak's Kiln in the following terms.

Post-Glacial	(Stiff brownish-grey loam, greenish at bottom, 14 feet.
	(
Drift	(Black peat with bits of wood, 2 or 3 feet, not bottomed.

The related Old Series map (51 S.E.) shows at the site a small area of brickearth almost surrounded by boulder clay.

RESULTS

A re-investigation of the site was made in 1952 and 1953, during a survey of possible interglacial sites in Suffolk. The pit, still present, lies on the 175-foot contour of the 1:25000 map. Two boreholes were made at that time with a single-spiral auger, one (1 below) at the base of a section at the north-west side of the pit, and the second (2 below) 20 m to the east. The results were as follows.

1.	Section	0 - 50 cm	Soil or disturbed
		50 - 350 cm	Red and grey mottled sandy clay and sand
	Borehole	0 - 50 cm	Disturbed
		50 - 100 cm	Red and grey mottled sandy clay with occasional flints
		100 - 130 cm	As above, with small chalk pebbles
		130 - 155 cm	Buff clay with a little sand, becoming dark towards 150 cm.
		155 - 165 cm	Dark brown clay-mud with a little sand

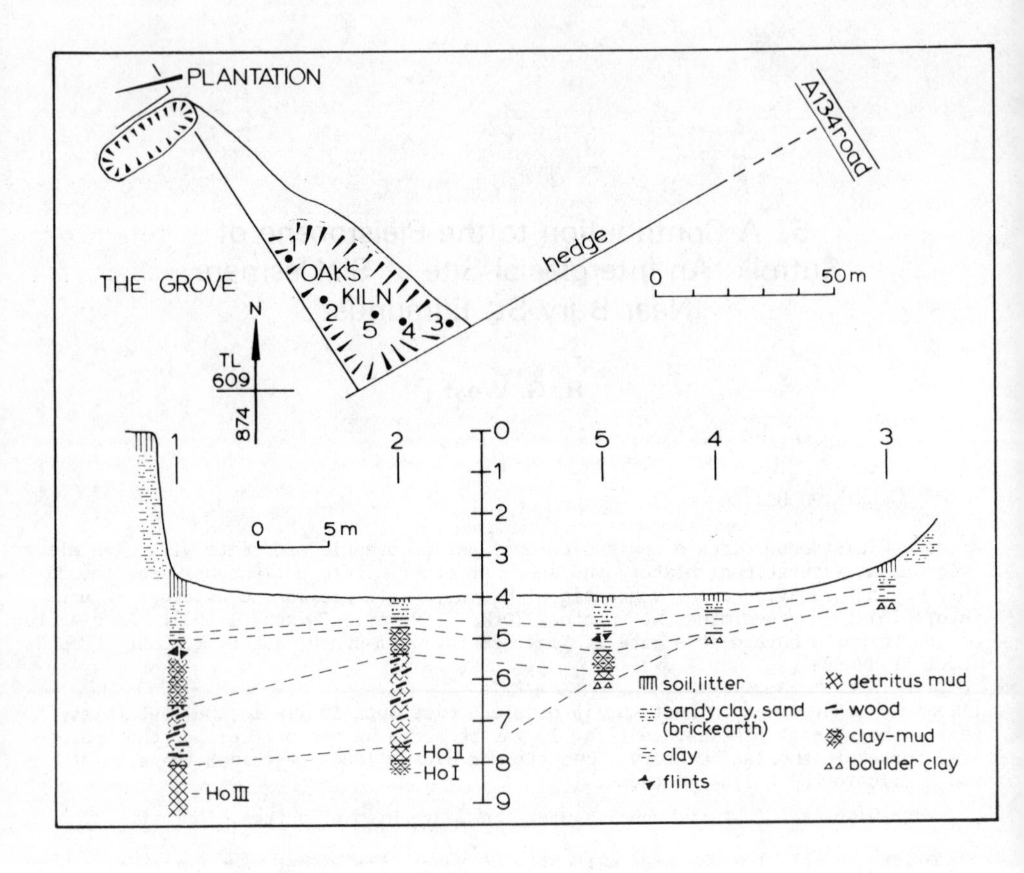

Fig. 1. Oak's Kiln, Sicklesmere and section through the deposits.

165 - 175 cm	Mottled grey and brown clay-mud with small flints and a little sand
175 - 200 cm	As above, becoming darker and more coarse towards the base
200 - 212 cm	Brown clay-mud with a little sand and plant remains
212 - 225 cm	As above, but getting greyer towards the base
225 - 280 cm	Grey brown clay-mud
280 - 285 cm	Red brown slightly sandy organic clay-mud with wood
285 - 305 cm	Lost from auger
305 - 450 cm	Medium coarse detritus mud, humified with wood
450 - 570 cm	Finer detritus mud with less wood

Boring stopped, auger not retaining sediment

2. Borehole

0 - 5 cm	Leaf litter and rubbish
5 - 25 cm	Red sandy clay
25 - 70 cm	Grey slightly sandy clay
70 - 125 cm	More organic and brown clay
125 - 135 cm	Transition
135 - 360 cm	Brown medium coarse detritus mud, humified with small pieces of wood

360 - 390 cm	Fine detritus mud
390 - 425 cm	Dark green slightly sandy clay-mud

Boring stopped, auger not retaining sediment

Three more single-spiral auger holes, nos. 3, 4 and 5, were made further east in 1980. The results of these were as follows.

3.	Borehole	0 - 30 cm	Leaf litter and soil
		30 - 55 cm	Red and grey sandy clay
		55 - 90 cm	Grey sandy clay
		90 - 135 cm	Grey-brown sandy clay with flints and chalk pebbles (boulder clay)
4.	Borehole	0 - 15 cm	Leaf litter and soil
		15 - 35 cm	Red and grey streaky sandy clay
		35 - 50 cm	Grey sandy clay, darker near base
		50 - 70 cm	Purple clay
		70 - 90 cm	Grey sandy clay
		90 - 120 cm	Grey-brown sandy clay with flints and chalk pebbles (boulder clay)
5.	Borehole	0 - 12 cm	Leaf litter and soil
		12 - 70 cm	Red and grey sandy clay
		70 - 80 cm	Similar but greyer
		80 - 90 cm	Purple clay
		90 - 100 cm	Grey clay with small flints
		100 - 130 cm	Grey-brown clay
		130 - 140 cm	Brown clay-mud
		140 - 180 cm	Grey-brown clay-mud
		180 - 195 cm	Dark brown wet crumbly mud
		195 - 220 cm	Dark grey sandy clay with flints and chalk pebbles (boulder clay).

Figure 1 shows a sketch section giving the location of the section and boreholes, and also the resulting stratigraphy. The stratigraphical section indicates the presence of a basin in the boulder clay deepening to the north-west. Towards the centre of the basin, the sequence of sediments is first a dark green clay-mud, then a humified detritus mud with wood fragments in its upper part, then a series of grey and brown clays with a mud component, and finally a red and grey sandy clay with flints, which is presumably the brickearth which was won from the pit in the last century. Towards the edge of the basin a mud directly overlies the boulder clay (borehole 5), with clays again overlying the mud. The whole sequence indicates filling of the basin by clay-mud then detritus mud, with a later resumption of clay deposition.

Three pollen analyses were made at the time of the investigation of samples taken from the auger, at a depth of 520 cm in borehole 1 and from depths of 375 cm and 425 cm in borehole 2. The results are given in Table 1. The upper organic sediments were very humified and this should be borne in mind in interpretation of the pollen spectra.

520 cm, borehole 1. Detritus mud with wood. The analysis shows a spectrum with *Pinus*, *Abies*, *Alnus*, *Tilia* and *Corylus* as the most important taxa represented. The high *Alnus* pollen frequencies, taken together with high frequencies of Filicales spores and the presence of wood in the detritus mud indicate the local presence of fen carr. The presence of *Abies* pollen, with *Corylus*, and low frequencies of *Ulmus* and *Quercus* pollen indicate a late temperate stage forest of an interglacial. Similar spectra have been recorded from the Hoxnian III substage, near the III a/b

TABLE 1 Pollen Analyses from Sicklesmere

	Borehole 1, 520 cm		Borehole 2, 375 cm		Borehole 2, 425 cm	
		% TLP*		% TLP		% TLP
Betula	6	3	25	10	120	76
Pinus	22	12	26	10	10	6
Ulmus	6	3	9	4	10	6
Quercus	3	2	20	8	7	4
Tilia	21	11	3	1	-	-
Alnus	84	46	64	25	2	1
Fraxinus	-	-	1	0.4	-	-
Picea	-	-	2	1	-	-
Abies	9	5	-	-	-	-
Ilex	-	-	1	0.4	-	-
Corylus	28	15	75	29	3	2
Salix	-	-	-	-	1	0.6
Type x	-	-	1	0.4	1	0.6
Hedera	3	2	-	-	-	-
Gramineae	-	-	19	7	1	0.6
Cyperaceae	-	-	8	3	1	0.6
Compositae Tubuliflorae	-	-	1	0.4	-	-
Plantago major/media	-	-	1	0.4	-	-
Ranunculus	-	-	-	-	1	0.6
	182		256		157	
		% TLP + aquatics		% TLP + aquatics		
Alisma	-	-	1	0.3	-	-
Azolla (massulae)	-	-	2	0.7	-	-
Typha latifolia	1	0.5	13	5	-	-
	1		16			
		% TLP + ferns		% TLP + ferns		% TLP + ferns
Filicales	103	36	321	55	11	6.5
Polypodium	-	-	1	0.2	-	-
Osmunda	3	1	1	0.2	-	-
	106		323		11	

* total land pollen

boundary, as at Marks Tey (Turner, 1970), except that the frequency of *Tilia* pollen is much higher, possibly related to the high degree of humification of the sediment. The only other forest assemblage with *Abies* in the East Anglian Pleistocene is in the Cromerian III b substage, where *Abies* is accompanied by substantial frequencies of *Picea* and *Carpinus* pollen, unlike the 425 cm analysis under discussion.

375 cm, borehole 2. Fine detritus mud. The analysis shows again a forest spectrum with *Betula*, *Pinus*, *Ulmus*, *Quercus*, *Abies* and *Corylus* as the main taxa, with lower frequencies of *Tilia* and the characteristic tricolpate type x pollen of the Hoxnian (Phillips, 1976). The pollen spectrum resembles those from Hoxnian II c at Hoxne (West, 1956) and Marks Tey (Turner, 1970), except for the absence of *Taxus*, not recognized at the time these spectra were counted. The abundance of Filicales spores and pollen of *Typha latifolia* indicates the nearness of fen carr and reed-swamp. Abundant massulae and megaspores of *Azolla filiculoides* were found in washing the sample, also indicating a shallow water environment. *A. filiculoides* is abundant at Hoxne in the later part of Hoxnian II c sediments. It is not recorded in the younger temperate stages. It is notable that 7% Gramineae pollen occurs in this sample. At Hoxne and Marks Tey such frequencies are not present in Hoxnian II c except in the deforestation phase towards the end of Hoxnian II c and it is possible that this pollen spectrum at Sicklesmere belongs to this episode.

425 cm, borehole 2. Dark green clay-mud. The analysis shows again the presence of forest, but with very high frequencies of *Betula* pollen. The assemblage is similar to those from the later part of Hoxnian I at Hoxne and Marks Tey, and indicates the presence of birch woodland at a time, as indicated by clay-mud sedimentation, when the water body was not so shallow as later in the temperate stage.

DISCUSSION

The pollen analyses thus indicate the presence of a Hoxnian forest sequence in sediment indicating progressive shallowing, with clay-mud in the older part of the temperate stage, then fine detritus mud, then coarser detritus mud with wood in the later part of the stage. The succeeding clays indicate a further change in the lake environment, their deposition possibly being related to a rise in water level and increased run-off from the surrounding boulder clay. The origin of the overlying 'brickearth' requires further study. An analysis of the Westley 'brickearth', occurring at about the same height 5.5 km to the north-west (Baden-Powell & Oakley, 1952) indicated an aeolian origin.

The position of this Hoxnian lake deposit in the local Pleistocene stratigraphy is as follows. It overlies a chalky boulder clay, dark grey in the centre of the basin, but grey-brown in the marginal parts where it is nearer the surface. This boulder clay must be correlated with the Lowestoft Till of East Anglia. The lake deposit underlies nearly 5 m of red and grey mottled sandy clay recorded by earlier geologists as 'brickearth'. Since this 'brickearth' appears to be post-Hoxnian, it is likely to be Wolstonian in age.

Further investigation of the site should reveal more detail of the relationship of the lake deposit to the 'brickearth', important because elsewhere in the neighbourhood the 'brickearth' contains Palaeolithic artefacts (Baden-Powell & Oakley, 1952), as well as detail of Hoxnian vegetational history and water level changes. Meanwhile, it is clear that the stratigraphical sequence at Sicklesmere is similar in outline to that at Hoxne (West, 1956) and extends our knowledge of the Anglian, Hoxnian and post-Hoxnian (with a Palaeolithic industry) stages into west Suffolk.

ACKNOWLEDGEMENTS

I am grateful to Mr O.R. Oakes of Breckey Ley House, Nowton, for access and to E. Dahl, F.H. Perring and A.G. Smith for field assistance.

REFERENCES

Baden-Powell, D.F.W. and K.P. Oakley (1952). Report on the re-investigation of the Westley (Bury St. Edmunds) skull site. *Proc. prehist. Soc. for 1952*, 1-20.

Bennett, F.J. and J.H. Blake (1886). *The geology of the country between and south of Bury St. Edmunds and Newmarket.* Mem. geol. Surv. Eng. Wales.

Penny, L.F. and P.F. Rawson (1969). Field meeting in East Yorkshire and North Lincolnshire. *Proc. Geol. Ass.*, 80, 193-218.

Phillips, L. (1976). Pleistocene vegetational history and geology in Norfolk. *Phil. Trans. R. Soc.*, B 275, 215-286.

Turner, C. (1970). The Middle Pleistocene deposits at Marks Tey, Essex. *Phil. Trans. R. Soc.*, B 257, 373-440.

West, R.G. (1956). The Quaternary deposits at Hoxne, Suffolk. *Phil. Trans. R. Soc.*, B 239, 265-356.

Professor R.G. West, F.R.S.,
Sub-department of Quaternary Research,
University of Cambridge,
Downing Street,
Cambridge. CB2 3EA.

6. The Devensian Glaciation on the North Welsh Border

D. S. Peake

INTRODUCTION

On the north Welsh border the Devensian Glaciation had been preceded by two earlier and greater glaciations in which controlling conditions differed from those of the last glaciation to the extent that north Welsh ice had had complete ascendancy each time in at least the southern part of the Cheshire basin, for the glaciers which then over-rode the Wenlock-Pennine watershed and advanced far down the present middle Severn valley were both north Welsh in type (Wills, 1937). Contemporary northern ice presumably accumulated in the Irish Sea basin, and may well have commenced encroachment on the Cheshire plain. The pressure of this ice could have been responsible for the south eastward deflection of either of these north Welsh ice sheets. There is no proof of this, however, except for scant Irish Sea erratics in the Midlands older drift, south of the limit reached in the Devensian Glaciation (Wills, 1937; Morgan, 1973).

Evidence of the earlier Welsh advances in the Cheshire basin is also lacking, but a recent geophysical survey in the Shrewsbury area (Finch, 1976) has shown the existence there of a deep linear sub-drift hollow. Reaching over 50 m below sea level in relatively soft Triassic sandstone at its western end and trending east south eastwards, this trough, termed by the author the Severn Trench (Fig. 1), is considered by him to be a glacially over-deepened feature. Even if scoured along a pre-existing valley line the direction strongly suggests that Welsh ice rather than Irish Sea ice was in operation. The most likely gouging force to have been in action would have been one or both of the two earlier great Welsh glaciers issuing from the upper Severn mountain valley system and deflected eastward between the Breiddens and the Nesscliff scarp (384193). At best this view can be only conjectural, but the postulation of possible pre-Devensian deposits south of the Severn Trench (Shotton, 1962) has already expressed with good reason an idea along these lines. Original deposits from the earlier glaciations may still be preserved at the base of the 100 m infilling in the deeper parts of the Severn Trench.

On the other hand the undulating floor of the Dee Trench, the very long north-south groove excavated mainly in Triassic sandstone along the north Welsh border (Fig. 1) may well have been deepened locally by all the Irish Sea glaciers in their uphill southward advance. A geophysical survey (Bilbo, 1975) has extended southward beyond Ellesmere the part of the trench floor estimated to be below sea level by Wills (1912), but it does not substantiate the suggested great curve of a pre-glacial Dee valley between Chirk and Ellesmere. Wills carefully pointed out that this unproven

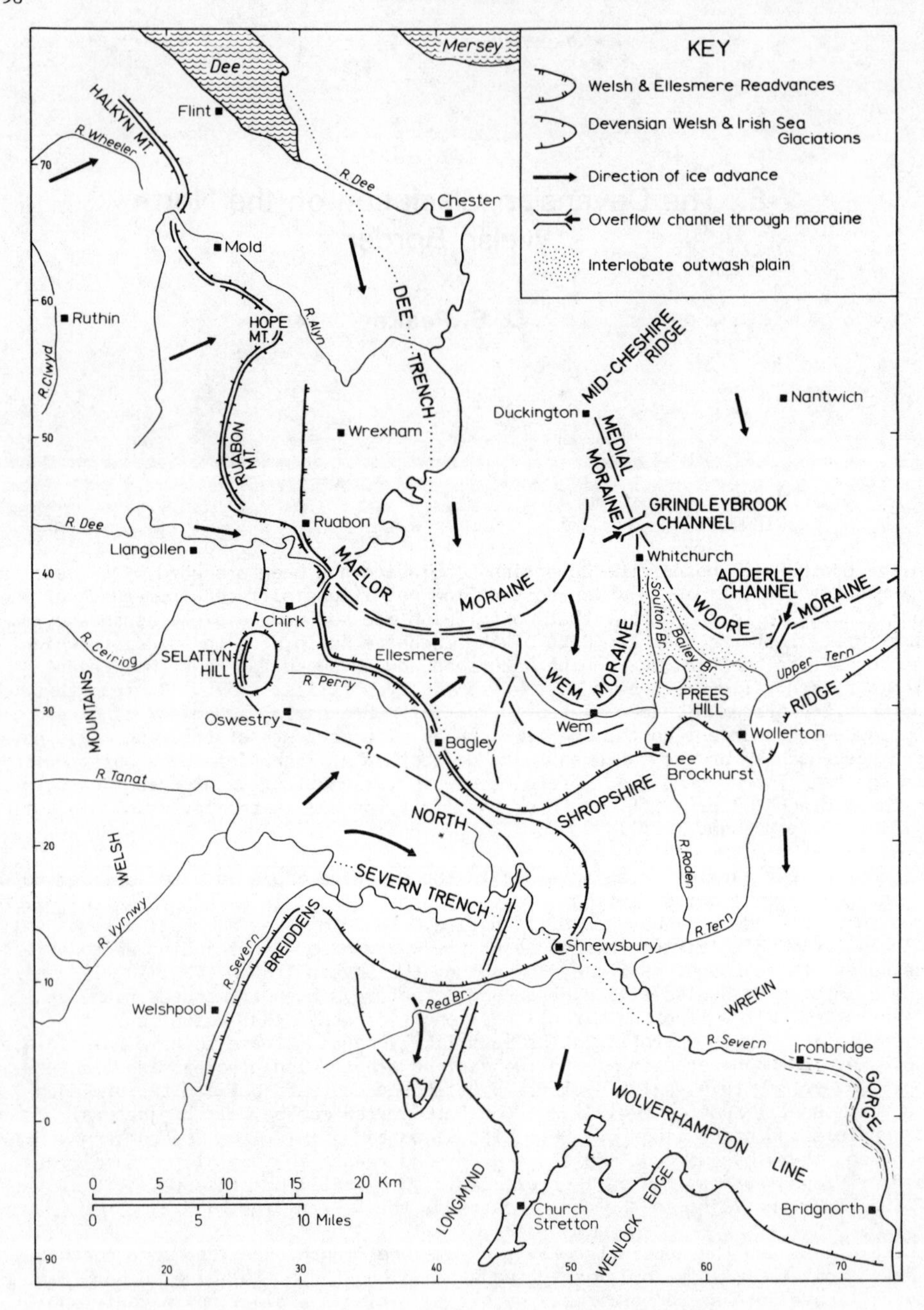

Fig. 1. The Devensian Glaciation on the North Welsh Border.

line was merely hypothetical, and it now seems likely that the buried valley which he traced southward through Chirk resulted from a temporary glacial diversion of the Dee along the foot of the hills (Peake, 1979).

THE DEVENSIAN MAIN GLACIATION

It was Lamplugh who named the last major glaciation the 'Irish Sea Glaciation' in recognition in the British Isles of the great central reservoir of ice collecting in the Irish Sea basin. Fed by ice from the mountain massifs to the north the ice accumulated to such a thickness that in its southward advance between Britain and Ireland it had the power to move uphill across surrounding lowland tracts. In thrusting against the ice-covered mountain barrier of north Wales an eastern lobe parted from the main glacier in the vicinity of the Great Orme and pressed round the coastal area of north east Wales into the Cheshire basin. West of the Vale of Clwyd its tills are thickly plastered against the ancient cliff line in Palaeozoic rocks along the coast, burying the rock platform, probably of marine origin, at its foot (Strahan, 1885). East of the vale at the seaward end of the Clwydian Range and the Lower Carboniferous tract the cliffs protrude through the drift for a short distance only; from near Gronant (092832), where the Cefn-y-fedw Sandstone bluffs curve to a south easterly trend along the Dee estuary, the cliff line is completely covered by the deep tills of the Irish Sea ice. Cut into Coal Measures here this older buried coastline, partly degraded, continues south eastward below the slopes of Halkyn Mountain through Mostyn (Wedd, 1912) at least to the neighbourhood of Flint (J. Everett, pers. comm.).

The Irish Sea drifts are characterized by the marine shells from the sea bed which were carried inland by the encroaching ice, as well as by northern erratics. Advancing up the Cheshire plain the glacier's western edge moved roughly along the strike of the solid rocks dipping eastward from north east Wales. In the central section of the north Welsh border, where the Carboniferous presented long easily mounted dipslopes to the oncoming ice, the presence of northern erratics and marine shells in sands and gravels high on the cuestas indicates the great altitudes reached on them by the Irish Sea ice. The eastern dipslopes of the two classic examples of this, Ruabon Mountain (Wedd, Smith and Wills, 1928) and Selattyn Hill (Wedd and others, 1929) (Fig. 1), are masked by thin Irish Sea till. Both cuestas are well above mountain height, the resistant Cefn-y-fedw Sandstone forming the crests of their westward facing escarpments. Welsh ice sheets moving eastward from the Lower Palaeozoic mountains appear to have met the Irish Sea ice along either ridge-top, for their tills are mutually exclusive. On the north Welsh border Welsh erratics, carried downhill, may occur in varying degree in the Irish Sea drifts, but there are no records of northern erratics and marine shells to the west of the main Carboniferous crests.

A difference in the division between the areas of influence of the opposing ice sheets is evident farther north beyond the Llanelidan Fault belt, due to a variation in topography. Well to the east of the main Carboniferous escarpment here, two long anticlinal limestone and grit ridges, Halkyn Mountain and Hope Mountain (Fig. 1), form barriers which prevented westward penetration of the Irish Sea ice. The Welsh ice, having over-ridden the lengthy main escarpment, appears to have filled the dipslope to a depth sufficient to debar the Irish Sea ice in general from all but the outer slopes of the high ground of this northern part of north east Wales (Wedd and King, 1924).

In the southern section of the north Welsh border also it seems likely that the main Carboniferous escarpment was over-ridden by contemporaneous Welsh ice, for south of the general latitude of Oswestry there are no records of Irish Sea drift on the most southerly cuestas which end in Llanymynech Hill (262222), the latter being well below mountain height due to the absence here of the Cefn-y-fedw Sandstone. The escarpment dies away where the three great mountain valleys of the

Severn and its tributaries the Vyrnwy and Tanat open to the piedmont plain. Their powerful glaciers coalesced into one great sheet, in all probability the main thrust initially being north of east from the deep wide upper Severn valley, directly opposing the Irish Sea ice front advancing from the north (Fig. 1). Despite the local obstacle to progress suggested by the distribution of Welsh till (Wedd and others, 1929; Pocock and others, 1938) the Irish Sea ice sheet reached the irregular southern rim of the Cheshire basin and mounted its slopes to heights of 350 m on the Longmynd (Wright, 1968) and 250 m on Wenlock Edge (Pocock and others, 1938). These elevations, comparable to those attained along the Welsh border, demonstrate the great thickness of the ice lobe which covered the Cheshire plain. Surmounting the Wenlock-Pennine watershed to a limit known as the Wolverhampton Line (Fig. 1), its outwash down the valleys south of the divide built up valley trains designated as the Main Terrace of the Severn (Wills, 1938). On the retreat of the ice sheet from the watershed the overflow from meltwater impounded by the ice along the northern slopes cut the Ironbridge gorge to a depth sufficient to retain permanently the upper Severn drainage.

THE LATE DEVENSIAN READVANCES

Although it is now generally accepted that the major ice advance in the last glaciation took place in the Late Devensian, Shotton (1977a) has argued the probability that advances and retreats of the glaciers in this period were not necessarily synchronised throughout their domain. It is also unlikely that in coalescing and advancing to their maxima ice sheets of the volume and power of the main Devensian glaciers would have waxed and waned without fluctuation.

In the southern section of the north Welsh border, in an area within the Severn catchment north west of Shrewsbury, it is well established that Welsh drift locally overlies Irish Sea drift (Pocock and Wray, 1925; Wedd and others, 1929; Pocock and others, 1938). Recording the superimposition of Welsh till on outwash gravels attributable to the main Devensian ice retreat, Whitehead (in Pocock and others, 1938) wrote of the inferred glacier advance as a Welsh readvance, regarding it as a late fluctuation of ice from the valleys of the Severn, Vyrnwy and Tanat. The now adopted term Welsh Readvance (Fig. 1) is preferable to the name Little Welsh Glaciation used by Wills (1938) at the time when the status of the preceding interstadial was still indeterminable, well aware though he was of the likelihood of its not being an interglacial.

The Welsh Readvance reached a line extending in general south eastward from Ruabon to Shrewsbury (Fig. 1). Acceptance of its status raises the possibility of contemporaneous reactivation of the Irish Sea ice, with the probability of readvance from the north as suggested on the map, or at least a prolonged pause in northward retreat. This conclusion is further supported by the mixed nature of the surface tills in a zone along the depicted line of contact (Fig. 1), with a north west to south east ridge of till at Bagley. Pocock and Wray (1925) claim a pronounced drift ridge between Ellesmere and Whitchurch to have been 'a terminal moraine thrown down during a pause in the retreat', and amongst other authors of similar views Wills (1948) named an Ellesmere moraine as part of a second glacial maximum on his map of the Newer Drifts. Many writers have correlated the great drift ridge with a similar drift belt east of Whitchurch, termed the Woore moraine (Fig. 1). An examination of the main features of the border landscape has some bearing on an interpretation of the drift sequence in the region.

Some authors, notably Poole and Whiteman (1966) and Worsley (1970), have given prominence to the effect of the mid-Cheshire Triassic ridge on the retreat of the Irish Sea ice, few references having been made to its dominant role in ice advances. Although oblique at its southern end to the main direction of ice movement southward, initially it divided encroaching ice fronts into twin east and west lobes. From near Duckington (Fig. 1) at its southern extremity the resultant undulating medial moraine formed between the two advancing ice lobes now stretches south south east-wards to the vicinity of Whitchurch, filling in the upper end of a subdrift valley falling eastwards (Yates and Moseley, 1967). Doubtless the mid-Cheshire ridge was entirely over-ridden by ice many times in the Pleistocene glaciations, the medial moraine persisting subglacially. It is no accident that over 30 m of Middle Lias (Pocock and Wray, 1925), including the youngest strata to have escaped destruction by glacial scour in the main Cheshire syncline, form the small but conspicuous conical outlier of Prees Hill, south south east of Whitchurch on the medial line (Fig. 1).

Fewer writers have given due prominence to the Triassic anticlinal ridge trending from south west to north east across the north Shropshire plain into Staffordshire, and which was probably the physical barrier against which the Irish Sea readvance ended (Fig. 1). Across the ridge only one low col, now breached by the Tern, existed to the east of the medial line, at Wollerton (625298). The eastern lobe of Irish Sea ice, arrested by the rising ground, pushed a tongue of ice southward over the col, its outwash pouring southward towards the Severn. Correlation of this valley train with the Severn terrace system will be discussed later.

As retreat from the barrier commenced, the northward withdrawal of the ice front resulted in channelling of meltwater into the marginal gap between the ice and the northern slopes. Denuding the latter of most of its drift and establishing a deep valley along the hillside from the north east (Gibson, Wedd and Scott, 1925), the drainage escaped southward over the col at Wollerton at the proto-Tern. At this stage the front of the eastern ice lobe, standing parallel to the ridge and to Prees Hill with the emerging medial moraine, began to assume the form of a convex arc from Bar Hill (763437) to Whitchurch, its asymmetrical curve reflecting the trend of the slopes from which it was retreating (Fig. 1). Yates and Moseley (1967) described in detail the composite drift belt, the Woore moraine, deposited at this ice front, putting forward strong evidence that it is not the major retreat feature of an earlier glaciation as maintained by Poole and Whiteman (1966), but a Devensian moraine marking the terminal position of an ice advance. Their account of the diversity of the sequence within the moraine and of the number of minor east-west crests on its northern flank suggests alternatively that the drift belt as a whole might mark a pause in general northward withdrawal from the ridge, characterized by a period of limited fluctuation. Such activation of the eastern ice lobe along the southern perimeter of the east Cheshire Keuper and Lias basin would result in the linear accumulation of an exceptional thickness of sand, gravel and till in very complex form.

On the other side of the Whitchurch medial line the southward progress of the west-ern Irish Sea ice lobe was severely hampered on the west flank by its meeting with the Welsh Readvance. Issuing from the Severn, Vyrnwy and Tanat valleys the Welsh ice had again filled the piedmont plain, the distribution of Welsh till suggesting coalescence this time with glaciers from the Dee and Ceiriog valleys. Preservation of Irish Sea drift and morainic forms on Selattyn Hill (Wedd and others, 1929) suggests that there the Carboniferous crest may have remained a nunatak above the level of the readvancing Welsh ice (Fig. 1), as had Ruabon Mountain farther north between the two opposing ice sheets.

The wide zone of mixed Irish Sea and Welsh drift forming the divide between the spheres of influence of the readvancing ice was discussed earlier (Peake, 1961). At Wood Lane quarry (422325) south east of Ellesmere the strongly disturbed nature of the mixed sands, gravels and tills may be the result of interplay between opposing ice fronts (Worsley, 1970; Shotton, 1977b). Poole and Whiteman (1966) give significant evidence of local concentrations of Welsh erratics in Irish Sea till at Penley (419397) and Gredington (446385), 6 km north east of Ellesmere.

Indisputable evidence of the incursion of Welsh ice half way across the Irish Sea western lobe is provided by the Soil Survey's map and bulletin of the soils of the Wem district (Crompton and Osmond, 1954). They show that the Baschurch series, a soil developed on mixed Palaeozoic and Triassic till, interpreted by the authors as respectively Welsh and Irish Sea till, extends in an eastward lobe 8 km beyond the medial Bagley moraine through the Ellesmere locality to Welshampton (434350) and Bettisfield (460354). Hindered in its passage up the border piedmont plain by the buttress of Welsh ice, a narrower lobe of Irish Sea ice curved round the protruding obstacle to fill the remaining unoccupied lowland to the east (Fig. 1).

In the quadrant south east of Ellesmere the terrain in the few square kilometres immediate to the town is a classic example of tumultuous kettledrift topography. The hillocks and hollows, many of the latter holding meres, show a radial alignment from the north west. The likely explanation is apparent in Fig. 1. At the forefront of the Welsh Readvance, the area was the pivot round which the rotational forward movement of the Irish Sea ice was directed. At best the map can give only an indication of the overall forces at work in the area. No doubt the contest between the ice fronts continued for some considerable time with varying impetus on either side, and with local stress and disturbance as evidenced at Wood Lane quarry. In this respect it is of interest that the quarry is situated vertically above the south east rim of a circular hollow indicated in the sub-drift solid floor by geophysical surveying (Bilbo, 1975). The depression, 2 km in diameter and reaching more than 10 m below sea level, appears to be the only one of its type in the vicinity. Gouged out of the Keuper Marl and adjoining the east side of the Dee Trench, could it be the result of rotational pressurized ice scour at this locality?

Other forces may have been at work, for a careful study by Francis (1978) attributes much of the disturbance within the kettledrift to differential subsidence in ice melting. Both ice sheets were heavily charged with sand and silt from their passage over the border Triassic sandstones and with gravel and heavier erratics, for a great deal of unconsolidated material would have been incorporated from the region's earlier drift cover. At the centre of the milling ice large masses were thus very rapidly buried by meltwater debris, their eventual wastage resulting in kettleholes of every size and shape. The finest example is perhaps that holding the twin lakes of Blake Mere and Kettle Mere (416340), well over 30 m in depth. Other lakes such as Cole Mere (433333) appear to fill hollows between the sandy ridges.

Along the fringe of the thrusting Welsh ice front the southward progress of the diminished Irish Sea western ice lobe was halted against the slopes of the north Shropshire ridge. Its ponded meltwaters were responsible for the long poorly drained belt of mixed lacustrine clays and till, the Crewe series, mapped by the Soil Survey (Crompton and Osmond, 1954) along the northern flank of the hills. The main drainage outlet was a col across the ridge at Lee Brockhurst (548270) through which sand and gravel poured southward towards the Severn down the Roden valley in a manner similar to that of the Tern, but fanning more widely over the gentle till slopes around Shawbury (558213) and Rodington (589144), less restricting than the rock knolls of the Tern valley to the east. With the exception of the gravel trains of the Roden and Tern beyond the ridge, the formation of an extensive outwash plain from the Irish Sea ice front was not possible in the confined space north of the ridge when retreat from the rising slopes began.

In 1961 I described very briefly the retreat of the Irish Sea western ice lobe from the ridge and the Bagley moraine, punctuated by two minor readvances which are marked by crescentic moraines with sandy outwash (Pocock and Wray, 1925). The second of these, reaching almost to Wem, can be correlated very convincingly with the Woore moraine (Fig. 1). Rising in height and in bulk towards Whitchurch the curving feature merges with the latter north of the town, combining with it to extend north north westwards to the mid-Cheshire ridge as the Whitchurch medial moraine. Southwards the formation of the interlobate outwash plain of Whitchurch Heath and Prees Higher Heath demonstrates the contemporaneity of the diverging pair of moraines. Commencing at 100 m A.O.D. near Mossfield (545403), 1 km south south east of Whitchurch, the widening flat has an imperceptible fall of little more than 10 m in the 6 km that it extends south south east to Prees Hill (Fig. 1). At either edge of the plain the sand and gravel slopes of the high flanking moraines grade to its level surface, evidence of the dual aggradation that built it up. Down these slopes the shallow valleys of streamlets, many now dry, fall obliquely southwards to the plain. On the east side deeper incision above Twemlows Woods (570360) has cut the Woore moraine into sloping ridges of gravel; along the edge of the Wem moraine on the west side the sands and gravels in the neighbourhood of Steel Grange (550367) form strings of mounds and hollows trending in a southerly direction into the plain. Contemporaneous braided streams from either direction levelled emerging morainic material on the medial line, replenishing it with a sand and gravel infill of considerable thickness, but thinning southward. At least 21 m of sand and gravel were encountered at Tilstock airfield (557385) in the north (Poole and Whiteman, 1966), a depth of 12 m at the Sanatorium (565352) 1 km north of Prees Hill, and shallower deposits farther south (Pocock and Wray, 1925).

The wedge-shaped plain divides to continue southwards on either side of the rounded Middle Lias hill rising somewhat abruptly above its flat surface. Shallow channels were developed in the outwash deposits along these lateral drainage lines, that of Bailey Brook to the east, Soulton Brook to the west. South west of Prees Hill the Wem moraine is much diminished, the space in the wide angle between the hill and the oblique north Shropshire ridge being too great to be evenly filled by its decreasing outwash. Thus the western branch of the interlobate outwash plain ends here at Lacon (540326), the sands and gravels breaking into low mounds rising through black peaty alluvium, poorly drained by the network of canalized channels of Soulton Brook.

In the more restricted space to the east of Prees Hill the plentiful supply of outwash from the Woore moraine fills the valley from side to side, confining Bailey Brook to the south side of this marginal hollow between the moraine and the north Shropshire ridge. Well developed ridges of sand and gravel slope obliquely downstream along the brook, road widening across them exposing deposits of fine evenly bedded clean red sands dipping parallel to slope direction. Deeply incised, the main channel and those of the lateral streams from the north have floor widths suggesting a flow of water greatly in excess of today's meagre drainage. Whereas the retreat of the smaller western ice lobe from the north Shropshire ridge to the line of the Wem moraine was effected in two stages, the more entrenched but fluctuating eastern ice lobe maintained a strong steady outwash all along its distal southern slopes, both east and west of the Wollerton gap. This continued until an effectual northward ice retreat caused ponding of meltwater along the north side of the Woore moraine, its overflow southward to the Tern initiating the Adderley channel through the drift belt (Fig. 1).

West of the Whitchurch medial line the turbulent period in the Ellesmere pivotal area ended with this northward retreat of the Irish Sea ice, to be followed here by a further fluctuating readvance which deposited a complex belt of drift comparable in width and diversity to the Woore moraine. From topographical evidence the ice appears to have pushed into the northern fringe of the kettledrift on a curving front, and its progress to have been checked by the latter, or by persisting Welsh ice, or by the lack of impetus for further advance. In 1961 I referred to this wide drift belt as the Flintshire Maelor moraine, but following the establishment of new Welsh counties south eastern Clwyd has been given the name of Wrexham Maelor. For spatial reasons the single name Maelor is therefore appropriate for this complex moraine, extending westward as it does beyond the Dee to join the mountain dipslope south of Wrexham. East of the Ellesmere kettledrift the outwash plain from the moraine grades across to the base of the steep proximal slope of the earlier Wem moraine, a shallow linear peat-filled hollow on its flat surface, Fenn's Moss (490370) originally having held a large lake. Ponded meltwater from the Maelor moraine's eastern extremity overflowed across the Whitchurch medial moraine, initiating the Grindleybrook channel (Fig. 1), further indication that the Maelor moraine is not a true correlative of the Woore moraine. It may correlate with ice stands north of the Woore moraine described by Poole and Whiteman (1966). Many of the glacial features mapped by these authors show similarities to those depicted in Fig. 1, but in chronology we are not in agreement. Everything pertaining to the glacial sequence which I have attempted to describe points to its derivation from the final retreat of the main Devensian ice sheet from the Cheshire basin, with one major readvance, for which in 1961 I suggested the name Ellesmere Readvance to distinguish it from the contemporaneous Welsh Readvance, but also in recognition of the singularly impressive glacial topography of that locality.

THE RIVER TERRACES OF THE SEVERN SYSTEM

Pocock and others (1938) mapped three Severn river terraces above the Ironbridge gorge, in descending order the Uffington, Cressage and Atcham terraces. Their distribution shows that aggradation post-dated the downcut of the gorge across the Wenlock-Pennine watershed, with the draining of ice or drift ponded water from the Shrewsbury area. The gravels of the Uffington terrace, correlated by Wills (1938) with the Worcester Terrace below the Ironbridge gorge, commence near Shrewsbury along both banks of the main river and in the valley of its southern tributary the Rea Brook, immediately downstream from the limit of attenuated tills of the Welsh Readvance. They appear to have no correlatives in the valleys of the local northern Severn tributaries, the Perry, Roden and Tern. The extent of Welsh surface till (Pocock and Wray, 1925; Crompton and Osmond, 1954) implies that the south western end of the north Shropshire ridge and the entire Perry valley were overwhelmed by the Welsh Readvance, whereas the Roden and Tern drained into the Severn valley respectively, as suggested earlier, from the western and eastern twin lobes of the Ellesmere (Irish Sea) Readvance (Fig. 1).

When the great thickness of the main Devensian ice sheets is borne in mind the duration of their ablation in the warmer climatic hiatus before the stadial of the readvances is the more impressive, although no organic deposits from this amelioration have been recognised on the north Welsh border. In the succeeding stadial the new Welsh glacier, moving downhill eastward from the three great mountain valleys may have reached its limit at Shrewsbury before the readvancing northern ice banked up against the north Shropshire ridge. No Welsh terminal moraine was deposited, the glacier advancing over gravels and till of the earlier glaciation and over its own outwash (Worsley 1977), downstream from which the third or Uffington terrace was built up.

The second or Cressage terrace occurs in patches along the winding course of the Severn from a point much farther to the west, with a small combined development at the Perry confluence, an indication that the Welsh ice front was already in retreat at the time of aggradation. In the Tern valley the train of coarse roughly abraded gravel and red sand mentioned earlier, mapped by Pocock and others (1938) as fluvio-glacial flood gravels, is best interpreted as outwash from the Ellesmere Readvance eastern ice lobe. It extends intermittently high on either side for 8 km, from above Waters Upton (635195) down to Isombridge (610137), where it passes into the head of the Tern's correlative of the Cressage terrace 8 km above the river's confluence with the Severn. In the Roden valley the correlative of the Cressage terrace is of short extent, but immediately downstream from the termination of the glacial sand and gravel spreads at Rodington it is well developed for about 2 km, merging imperceptibly with the Cressage terrace of the Tern at The Lees (592125). Once again these combined factors suggest that the maximum of the Ellesmere Readvance may have post-dated that of the Welsh Readvance, during which the higher Uffington terrace had been formed along the Severn.

The lowest Severn terrace, the first or Atcham terrace, rises little above the alluvial plain along the main river, similar low terraces being traceable upstream in the three great mountain valleys (Wedd and others, 1929). By the late glacial period, which probably they represent, overflow from narrow lakes impounded in the Perry and Roden valleys between moraines north of the north Shropshire ridge (Peake, 1961) had lowered the rock sills in the two transverse gorges and drained the lakes. The remaining two small misfit rivers had little power for further transport or erosion, only the Roden showing a minor development of a correlative of the Atcham terrace (Pocock and others, 1938).

To the east the Tern valley had a late glacial history of greater significance. The Woore moraine appears to have remained fairly intact as an unbroken barrier while outwash from the ice front filled the linear hollow along the north Shropshire ridge with sand and gravel and drained southward over the wide Wollerton sill in Keuper Marl. When at length the northward retreat of the icefront began to impound melt-water behind the moraine, the main southward overspill initiated near Adderley descended the distal slope with considerable erosive power. Mapped by Pocock and Wray (1925) downstream from Longslow (655354), remnants of its sandy deposits commence high on the east side of the deep channel which was swiftly cut through the unconsolidated drift, about 20 m above its wide alluvial floor. Falling rapidly southwards to the Wollerton gap this terrace then becomes the dominant terrace of the Tern, its comparatively straight, narrow and incised course through the low Carboniferous and Triassic sandstone hills down to Longdon-upon-Tern (623155) being evidence of the strong constant flow of water from the Adderley channel. At this time the Wollerton col was more than 10 m higher than it is today, and across the Keuper Marl sill the torrent from the north excavated a trench with a flood plain over 1 km in width. As the volume of meltwater from the deepening Adderley channel rapidly increased, due to progressive enlargement of the proglacial lake north of the Woore moraine, a narrower and deeper channel was eroded across the Wollerton sill, exposing a 10 m step in Keuper Marl below the earlier entrenched level now left as first terrace remnants on either side (Pocock and Wray, 1925). Also bisected to expose the Keuper Marl below was the earlier sand and gravel outwash from the Woore moraine, both in the end of the Adderley channel and along Bailey Brook. Downcut was maintained until the level of the proglacial lake to the north had fallen below the level of the Adderley col, by which time an impressively wide straight-sided and flat floored channel had been established from this col for 11 km southwards through the Wollerton gap to the neighbourhood of Peplow (637247), at which locality the channel width is restricted downstream alongside the outcrop of more resistant Keele Beds.

Within the deep Adderley channel through the Woore moraine very low terraces rising on either side of the peat deposits on the alluvial floor were mapped by Pocock and Wray (1925) and Yates and Moseley (1967). A short distance downstream they end where the wide meltwater channel, now deserted, passes from the drift of the moraine on to the Keuper Marl of the Wollerton sill, which suggests that they may represent brief aggradation resulting from the restraint imposed on downcutting in drift when this rapid erosion reached the base level of the solid sill.

Below Wollerton the height of the Tern's first terrace decreases steadily until at Stoke-upon-Tern (640280), where maximum width is reached, it rises little above the level of the present flood plain, a general elevation maintained for 22 km downstream until it grades insensibly into the Severn's Atcham terrace near Atcham (540093). Throughout its aggradation the Tern's main source in a proglacial lake suggests that in the main the Atcham terrace, like the Cressage and Uffington terraces, belongs to the late glacial rather than to the post-glacial period.

CONCLUSIONS

In the Triassic sandstone of the north Welsh border the excavation of two deep trenches by ice was most probably initiated in pre-Devensian glaciations. The Late Devensian main Irish Sea glacier filled the Cheshire basin, mounting the western and south western perimeter to altitudes well above mountain height; further penetration of the uplands was prevented by contemporaneous Welsh glaciers. The ensuing retreat was followed by ice readvance from both sources to a diagonal front on the north Shropshire plain, the late glacial drainage forming a descending series of terraces along the Severn and its tributaries.

ACKNOWLEDGEMENTS

I would like to thank Mr and Mrs J.K. Shanklin, Mr J.R. Riden of the Welsh Water Authority and my husband for their assistance. I am indebted to the Severn Trent Water Authority for access to information contained in the Shropshire Groundwater Investigation reports.

REFERENCES

Bilbo, M. (1975). The geology of the Perry pilot area. Severn Trent Water Authority. *Shropshire groundwater investigation (3rd Report)*, 81 pp.
Crompton, E. and D.A. Osmond (1954). The soils of the Wem district of Shropshire. *Mem. Soil Surv.* vii + 79 pp.
Finch, J. (1976). Geophysical survey of the Severn Trench. Severn Trent Water Authority. *Shropshire groundwater investigation (4th Report)*, 93 pp.
Francis, E.A. (1978). *Field handbook, Keele meeting. Q.R.A.*, 109 pp.
Gibson, W., C.B. Wedd, and A. Scott (1925). *The geology of the country around Stoke-upon-Trent.* Mem. geol. Surv. Eng., 3rd edn. xv + 112 pp.
Morgan, A.V. (1973). The Pleistocene geology of the area north and west of Wolverhampton, Staffordshire, England. *Phil. Trans. R. Soc.*, B 265, 233-297.
Peake, D.S. (1961). Glacial changes in the Alyn river system and their significance in the glaciology of the north Welsh border. *Q. Jl geol. Soc. Lond.*, 117, 335-366.
Peake, D.S. (1979). The limit of the Devensian Irish Sea ice sheet on the north Welsh border. *Quaternary Newsl.*, 27, 1-4.
Pocock, R.W., T.H. Whitehead, C.B. Wedd and T. Robertson (1938). *The geology of the country around Shrewsbury.* Mem. geol. Surv. Gt Br.. xii + 297 pp.
Pocock, R.W., and D.A. Wray (1925). *The geology of the country around Wem.* Mem. geol. Surv. Eng. Wales. vi + 125 pp.
Poole, E.G. and A.J. Whiteman (1966). *The geology of the country around Nantwich and Whitchurch.* Mem. geol. Surv. Gt Br.. viii + 154 pp.
Shotton, F.W. (1962). A borehole at Conduit Head, Shrewsbury. *Trans. Caradoc Severn Vall. Fld Club*, 15, 1-4.
Shotton, F.W. (1977a). The Devensian Stage: its development, limits and substages. *Phil. Trans. R. Soc.*, B 280, 107-118.
Shotton, F.W. (1977b). The English Midlands. *Guide book for Excursion. X INQUA Congress*, 51 pp.
Strahan, A. (1885). *The geology of the country around Rhyl, Abergele and Colwyn Bay.* Mem. geol. Surv. Eng. Wales. vi + 77 pp.
Wedd, C.B. (1912). In G.W. Lamplugh and others, 11. Field Work (A.) England and Wales, 1. — Denbighshire District. *Summ. Prog. for 1911, Mem. geol. Surv., Gt Br..* pp. 6-20.
Wedd, C.B. and W.B.R. King (1924). *The geology of the country around Flint, Hawarden and Caergwrle.* Mem. geol. Surv. Eng. Wales. viii + 222 pp.
Wedd, C.B., B. Smith, W.B.R. King and D.A. Wray (1929). *The geology of the country around Oswestry.* Mem. geol. Surv. Eng. Wales. xix + 234 pp.

Wedd, C.B., B. Smith and L.J. Wills (1928). *The geology of the country around Wrexham (2)*. Mem. geol. Surv. Eng. Wales. xvii + 237 pp.

Wills, L.J. (1912). Late-Glacial and Post-Glacial changes in the lower Dee valley. *Q. Jl geol. Soc. Lond.*, 68, 180-198.

Wills, L.J. (1937). The Pleistocene history of the west Midlands. *Rept Br. Ass. (Nottingham)*, 71-97.

Wills, L.J. (1938). The Pleistocene development of the Severn from Bridgnorth to the sea. *Q. Jl geol. Soc. Lond.*, 94, 161-242.

Wills, L.J. (1948). *The palaeogeography of the Midlands*. Liverpool Univ. Press, Hodder and Stoughton, Ltd., London. vii + 144 p.

Worsley, P. (1970). The Cheshire-Shropshire lowlands. In C.A. Lewis (Ed.), *The glaciations of Wales and adjoining regions*. Longman, London, pp. 83-106.

Worsley, P. (1977). The Cheshire-Shropshire plain. In D.Q. Bowen (Ed.), Wales and the Cheshire-Shropshire lowland: *Guidebook for Excursion. X INQUA Congress*, 64 pp.

Wright, J.E. (1968). *The geology of the Church Stretton area*. H.M.S.O. 87 pp.

Yates, E.M. and F. Moseley (1967). A contribution to the glacial geomorphology of the Cheshire plain. *Trans. Inst. Br. Geogr.*, 42, 107-125.

Mrs D.S. Peake,
Rosewall,
Portley Wood Road,
Whyteleafe,
Surrey. CR3 OBP.

7. The 'South Wales End-Moraine': Fifty Years After

D. Q. Bowen

INTRODUCTION

J.K. Charlesworth's paper on 'The South Wales End-Moraine', outlining the limit of what would now be termed the Late Devensian Ice Sheet (Fig. 1), was read to the Geological Society of London on January 23rd 1929. There, it drew an exceedingly critical response from O.T. Jones, a harbinger of many attempts at modification of Charlesworth's line (Fig. 2). Many of these were tacitly based on the assumption that the Quaternary is different from other periods, hence special approaches are necessary. Others coupled this with 'count from the top' methods (below). Such *ad hoc* methods have been characteristic for most of the fifty years since Charlesworth, unlike the lithostratigraphic approach of recent years.

CHARLESWORTH'S LINE

Maps and memoirs of the Geological Survey were only available south of a line drawn through approximately Carmarthen. North of this Charlesworth was only able to base his work on papers by Jehu (1904), on Preseli (north Pembrokeshire), and Williams (1927), on south Ceredigion (Cardiganshire). Neither of these had produced geological maps of the Pleistocene deposits. Lacking such fundamental evidence, Charlesworth argued that the Irish Sea Ice Sheet would have blocked the mouths of local streams, causing proglacial lakes to develop. Overflow channels, which drained the lakes, would therefore serve to indicate the ice-margin. Altogether he described 24 overflows, designated 'marginal drainage features' (Charlesworth, 1929 p.342). Yet, of these, 7 are 'direct overflows' (in the sense of Kendall, 1902), and only 17 are 'marginal' in the sense that they were deemed to lie parallel or sub-parallel to the ice-margin. This was the control for reconstructing some 72 km of the Irish Sea Ice Sheet margin, out of a total length of some 96 km. The other 24 km was indicated by 'moraines', described as 'discontinuous', 'moderate in dimension' and 'rarely conspicuous'; characteristics attributed to a paucity of englacial debris in the ice, the steep coastal slope having prevented debris-rich basal ice layers from penetrating inland. Two 'moraines' are large features by British standards: the Monington esker (Gregory and Bowen, 1966), and Banc-y-warren delta (Jones, 1965; Helm and Roberts, 1975). Four major proglacial lakes (Manorowen, Nevern, Teifi - 54 km from Cardigan to Lampeter and Aeron), and six minor ones, were postulated on the basis of the ice-margin inferred from moraines and overflows. Only in the case of Lake Teifi, the sole instance to have withstood modern criticism (Bowen and Lear, 1980), was independent evidence cited: the Llanwer delta, and deltaic deposits upstream (Dulas Valley) and downstream from Lampeter.

Evidence of variable quality was adduced to establish the margin of the coeval Welsh Ice Sheet (Fig. 1). East of Swansea, the limit was drawn at the margin of continuous glacial deposits mapped by the Geological Survey. Some uncertainty obtained in the eastern coalfield valleys, and those north of Port Talbot. West of Swansea, however, the precept followed farther east was disregarded, and an arbitrary line, admittedly conjectural, was drawn north to Tregaron, where a prominent end-moraine occurs.

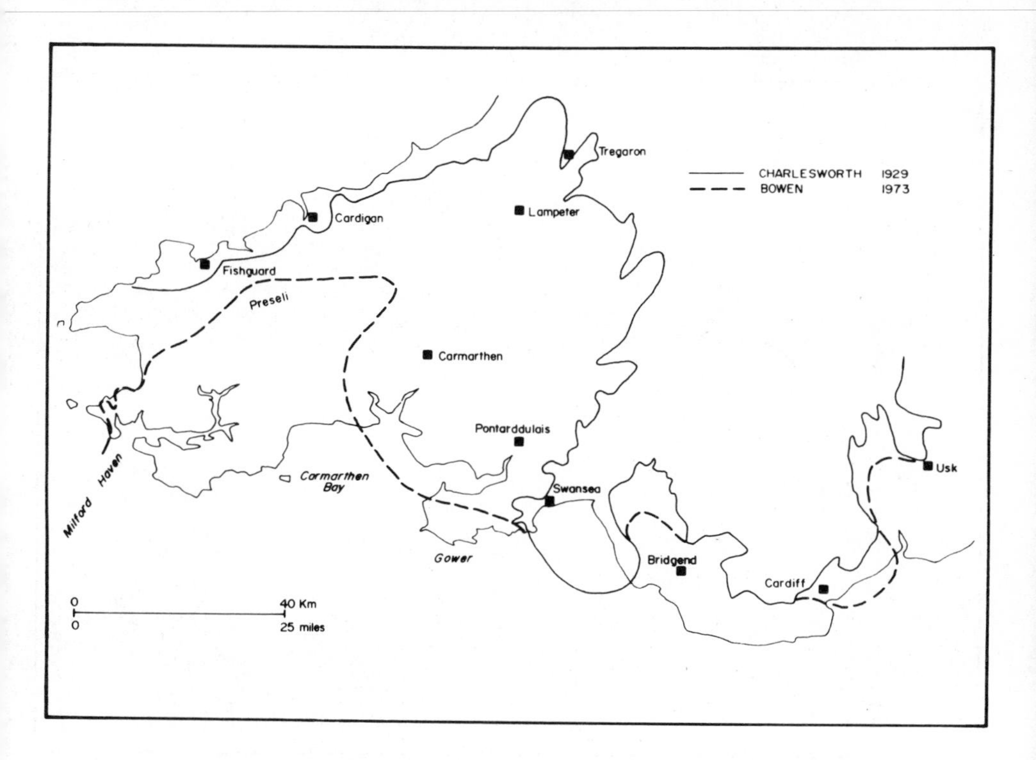

Fig. 1. J.K. Charlesworth's (1929) South Wales End-Moraine and the Late Devensian Limit recognised fifty years after (Bowen, 1973).

The 'newer drift' glaciation, of which the South Wales End-Moraine marked the margin, was dated as early Magdalenian (Late Devensian), because its deposits sealed Aurignacian artefacts inside the Cae Gwyn Cave in the Vale of Clwyd, North Wales.

LOCAL ELABORATION

T.N. George's (1932, 1933) work on the glacial and coastal Pleistocene deposits of Gower confirmed Charlesworth's limit on the western side of Swansea Bay. But Griffiths (1937, 1939, 1940), in examining the mineralogy and erratic content of glacial deposits west of Cardiff, confirmed Jones's criticisms by showing that a continuous formation crossed Charlesworth's line north of Swansea (Fig. 2) (cf. also Archer, 1968). The ice-margin for this formation was placed far outside

Charlesworth's (Fig. 2), who had, however, thought it conceivable that deposits near Carmarthen and Pontarddulais might be 'newer drift' in age.

COUNT FROM THE TOP CLASSIFICATION

'Count from the top' methods (Kukla, 1977) are those which seek to classify the history of a region according to a model of Pleistocene history inferred elsewhere. In other words, local evidence is ascribed to 'pigeon hole' stages of a standard

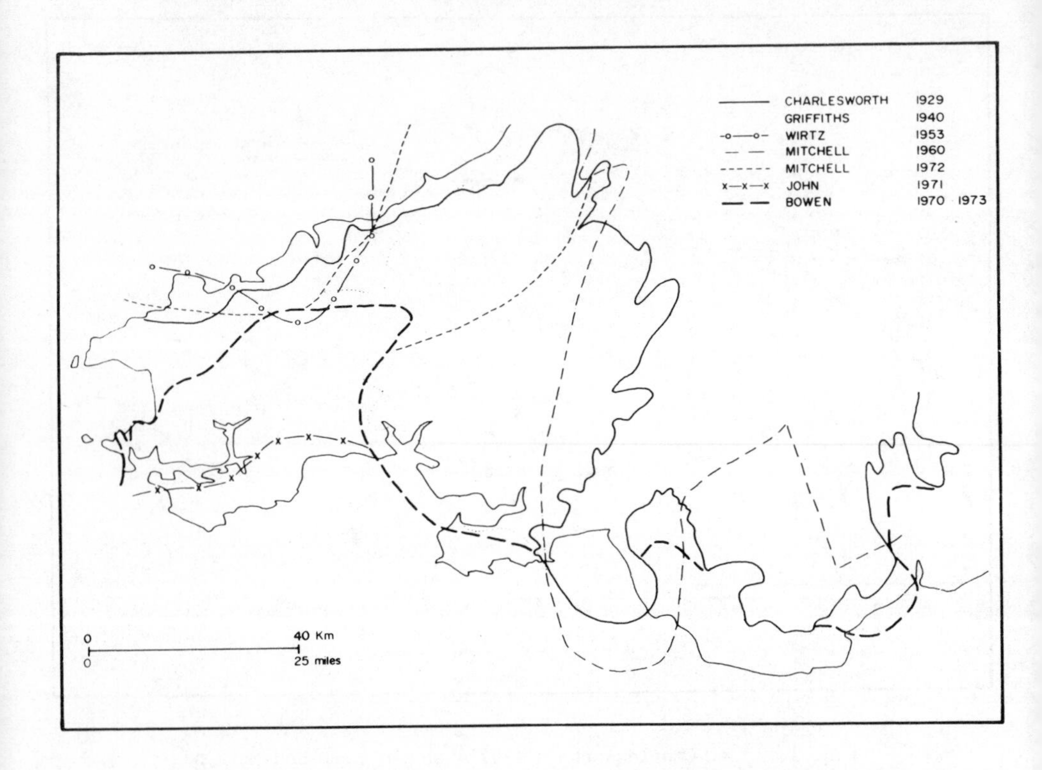

Fig. 2. Attempts to define the Last Glaciation extent in South Wales.

scheme. Usually lithostratigraphic units are accorded chronostratigraphic status, on a one to one basis. The model applied to South Wales was that for south-east Ireland. Two particular factors are of some importance. First, the belief (later modified) that cryoturbation and ice-wedge pseudomorphs (and later remains of pingos) only occur on glacial deposits older than the Last Glaciation (e.g., Synge, 1964); and second, the belief that the raised beaches of South Wales and southern Ireland are older than the Last Interglacial.

Daniel Wirtz (1953) visited South Wales briefly during his tour of the entire Irish Sea basin, and suggested that only limited local valley glaciation occurred during the Last Glaciation, with the Irish Sea Ice Sheet impinging locally around Cardigan and Fishguard (Fig. 2). Cardigan Bay was deemed to be ice-free at this time because of the cryoturbation structures in local glacial drifts. Although he used the Irish model as a basis for interpretation, the labels applied were those of the

European Alpine scheme. Premature though this may seem, it was no less so than the application of East Anglian labels some years later when much the same approach was followed by Mitchell (1960). He did, however, delimit the margin of a Welsh Ice Sheet to the north of the Llŷn peninsula. Following Wirtz (1953) Cardigan Bay was deemed to have been ice-free due to the cryoturbation structures in its drifts. A decade and more later (Mitchell, 1972), he revised his interpretation slightly by suggesting that till in the Teifi estuary was Late Devensian in age. He thus inserted a lobe of Irish Sea Ice across Cardigan Bay, much as Wirtz (1953) had done earlier.

Some support for these ideas came from Watson's (1968) work in mid-Wales. He concluded that the periglacial deposits of the region showed it had been ice-free during the Last Glaciation: later he used the remains of pingos to indicate those ice-free areas (Watson, 1972).

LITHOSTRATIGRAPHIC ASSESSMENT

This phase commenced in the early 1960's. It involved detailed local investigations which culminated in *The Glaciations of Wales* (Ed. Lewis, 1970). In south-east and central South Wales (Bowen, 1970), the Pleistocene geology was related to the lithostratigratigraphy of Gower, and the extent of Late Devensian Welsh ice shown to be more extensive than envisaged by Charlesworth (1929), and marginally more so than by Griffiths (1940).

In Preseli, the coastal lithostratigraphic succession was shown to be identical on both sides of Charlesworth's (1929) line, thus rendering it redundant as a significant ice-margin (John, 1970). This confirmed the interpretation of the Fishguard meltwater channels as subglacial in origin (Bowen and Gregory, 1965), and the Cregiau Cemmaes 'moraines' (Charlesworth, 1929) as fluvioglacial dead ice features (Gregory and Bowen, 1966), both cases necessitating a more extensive ice-cover than envisaged by Charlesworth (1929).

By correlating the Preseli lithostratigraphy with that in Devon and the Isles of Scilly, John (1970) suggested that Late Devensian ice covered Pembrokeshire and the Bristol Channel. The sequence was: (1) raised beach, (2) head, (3) till, (4) head. But only head overlies the beach in South Pembrokeshire, so he suggested that the glaciation was lost in unconformity between a cemented and uncemented head unit (John, 1970 p. 240). Subsequently it was shown that the 'till' unit of South West England was a lithofacies of the head, being recycled older glacial drift (Bowen, 1969). John (1971) subsequently revised his ice limit, placing it immediately south of Milford Haven. This line was flanked on both sides by an identical lithostratigraphy (head on beach), which had been the method by which he had previously invalidated Charlesworth's line in Preseli (John, 1970)! Subsequently, in the belief that human occupation of south Pembrokeshire had occurred 'at approximately the coldest part of the Last Glaciation', he maintained that 'Irish Sea Ice did not extend further south than Milford Haven' (John, 1974, p. 67).

Lithostratigraphic documentation of South Wales as a whole shortly followed (Bowen, 1973a, 1973b, 1974, 1977). It showed that a glaciation ('older drift') had antedated the raised beach interglacial, so-called 'older drift' overlying the raised beach being recycled deposits and as such technically head; and that the 'newer drift' glaciation (South Wales End Moraine) was the sole glacial event subsequent to that interglacial. Thus the raised beach emerged as a stratigraphic marker of some importance, as it is in South West England (Mitchell and Orme, 1967; Bowen, 1969), and Ireland (Bowen, 1973b, Warren in press). Those who would argue that two glacial events postdate the raised beach episode in Ireland (e.g. Mitchell, 1972), implicitly subdivide the post-beach period using the South Irish End Moraine as a stratigraphic artefact (notwithstanding the Shortalstown site: Colhoun and

Mitchell 1971, Bowen, 1973b). In this way, the South Irish End Moraine of Charlesworth (1928) becomes a gigantic circular argument quite devoid of any kind of stratigraphic validation. Some Irish workers have accepted that only one glacial event, the Late Devensian, postdated the raised beach in Ireland (Synge, 1977, Warren in press).

Ascription of the raised beaches to the last Interglacial (Oxygen Isotope Stage 5e) allows the identification of areas glaciated before and after that event, on straightforward lithostratigraphic grounds (Bowen, 1973a, 1973b, 1974), (Figs. 1 and 2): i.e., head on beach (unglaciated) or head and glacial beds on beach (glaciated) respectively. In this way the limit can be fixed with some accuracy along the coastline. Inland the margin of the Late Devensian glaciation is placed at the limit of continuous, as opposed to greatly dissected and highly fragmentary drift (see map in Bowen, 1970).

FIFTY YEARS AFTER

Several problems remain. The age of the pre-raised beach glaciation is unknown. In West Gower comparatively undissected drift (George, 1933) stands in great contrast to other drift covered areas outside the Late Devensian limit. Does this mixed drift, of mixed Irish Sea and Welsh provenance signify a second, and younger 'older drift'? It cannot be Devensian, for its soliflucted facies is part of the Devensian head formation at Eastern and Western Slades (Bowen, 1971).

Two equivocal sites occur in Milford Haven. Near the entrance, at West Angle Bay, occurs a sequence of: till (Dixon, 1921), loams with interglacial peat (John, 1970), and head (Bowen, 1974). John (1970) claims the basal till does not exist, and that the unit interpreted as head is in fact till (John, 1970). Some doubt occurs as to whether the loam unit is marine or fluviatile (Bowen, 1977). Recent ascription of the interglacial to the Hoxnian is an unfortunate revival of 'count from the top' procedures: that term has no meaning in Wales. Upstream, at Landshipping, Geological Survey Officers described glacial drift overlying raised beach. This site lies well outside the Late Devensian Ice Margin. While there is no reason why an older raised beach could not be present, it is just as likely that the glacial unit is a proglacial sediment formed on the main meltwater route from the Irish Sea Ice Sheet in Preseli, supplemented by overspill of Lake Teifi (Bowen, 1980), for in such environments a wide range of sediments can be deposited.

It is possible that interglacial beaches of more than one stage are preserved along the coastline. But recent work in Gower tends to confirm the unitary nature, and Last Interglacial age, of the raised beach adduced in this account as a critical stratigraphic marker. At Minchin Hole Cave, two marine units occur (Bowen, 1977): an inner beach (amino-acid ratio on *Patella vulgata*: 1.40 unpublished) and an outer beach (amino-acid ratio: .099). The outer (*Patella*) beach correlates, by aminostratigraphy, with other Last Interglacial beaches in South West England (Andrews, Bowen and Kidson, 1978). Furthermore, an $^{234}U/^{230}Th$ date on stalagmite overlying the beach at nearby Bacon Hole Cave (Stringer, personal communication), tends to confirm its 125 ka age. An attempt to recognise two interglacial beaches elsewhere (Mitchell, 1972) is invalid: in Gower, because the evidence is incorrectly described, and the method stratigraphically inadmissible (Bowen, 1973a, 1973b); and on St. Martin's, Isles of Scilly, because the critical stratigraphy is inferred, not superposed, and granite core-stones have been interpreted as a marine deposit (Bowen, in preparation).

After just over fifty years, J.K. Charlesworth's (1929) line has survived with only minor modification east of Swansea Bay. Elsewhere, an extension of some 24 km has been necessitated on the west coast of Pembrokeshire, but with rather more drastic modification in south-east Dyfed (S.E. Carmarthenshire) (Figs. 1 and 2). The modern version(Figs. 1 and 2) is based on lithostratigraphy and the distribution of glacial drifts, not on geomorphology (Bowen 1973a, I.G.S. Quaternary Map of the British Isles, 1977). Its age, anticipated in a penetrating and incisive review of the Last Glaciation in Britain by L.F. Penny (1964), rests on the position of glacial drift in sequences overlying the Last Interglacial raised beach, and also on minimum dates established from pollen analysis of kettle hole fills on its surface (Bowen, 1974). The significance of this glacial limit far transcends local considerations. Any relegation of it to merely local circumstances fails to appreciate the cardinal role it, and its alleged counterpart in Ireland (Charlesworth, 1928), have played in the elucidation of Pleistocene history on the maritime margins of Europe, as well as the stratigraphic philosophy behind its modern location which deems that Pleistocene rocks are no different from earlier ones, hence should be classified similarly. While there must always remain a possibility of homotaxial error when dealing with such lithostratigraphic successions, the current conclusion that identical successions represent identical things, is one based on an economy of hypotheses. At the present time there is no justification in complicating that hypothesis in the advance of any contrary evidence.

ACKNOWLEDGEMENTS

The author thanks the editors for inviting him to contribute to this volume for Lewis Penny whose achievements in Quaternary Research, and its organisation, are well known. Not so well known, perhaps, is the enormous contribution of his wife, Mary Penny, to the X INQUA Congress in 1977. This essay is a salute to both of them. I thank Jenny Penman, Tish Gorton and Cheryl Pemberton of the University of Canterbury, Christchurch, New Zealand (where this essay was written) who drew the maps and typed the manuscript.

REFERENCES

Andrews, J.T., D.Q. Bowen and C. Kidson (1978). Amino acid ratios and the correlation of raised beach deposits in south-west England and Wales. *Nature, Lond.*, 281, 556-558.

Archer, A.A. (1968). *Geology of the Gwendraeth Valley and adjoining areas.* Mem. geol. Surv. Gt Br.. xi + 216 pp.

Bowen, D.Q. (1969). A new interpretation of the Pleistocene deposits of the Bristol Channel. *Proc. Ussher Soc.*, 8, 89.

Bowen, D.Q. (1970). South-east and central South Wales. In C.A. Lewis (Ed.), *The Glaciations of Wales and adjoining regions*, London. pp. 197-227.

Bowen, D.Q. (1971). The Quaternary Succession of South Gower. In D.A. Bassett and M.G. Bassett (Eds), *Geological Excursions in South Wales and the Forest of Dean*, Cardiff. pp. 135-142.

Bowen, D.Q. (1973a). The Pleistocene history of Wales and the borderland. *Geol. J.*, 8, 207-224.

Bowen, D.Q. (1973b). The Pleistocene succession of the Irish Sea. *Proc. Geol. Ass.*, 84, 249-272.

Bowen, D.Q. (1974). The Quaternary of Wales. In T.R. Owen (Ed.), *The upper Palaeozoic and post-Palaeozoic Rocks of Wales*, Cardiff. pp. 373-426.

Bowen, D.Q. (1977). The coast of Wales. In C. Kidson and M.J. Tooley (Eds), *The Quaternary History of the Irish Sea.* Geol. J. Spec. Issue, No. 7, pp. 223-256.

Bowen, D.Q. (1980). Pleistocene deposits and Fluvioglacial Landforms of North Preseli. In M.G. Bassett (Ed.), *Geological Excursions in Dyfed*, Cardiff.

Bowen, D.Q. and K.J. Gregory (1965). A glacial drainage system near Fishguard, Pembrokeshire. *Proc. Geol. Ass.*, 74, 275-282.

Bowen, D.Q. and D.L. Lear, (1980). The Quaternary Geology of the Lower Teifi Valley. In M.G. Bassett (Ed.), *Geological Excursions in Dyfed*, Cardiff.

Charlesworth, J.K. (1928). The glacial retreat from central and southern Ireland. *Q. Jl geol. Soc. Lond.*, 84, 295-300.

Charlesworth, J.K. (1929). The South Wales End-Moraine. *Q. Jl geol. Soc. Lond.*, 85, 335-358.

Colhoun, E.A. and G.F. Mitchell (1971). Interglacial marine formation and lateglacial freshwater formation in Shortalstown townland, Co. Wexford. *Proc. R. Ir. Acad.*, B 71, 211-245.

Dixon, E.E.L. (1921). *The geology of the country around Pembroke and Tenby.* Mem. geol. Surv. Eng. Wales. vi + 220 pp.

George, T.N. (1932). The Quaternary Beaches of Gower. *Proc. Geol. Ass.*, 43, 291-324.

George, T.N. (1933). The glacial deposits of Gower. *Geol. Mag.*, 70, 208-232.

Gregory, K.J. and D.Q. Bowen (1966). Fluvioglacial deposits between Newport, Pembrokeshire and Cardigan. In R.J. Price (Ed.), *Deglaciation.* Br. Geomorph. Res. Gp., pp. 25-28.

Griffiths, J.C. (1937). *The glacial deposits between the River Tawe and River Towy.* Unpublished Ph.D. Thesis, University of Wales.

Griffiths, J.C. (1939). The mineralogy of the glacial deposits of the region betwee the rivers Neath and Towy, South Wales. *Proc. Geol. Ass.*, 50, 433-462.

Griffiths, J.C. (1940). *The glacial deposits west of the Taff.* Unpublished Ph.D. Thesis, University of London.

Helm, D.G. and B. Roberts (1975). A re-interpretation of the origins of sands and gravels around Banc-y-Warren, near Cardigan. *Geol. J.*, 10, 131-146.

Jehu, T.J. (1904). The glacial deposits of northern Pembrokeshire. *Trans. R. Soc. Edinb.*, 41, 53-87.

John, B.S. (1970). Pembrokeshire. In C.A. Lewis (Ed.), *The Glaciations of Wales and adjoining regions*, London. pp. 229-265.

John, B.S. (1971). Glaciation and the West Wales landscape. *Nature in Wales*, 12, 1-18.

John, B.S. (1974). Ice Age events in South Pembrokeshire. *Nature in Wales*, 15, 66-68.

Jones, O.T. (1965). The glacial and post-glacial history of the lower Teifi Valley. *Q. Jl geol. Soc. Lond.*, 121, 247-281.

Kendall, P.F. (1902). A system of glacier-lakes in the Cleveland Hills. *Q. Jl geol. Soc. Lond.*, 58, 471-571.

Kukla, G.J. (1977). Pleistocene Land Sea Correlations. I. Europe. *Earth Sci. Rev.*, 13, 307-374.

Lewis, C.A. (Ed.) (1970). *The Glaciations of Wales and adjoining regions*, London.

Mitchell, G.F. (1960). The Pleistocene history of the Irish Sea. *Advanc. Sci. Lond.*, 17, 313-325.

Mitchell, G.F. (1972). The Pleistocene history of the Irish Sea: Second Approximation. *Scient. Proc. R. Dubl. Soc.*, A 4, 181-199.

Mitchell, G.F. and A.R. Orme (1967). The Pleistocene deposits of the Isles of Scilly. *Q. Jl geol. Soc. Lond.*, 123, 59-92.

Penny, L.F. (1964). A review of the Last Glaciation in Great Britain. *Proc. Yorks. geol. Soc.*, 34, 387-411.

Synge, F.M. (1964). The glacial succession in West Caernarvonshire. *Proc. Geol. Ass.*, 75, 431-444.

Synge, F.M. (1977). Records of sea levels during the Late Devensian. *Phil. Trans. R. Soc.*, B 280, 211-228.

Watson, E. (1968). The Periglacial landscape in the Aberystwyth region. In E.G. Bowen and others (Eds), *Geography at Aberystwyth*, Cardiff.

Watson, E. (1972). Pingos of Cardiganshire and the latest ice limit. *Nature, Lond.*, 238, 343-344.

Williams, K.E. (1927). The glacial drifts of Western Cardiganshire. *Geol. Mag.*, 64, 206-227.

Wirtz, D. (1953). Zur Stratigraphie des Pleistocans in Westen der Britischen Inseln. *Neues Jb. Geol. Paläont. Abh.*, 96, 267.

Dr. D.Q. Bowen,
Department of Geography,
University College of Wales,
Llandinam Building,
Penglais,
Aberystwyth,
Dyfed SY23 3DB

QIB - F

8. The Glaciation of Charnwood Forest, Leicestershire and its Geomorphological Significance

J. F. D. Bridger

INTRODUCTION

Charnwood Forest lies within the triangle made by the East Midland towns of Leicester, Loughborough and Coalville with the River Trent to the north and its tributary, the River Soar, to the east (Fig. 1). Topographically the locality forms a range of hills rising to over 270 m O.D. and aligned north-west to south-east to reflect the geological trend (Fig. 3).

The well-known, solid geology (Watts, 1947) is basically a south-easterly pitching anticline of late Precambrian age. Intrusive rocks also occur and these, together with the Charnian structure, are overlain by the remnants of a former cover of Triassic sediments. In contrast to the attention given to the ancient rocks of Charnwood forest the superficial deposits were neglected until the nineteen-sixties when a series of excavations, mainly connected with motorway construction, stimulated their study. Contradicting earlier views these investigations confirmed that the area had been extensively glaciated.

Drift in motorway cuttings has been briefly described by Poole (1968) while Bridger (1972, 1975) has, in addition to examining motorway sections in detail, mapped glacial deposits in other parts of Charnwood Forest. The relevance of the new appreciation of Charnwood Forest's glaciation to certain aspects of its morphology has been outlined by Ford (1967, 1968). Bridger (1968) has considered the relationship of glaciation to the origin of the area's discordant drainage.

The purpose of this article is to review the evidence for Pleistocene glaciation in Charnwood Forest and to discuss the geomorphological significance of the event with particular reference to the role of glacial deposition and meltwater erosion.

THE EVIDENCE FOR GLACIATION

Glacial deposits have been shown in and around Charnwood Forest on the Geological Survey's (now the Institute of Geological Sciences') one-inch drift sheets (Nos. 141 142, 155, 156) since the turn of the century. Spreads of sand and gravel are located mainly on the northern fringe forming a broad band running from Thringstone (SK 430180) to Shepshed (SK 480190) and at the edges of the valleys of the River Sence and Rothley Brook. Widespread boulder clay has been mapped outside the rim of Charnwood Forest's hills but within them its occurrence is sparse being mainly restricted to isolated patches on the higher ground, with pockets in some of the valleys. Compared with the extensive spreads plotted on the surrounding lowlands

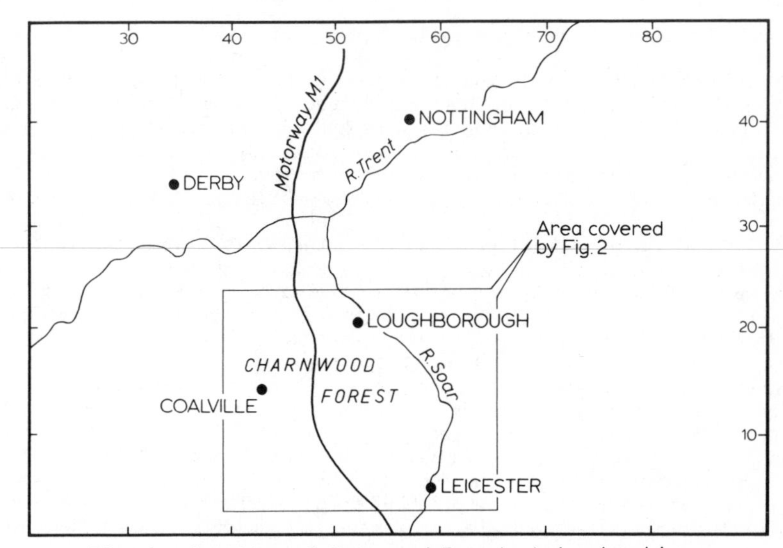

Fig. 1. Location of Charnwood Forest, Leicestershire.

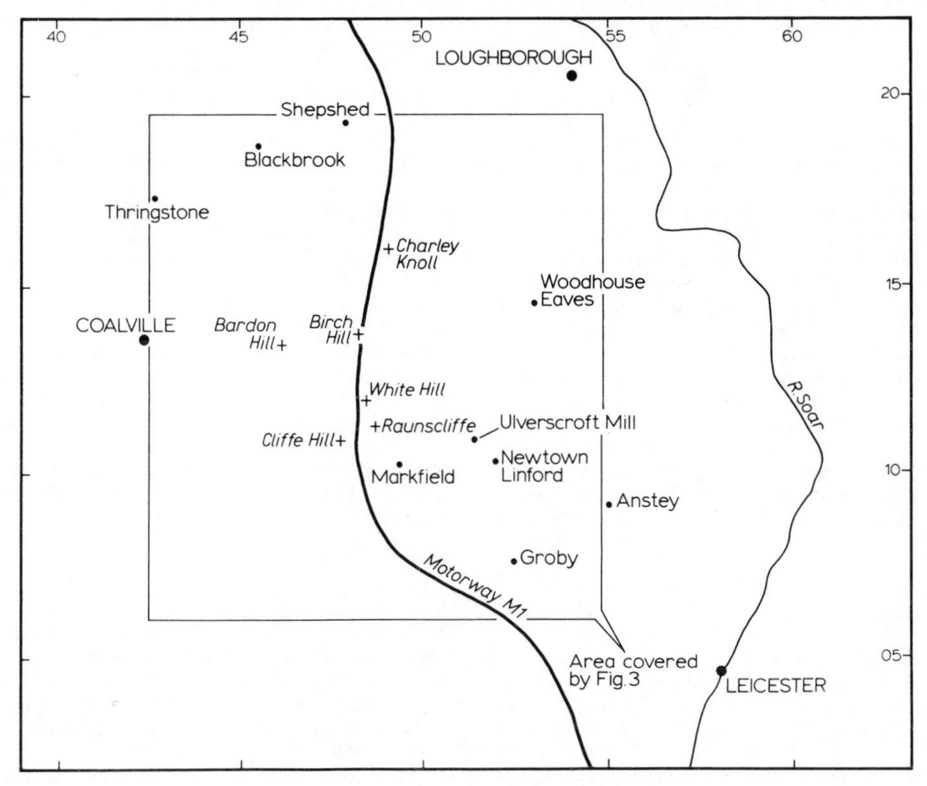

Fig. 2. Localities mentioned in text.

Charnwood Forest appears to be virtually free of drift.

Although it seems most unlikely that this impression alone led W.W. Watts, the foremost geological authority on Charnwood Forest throughout the first half of the twentieth century, to conclude that the area had escaped glaciation, it is hard to find alternative reasons for his opinion. While it is probable that the true cause of his oversight of the role of cold climate processes in Charnwood Forest will never be known with certainty, there is little doubt that his widely published belief, that the scenery represented an uninjured Triassic landscape (Watts, 1903, 1905, 1947), had a profound influence on the interpretation of local features by other workers, for example, Bennett (1929). Thus, in contradiction of the depositional evidence of boulder clay in the centre of Charnwood Forest, the accepted view was that the locality had stood as a giant nunatak above an ice-sheet which otherwise submerged the topography of the east Midlands.

The comparatively small area of boulder clay in Charnwood Forest plotted on the Institute of Geological Sciences' drift sheets has been shown by recent work (Bridger, 1972) to be an underrepresentation of its full extent. The cause of this discrepancy appears to have been the failure by the Officers of the Institute (formerly the Survey) always to succeed in separating Triassic bed rock from till, particularly when the latter is composed of local material together with relatively inconspicuous erratics from outside the area. Remapping of the drift in Charnwood Forest still awaits completion; nevertheless, sufficient has been completed to reveal several instances of such confusion, for example, at Newtown Lindford (SK 519099) where a Keuper - rich till with an assortment of Charnwood Forest, Bunter and Carboniferous erratics has been recorded as Keuper Marl. For mapping purposes, and also for understanding the local Pleistocene succession, it is fundamentally important to distinguish between Keuper Marl, Keuper-rich till, soliflucted Keuper Marl and soliflucted Keuper-rich till.

The drift of Charnwood Forest includes outwash and proglacial lacustrine sediments but, in fact, the most important evidence for glaciation is till of which three main types have been recognised. In order of decreasing age they comprise first a brownish, calcareous till carrying a predominance of Liassic rock. Second is a reddish usually non-calcareous, Keuper-rich till characteristically containing a proportion of Coal Measure erratics. The third is greyish with a distinctive component of chalk and flint.

The earliest of these, the Liassic-rich, Raunscliffe till of south Charnwood Forest has not been recorded outside of the Markfield by-pass cutting (Fig. 4). Reference to a map of the solid geology of the area (Fig. 5) shows that ice, carrying a dominantly Liassic suite of erratics, almost certainly moved across the county from the north-east because Liassic strata do not crop out west of the Soar valley. When the ice moved into south Charnwood Forest it ponded proglacial water to form a lake in the vicinity of White Hill (SK 482120). An altitudinal maximum for the lake sediments of 215 m O.D. implies that the impounding ice was of sufficient thickness to allow it to overwhelm the centre of Charnwood Forest.

Confirmation of such an early advance into the locality by ice of north-eastern provenance occurs at Birch Hill (SK 478135) in the form of 6 m of banded till underlying the northward extension of the White Hill lacustrine clays and silts. Unlike the Raunscliffe till which has lodgement features, the Birch Hill till is stratified and displays alternating red, grey and brown bands of silty clay usually less than 20 mm in thickness. The grey and brown colours are associated with a calcareous matrix containing Liassic rock fragments and micro-fossils. In contrast, the red bands are virtually free of $CaCO_3$ with Keuper Marl and Charnian rock particles as the main constituents. There are also a few beds of intermediate shades reflecting a mixture of calcareous and non-calcareous elements.

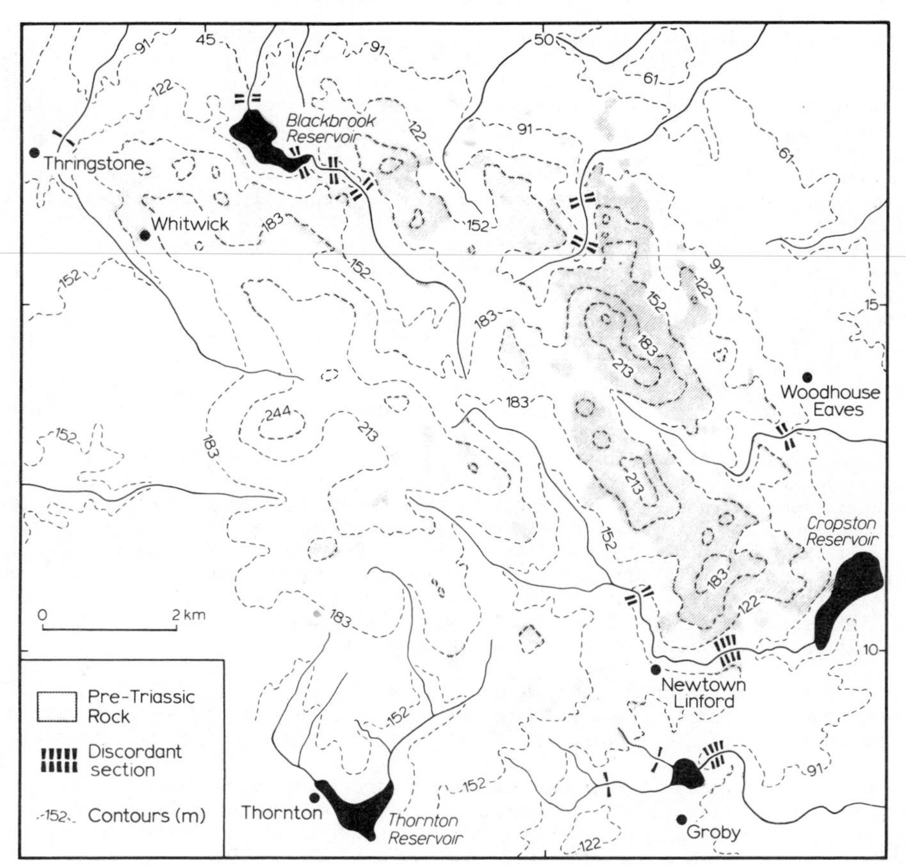

Fig. 3. Relief and drainage of Charnwood Forest.

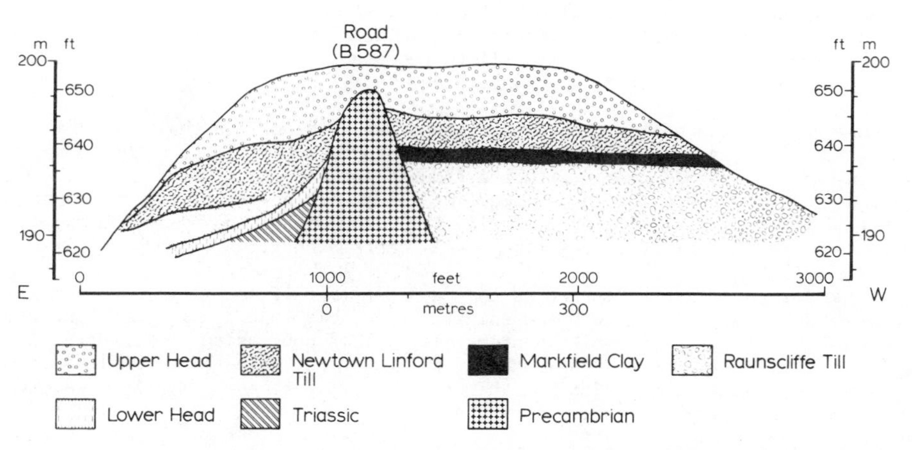

Fig. 4. Generalised section of the Markfield By-pass cutting.

The mechanism responsible for such a till is not fully understood but it is suggested that local topographical conditions, possibly near the ice-front, caused glacial shearing to introduce Charnwood Forest bed-rock into an ice-sheet otherwise loaded with Liassic debris. Subsequent melting of the ice, in environments similar to those postulated by Boulton (1968) for the formation of flow-tills, would explain many of the properties displayed by the deposit.

Acceptance of a correlation of the banded till at Birch Hill with the Raunscliffe till implies that the ice, associated with the two deposits, succeeded in over-running the slopes of Charnwood Forest to advance at least as far as Birch Hill However, other local geographical aspects of this glaciation remain in the realm of speculation. There is a better understanding of the glaciation connected with the second of the main types of till in Charnwood Forest, the red, Keuper-rich deposit with Coal Measure erratics. This till has been proven at intervals along the route of the M1 by both Poole (1968) and Bridger (1972) while the latter has observed it elsewhere in the area, particularly in the south, where, as the Newtown Linford till, it may incorporate a significant proportion of Charnwood Forest rock (Fig. 6).

In the motorway cutting at Charley Knoll (SK 486158), Poole noted over 7 m of Keuper-rich till with Coal Measure rocks overlying outwash gravel of similar composition. However, analysis of the gravel matrix and the lowest metre of till gave $CaCO_3$ contents of up to 15% for both deposits. Furthermore, the till, which was in part stratified, contained a small percentage of Cretaceous and Liassic material mainly as micro-fossils and clasts of limestone. Isolated examples of a calcareous red till have also been observed in other parts of the area and it therefore appears that in certain localities eastern material may be present in addition to the characteristic component of Triassic and Carboniferous rocks.

Further reference to the solid geology map (Fig. 5) shows that ice, carrying Coal Measure rocks, probably travelled from the north-west across the exposed portion of the Leicestershire coalfield as it moved towards Charnwood Forest. The small but diagnostic presence of Cretaceous and Jurassic rocks, notably at Charley Knoll, attests to the additional regional presence of ice from the north-east. That ice from the north-west ultimately became ascendant and gained sufficient height to surmount the greater part, if not all, of Charnwood Forest is demonstrated by the location of north-western till at altitudes of over 200 m O.D.

The chalky till, forming the uppermost and third of the types found in Charnwood Forest, typically contains a mixture of Upper Cretaceous and Liassic rocks. Observations of this type of till in the centre of the area are comparatively few but it is possible that a red-brown till with flints recorded in quarries (Ford, 1967) may represent a highly weathered and decalcified chalky boulder clay.

With one possible exception, the highest recorded altitude for chalky till in Charnwood Forest is just under 190 m O.D., a datum which at first sight is surprisingly low when compared with the figures for the other types of till. However, Ford (1968) has noted the presence of flint in a solifluction layer near the summit of Bardon Hill (278 m O.D.) (SK 460132), the highest point in Charnwood Forest, and believes that at one time a till containing flint capped the hill.

Since flint has not been observed in the Raunscliffe till and is rare in the red, north-western tills of Charnwood Forest it is probable that the deposit formerly capping Bardon Hill was a chalky boulder clay. This conclusion supports the contention that even the highest parts of Charnwood Forest were overwhelmed in the final phase of glaciation which, on the evidence of Cretaceous erratics, involved ice movement into Leicestershire from the north-east.

The absence of chalky till from much of Charnwood Forest is to be anticipated for, with a stratigraphical position as the uppermost of the three main types of till, it would be the first to be exposed to post-glaciation denudation. The former extensiveness of Cretaceous-rich till is confirmed by the scatter of flint, noted by Lucy (1870), over wide areas of Charnwood Forest.

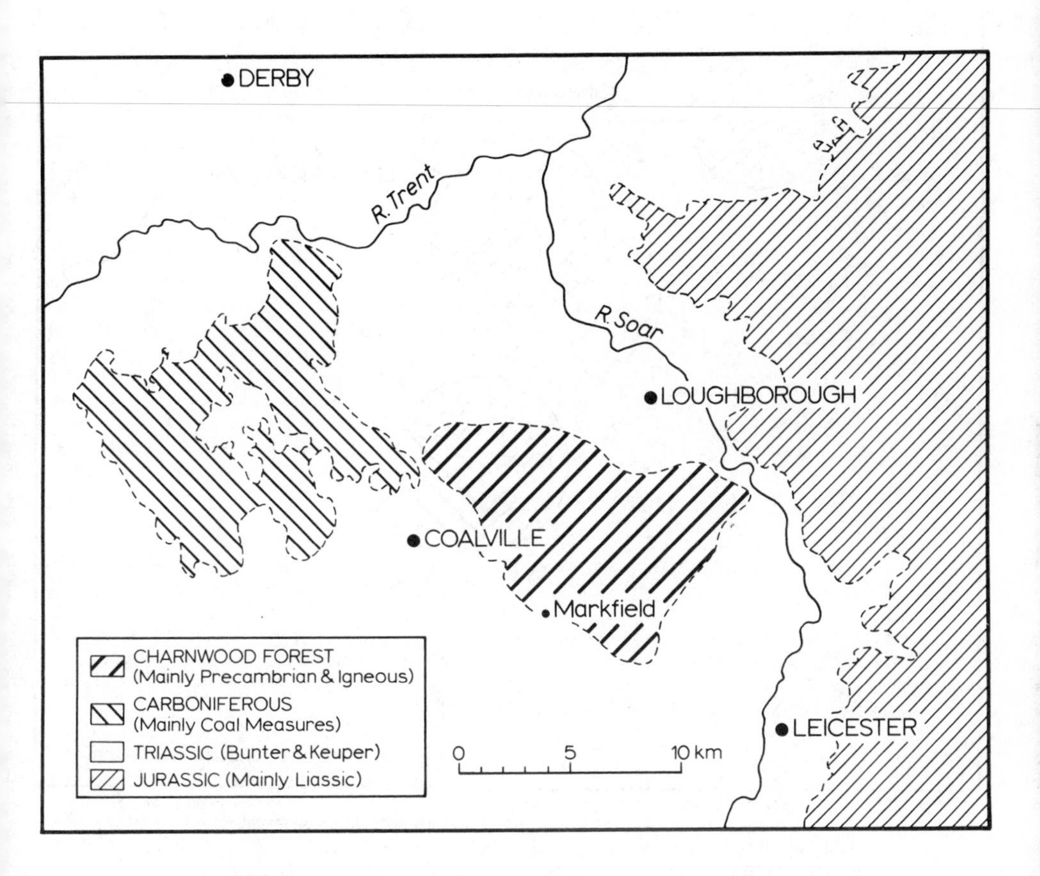

Fig. 5. Solid Geology of Leicestershire and Adjoining Counties

Mainly on lithological grounds both the red till and the chalky till of Charnwood Forest may be correlated with members of the Wolstonian succession elsewhere in the Midlands associated by Shotton (1953, 1976) and Rice (1968) with the formation of Lake Harrison. However, the dating of the lowest of the Charnwood Forest tills presents difficulties. The possibility of the Liassic-rich deposit belonging to the Anglian stage is not supported by evidence from other parts of the Midlands where Anglian tills are generally accepted as either north-western or Welsh in origin (Shotton, 1953, 1973).

An alternative is to position the till in a pre-Lake Harrison phase of the Wolstonian. A major problem here is the absence of a place for the required advance of early north-eastern ice in existing schemes for the Wolstonian glaciation of the Midlands. In this respect the uncertainty, expressed by Shotton (1968) of the precise position within the Wolstonian for the formation of Lake Harrison must be

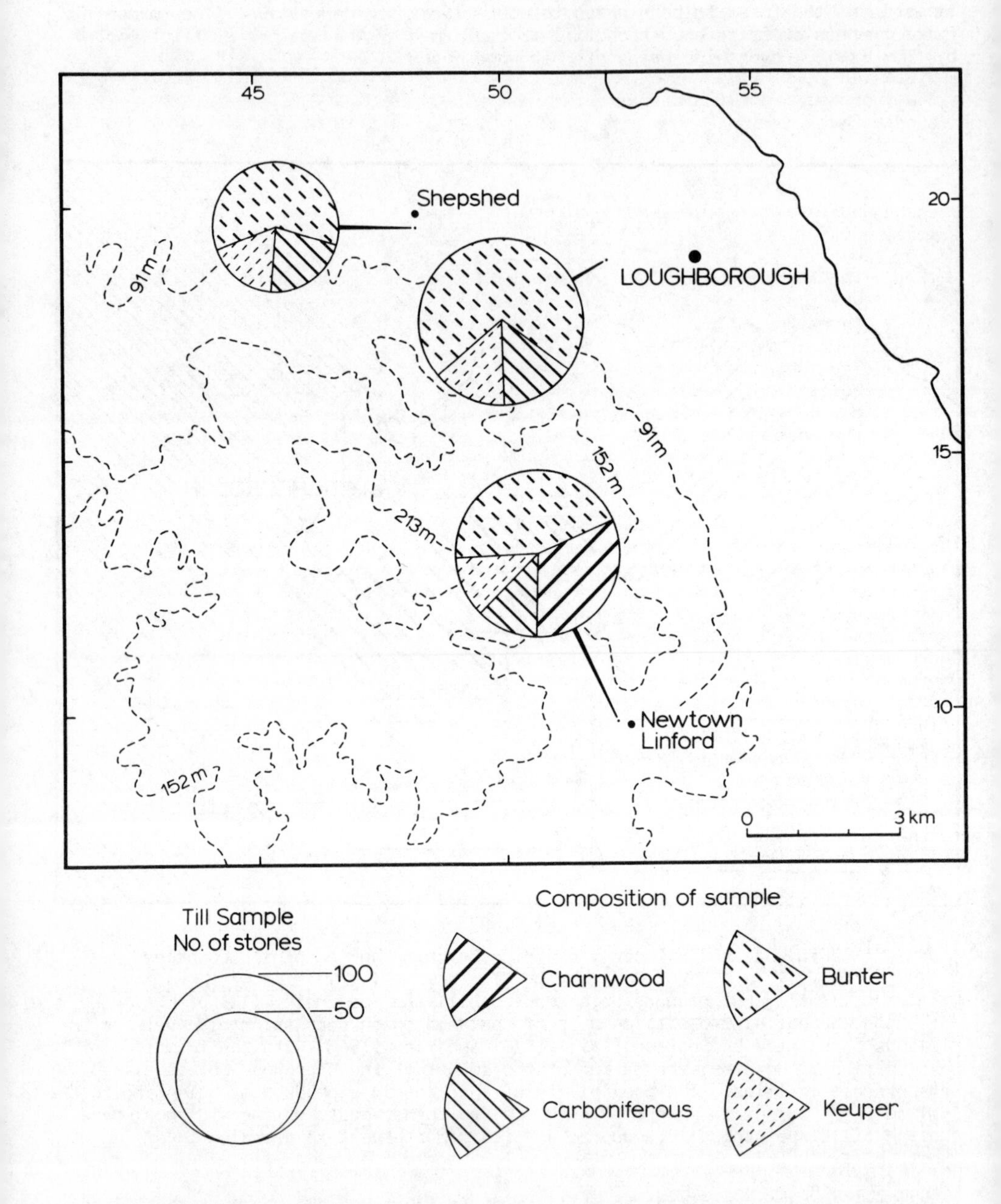

Fig. 6. Erratic content of tills of north western derivation.

kept in mind. Additonally, the work in the south Midlands by Bishop (1957) should not be overlooked for, in the Itchen Valley, he recorded a till with eastern erratics underlying the Wolstonian Series. Bishop explained the till as the deposit from a Wolstonian ice-lobe which retreated before Lake Harrison was created. These considerations have led to the tentative proposal of an early Wolstonian age for the Liassic-rich till of Charnwood Forest. Support for an early advance of eastern ice into western Leicestershire has also recently been presented by Douglas (1980) who gave a Wolstonian age to pre-Lake Harrison Basal Till from the area around Market Bosworth. The till has two facies and although the most widespread contains mainly Triassic erratics there is, significantly, also a very restricted occurrence of a deposit with chalk fragments.

Present understanding of the pattern of glaciation in Charnwood Forest may be summarised as follows:

1. The first recorded glaciation, provisionally dated as early Wolstonian, was by ice of north-eastern provenance which succeeded in advancing into Charnwood Forest to a minimum height of 215 m O.D.

2. The subsequent glaciation, placed in the Lake Harrison phase of the Wolstonian, completely over-ran Charnwood Forest and involved ice moving into Leicestershire first from the north-west, and later from the north-east.

THE GEOMORPHOLOGICAL IMPACT OF GLACIATION

The following discussion of the role of glaciation in the geomorphological evolution of Charnwood Forest is treated under the headings of glacial deposition and melt-water erosion. The omission of glacial erosion does not spring from the view that this process was locally unimportant, rather it reflects the failure of recent research to advance little from the position of recognising that the density of Charnian erratics in the Midlands implies considerable glacial scouring within Charnwood Forest. The only recent, specific, geomorphological reference to glacial erosion in the area has been by Ford (1967 and 1968) who, in brief discussion of the origin of local tors, concluded that their evolution included a phase of glacial sculpturing.

There is little doubt that glacial erosion has been of considerable geomorphological importance in Charnwood Forest but a precise appreciation of its significance awaits further investigation.

GLACIAL DEPOSITION

It has already been acknowledged that much of our understanding of glacial deposition in Charnwood Forest is based upon data derived from the construction of the M1 motor-way. Information from this source includes records of a resistivity traverse and bore-holes (kindly made available for publication by Sir Owen Williams and Sons) in addition to observations of temporary exposures. Collation of this data has allowed the compilation of a series of composite sections along the route of the motorway which runs north-south almost centrally through Charnwood Forest (Fig. 2).

Interpretation of the resistivity results has allowed the higher parts of the Precambrian surface to be identified; however, a limitation of the technique has been a failure to detect the boundary separating the Triassic from overlying Pleis-tocene deposits. Consequently, reliance has been placed upon temporary exposures and bore-holes to determine the Triassic-Pleistocene interface but, unfortunately, these lines of evidence have only permitted the plotting of the contact along a few short lengths of the route. Nevertheless the picture now available has clarified our understanding of the deposition of drift to the extent that a meaningful assessment of its geomorphological significance is possible.

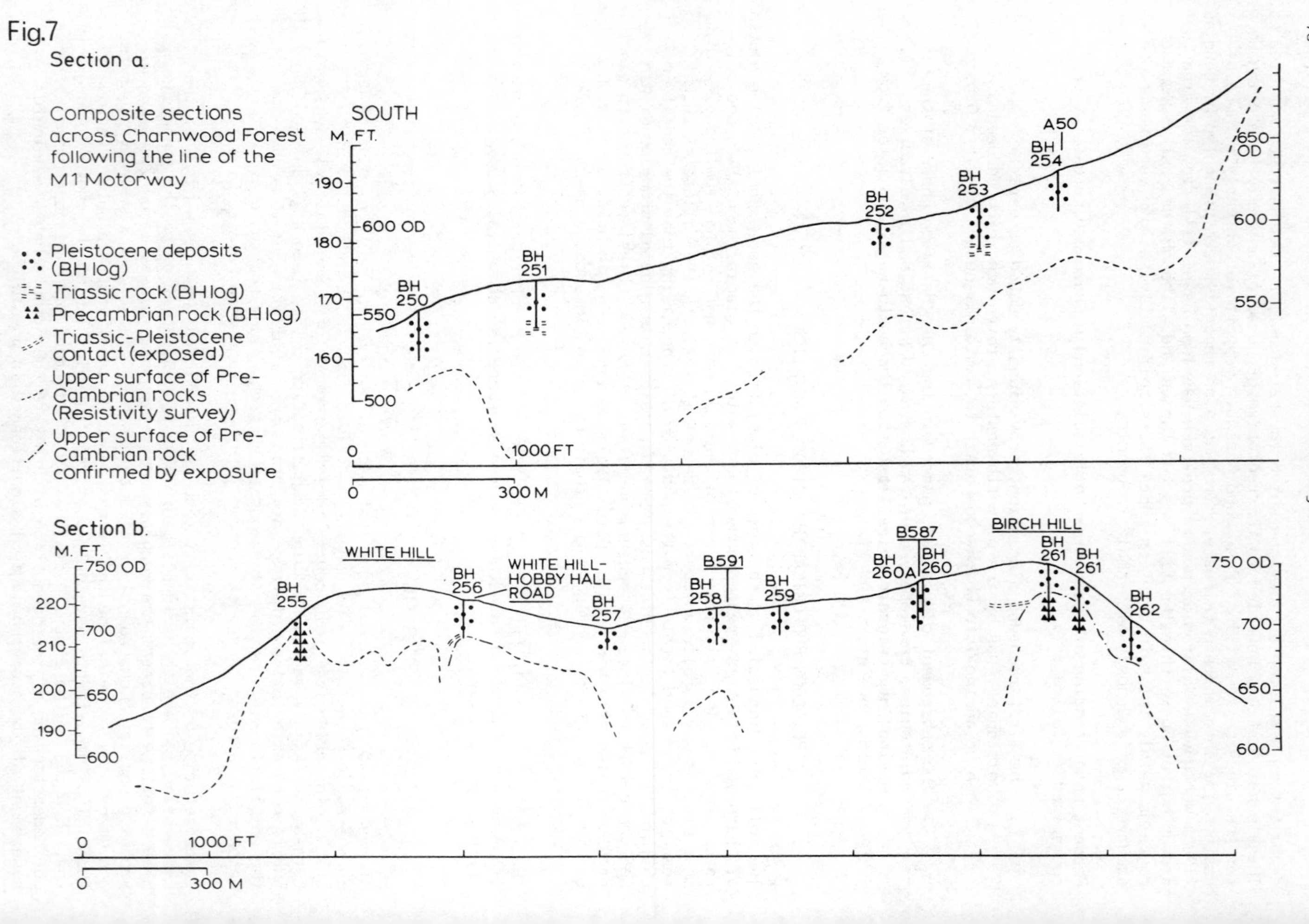
Fig.7
Section a.
Composite sections across Charnwood Forest following the line of the M1 Motorway
Pleistocene deposits (BH log)
Triassic rock (BH log)
Precambrian rock (BH log)
Triassic-Pleistocene contact (exposed)
Upper surface of Pre-Cambrian rocks (Resistivity survey)
Upper surface of Pre-Cambrian rock confirmed by exposure
SOUTH
M. FT.
190
180
170
160
600 OD
550
500
650 OD
600
550
BH 250
BH 251
BH 252
BH 253
BH 254
A50
0
1000 FT
0
300 M
Section b.
M. FT.
750 OD
700
650
600
220
210
200
190
WHITE HILL
BH 255
BH 256
WHITE HILL-HOBBY HALL ROAD
BH 257
BH 258
B591
BH 259
BH 260A
BH 260
B587
BIRCH HILL
BH 261
BH 261
BH 262
750 OD
700
650
600
0
1000 FT
0
300 M

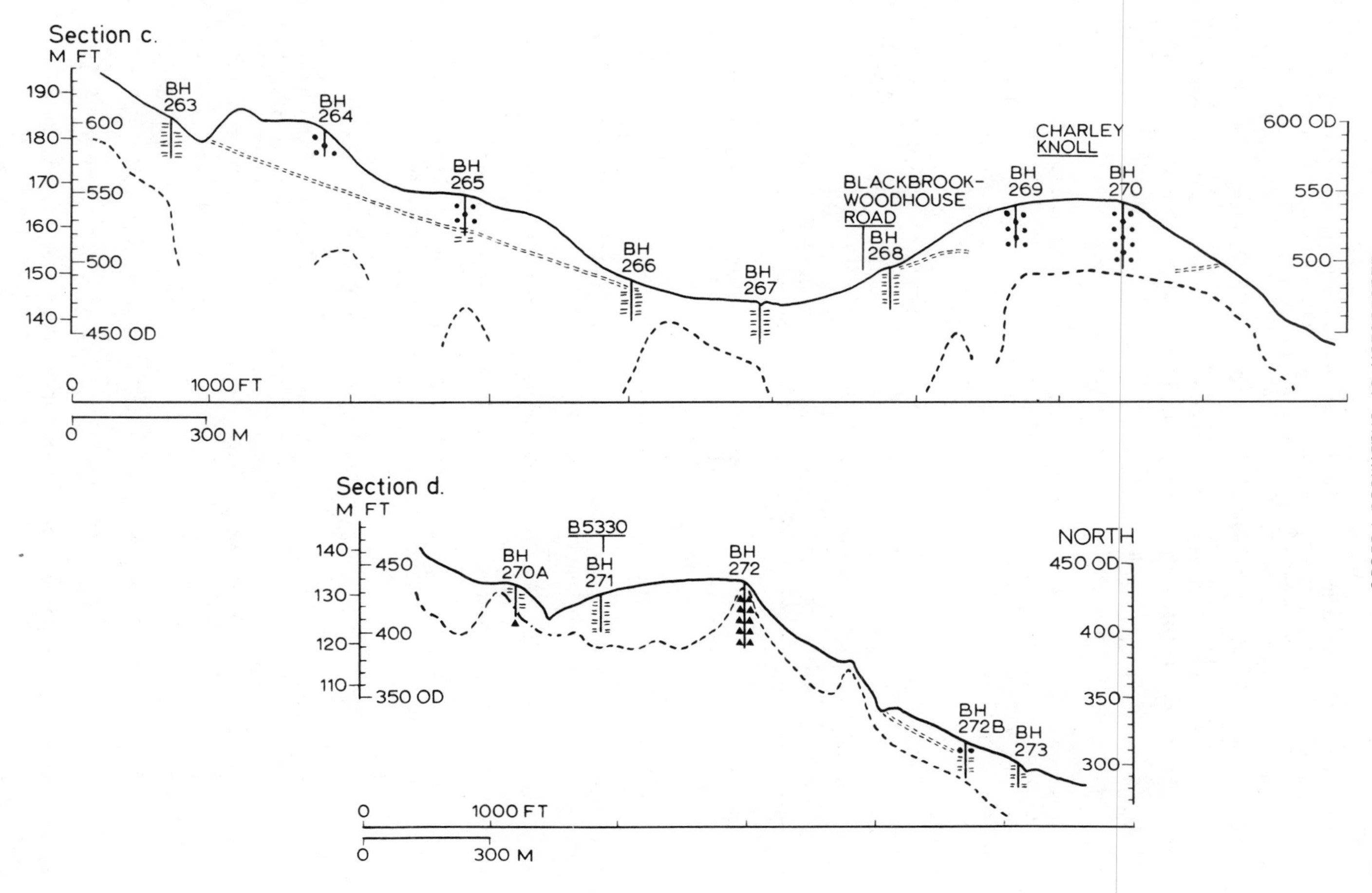
Section c.
M FT
190
180
170
160
150
140
600
550
500
450 OD
BH 263
BH 264
BH 265
BH 266
BH 267
BLACKBROOK-WOODHOUSE ROAD
BH 268
BH 269
CHARLEY KNOLL
BH 270
600 OD
550
500
0
1000 FT
0
300 M
Section d.
M FT
140
130
120
110
450
400
350 OD
BH 270A
B5330
BH 271
BH 272
BH 272B
BH 273
NORTH
450 OD
400
350
300
0
1000 FT
0
300 M

For this purpose it is convenient to divide the route of the M1 motorway in Charnwood Forest into four parts starting with the section showing the southern slopes (Fig. 7a). Here the Precambrian surface is seen to rise irregularly northward from around 150 m O.D. to over 210 m O.D.. To the south of the A50 junction Pleistocene deposits have a minimum thickness of 6 m while, to the north of it, meagre evidence suggests that the drift thins out to less than 3 m over the Precambrian at BH 255. Although the possibility of Triassic strata resting against the steeper faces of the Precambrian at depths below the base of the bore-holes makes the geomorphological impact of Pleistocene deposition on the southern flanks of Charnwood Forest difficult to assess, it is apparent that the presence of drift has undoubtedly acted to reduce the topographical expression of the Precambrian bedrock morphology.

The second section (Fig. 7b), between White Hill and Birch Hill, extends across some of the highest ground in Charnwood Forest. At a few points temporary excavations exposed the Precambrian and allowed comparisons of its surface with the record of the resistivity traverse. It was noted that, while the gradients shown on the resistivity profile were in close agreement with field measurements, adjustment had to be made for the lateral displacement of summits.

A notable feature of the resistivity profile is the 700 m wide valley in the Precambrian, lying to the south of Birch Hill, and having its base at not less than 180 m O.D.. The evidence of the trend of the Triassic-Pleistocene interface observed in local temporary exposures, together with recognition of the contact at the base of BH 260 A, indicates that at least the northern side of the valley is occupied by Triassic strata. Little is known about the possible occurrence of deposits of Triassic origin elsewhere along this length of the route but figures for minimum thicknesses of drift, in bore-hole records at five points between White Hill and Birch Hill, demonstrate that the infilling of the intervening depression has been largely accomplished by glacial deposition.

A clearer view of the position of Triassic rocks in the third section (Fig. 7c) indicates that they almost completely mask the Precambrian. The Triassic-Pleistocene interface is smooth and stream dissection of a former continuous cover of drift offers an acceptable explanation for the present amplitude of relief.

The fourth section (Fig. 7d) shows the northern edge of Charnwood Forest where a very irregular, Precambrian surface is seen to be almost totally hidden by Triassic strata. The virtual absence of glacial deposits in this area is in marked contrast to the situation on the southern slopes portrayed in the first section (Fig. 7a). The cause of the disparity is not thought to be the result of local inequalities in the thickness of glacial deposits but rather as a response to faster rates of post-glaciation, fluvial erosion in the north of Charnwood Forest.

The problem of an allowance for the amount of drift removed by post-glaciation denudation confounds attempts at reconstructing the topography of Charnwood Forest immediately following the melting of the ice. Notwithstanding this difficulty, the disposition of extensive areas of thick glacial deposits in addition to evidence of former till-cover on its highest summit, supports the notion that following glaciation Charnwood Forest had a blanket of drift giving it a form approximately that of an inverted saucer.

MELTWATER EROSION

The present drainage of Charnwood Forest (Fig. 3) is characterised by contrasts in stream pattern and valley morphology. The higher slopes have a near-radial stream network developing unexceptional valleys in both drift and Triassic rock. However, on the lower slopes, particularly at altitudes of around 120 m O.D., many streams are deflected from their radial orientations. In the northern and eastern areas of

Charnwood Forest these alterations of direction frequently take streams discordantly across outcrops of resistant rock, mainly of Charnian or igneous origin, where they have accomplished varying degrees of gorge development. The orientation of the deflected streams is not constant and minor gorges have been cut by streams showing no apparent deviation in course.

With knowledge of Charnwood Forest's glaciation it is no longer possible to accept the simple interpretation offered by Watts (1947) for the anomalous stream behaviour since it does not allow for disturbance of drainage evolution during the Pleistocene. Watts believed the pattern could be explained solely in terms of the superimposition of a Tertiary stream network developed upon a former cover of Mesozoic strata. The discordant gorges were thought by Watts to be largely the result of epigenesis on underlying ridges of 'ancient' rock.

An alternative view is put forward by Bridger (in press) who argues that a satisfactory explanation of the present drainage must take into account Pleistocene events with particular reference to the role of glacial meltwater, although it is acknowledged that preglacial discordant drainage probably developed in certain areas as demonstrated by a small drift-filled ravine at Ulverscroft Mill (SK 514108).

The details of the pre-glacial relief are believed to have had little effect on the general direction of meltwater-flow during the Wolstonian deglacial phase; the flow would, therefore, be controlled by glaciological factors and, in the envisaged conditions, would give rise to an essentially radial pattern.

The overall arrangement of the present drainage is thought to bear the imprint of a network which evolved in a Wolstonian setting of melting ice. An environment of deglaciation offers a variety of possibilities to account for the deflection of meltwater routes, including the influence of crevasses on englacial and subglacial streams during a phase of ice down-wasting (Clapperton, 1968). Crevasse development would presumably be at a maximum at the edges of Charnwood Forest where breaks of slope are most pronounced. Accordingly, meltwater streams entering the peripheral belt of fractured ice might be deflected, to a greater or lesser extent, from the radial arrangement they displayed on the upper slopes. Where such diversions became well established they would, in favourable conditions, be followed by streams after the ice had melted.

The discordant behaviour of streams may thus be seen as a further consequence of drainage evolution during deglaciation. The anomalous network in Charnwood Forest has frequently taken streams over ground where superimposition through ice on to underlying Charnian and intrusive strata would lead to discordance of drainage and, subsequently, to gorge development. Glaciological interpretations, involving glacial meltwater activity, have been put forward to explain gorge formation in other parts of Britain notably by Price (1960) and Sissons (1963). However the widespread cover of drift and also Triassic strata in Charnwood Forest has meant that the possibility of epigenesis need not necessarily cease at the end of the Wolstonian glaciation and some of the discordant stream sections manifesting little or no down cutting are likely to be of post-Wolstonian age.

The problem of the asymmetric distribution of discordant sections is probably as much concerned with the contrasting Upper Pleistocene histories of the valleys of the Proto-Trent and Proto-Soar, lying respectively to the north and south of Charnwood Forest, as with possible local differences in the conditions of deglaciation. The valley of the Proto-Soar has remained largely buried under drift (Shotton, 1953) whereas the valley of the middle Trent was probably re-established shortly after the melting of the Wolstonian ice (Straw, 1963). The latter event has given an important, erosional advantage to the tributaries of the Trent draining the northern and eastern slopes of Charnwood Forest and, accordingly, increased the chances of the formation of discordant sections by epigenesis through both Pleistocene and Triassic deposits in these localities.

CONCLUSION

Recent research has conclusively demonstrated that Charnwood Forest was glaciated during the Pleistocene. Two ice advances have been identified. The first was of north-eastern origin but there is doubt as to its age which is tentatively proposed as early Wolstonian and also the extent of its penetration into Charnwood Forest. The following glaciation, dated as Wolstonian, completely overran the area and involved ice-movements from the north-west as well as from the north-east.

Our knowledge of the geomorphological effect of glacial erosion in Charnwood Forest is far from complete but it is most unlikely that the local tors escaped a measure of glacial plucking. This point, together with indications of further modification by periglacial processes (Ford, 1967), challenges Watts' claim that the Charnwood Forest crags represent uninjured Triassic landforms.

A fuller comprehension of the role of glacial deposition indicates that, on the melting of the Wolstonian ice, a veneer of drift covered Charnwood Forest. Although subsequent denudation has eroded this cover, sufficient remains for it to continue to have considerable bearing on the morphology of the area.

The dissection of the drift by stream erosion has been associated with an anomalous, in part discordant, drainage pattern. Such stream behaviour is believed to be the imprint of a glacial meltwater network which has epigenetically produced discordant phenomena. In identifying superimposition as the process responsible for the discordant gorges of Charnwood Forest Watts was almost certainly correct but his failure to recognise the evidence for the area's glaciation precluded the possibility of his appreciating that present day features of the drainage may be of Upper Pleistocene origin.

ACKNOWLEDGEMENT

The author would like to thank Dr R.J. Rice for comments on this paper in draft form.

REFERENCES

Bennett, F.W. (1929). Remarkable features in the Ulverscroft valley. *Trans. Leicester lit. phil. Soc.*, 30, 40-45.

Bishop, W.W. (1957). Pleistocene geology and geomorphology of three gaps in the Middle Jurassic escarpment. *Phil. Trans. R. Soc.*, B. 241; 255-305.

Boulton, G.S. (1968). Flow tills and related deposits on some Vestspitsbergen glaciers. *J. Glaciol.*, 7, 391-412.

Bridger, J.F.D. (1968). Remarkable features in the Ulverscroft valley; a reappraisal of the drainage history. *Trans. Leicester lit. phil. Soc.*, 62, 73-77.

Bridger, J.F.D. (1972). *The Quaternary history of Charnwood Forest, Leics.* Unpub. MSc thesis, University of Leicester.

Bridger, J.F.D. (1975). The Pleistocene succession in the Southern part of Charnwood Forest. *Mercian Geologist*, 5, 189-203.

Bridger, J.F.D. (In press). The problem of discordant drainage in Charnwood Forest, Leicestershire. *Mercian Geologist.*

Clapperton, C.M. (1968). Channels formed by the superimposition of meltwater streams with special reference to the East Cheviot Hills, northeast England. *Geogr. Annlr.*, 50A, 207-220.

Douglas, T.D. (1980). The Quaternary deposits of western Leicestershire. *Phil. Trans. R. Soc.*, B. 288, 259-286.
Ford, T.D. (1967). Deep weathering, glaciation and tor formation in Charnwood Forest, Leicestershire. *Mercian Geologist*, 2, 3-14.
Ford, T.D. (1968). The Morphology of Charnwood Forest. In P.C. Sylvester-Bradley and T.D. Ford (Eds), *Geology of the East Midlands*. Leicester University Press.
Lucy, W.C. (1870). Notes upon the occurrence of the Post-Oligocene drift of Charnwood Forest. *Geol. Mag.*, 7, 497-499.
Poole, E.G. (1968). Some temporary sections seen during the construction of the M1 Motorway between Enderby and Shepshed, Leicestershire. *Bull. geol. Surv. Gt Br.*, 28, 137-151.
Price, R.J. (1960). Glacial meltwater channels in the upper Tweed drainage basin. *Geog. J.*, 126, 483-489.
Rice, R.J. (1968). The Quaternary deposits of Central Leicestershire. *Phil. Trans. R. Soc.* A. 262, 459-508.
Shotton, F.W. (1953). The Pleistocene deposits in the area between Coventry, Rugby and Leamington and their bearing on the topographical development of the Midlands. *Phil. Trans. R. Soc.*, B. 237, 209-260.
Shotton, F.W. (1968). The Pleistocene Succession around Brandon, Warwickshire. *Phil. Trans. R. Soc.*, B. 254, 387-400.
Shotton, F.W. (1973). A mammalian fauna from the Stretton Sand at Stretton-on-Fosse South Warwickshire. *Geol. Mag.*, 109, 473-476.
Shotton, F.W. (1976). Amplification of the Wolstonian Stage of the British Pleistocene. *Geol. Mag.*, 113, 241-250.
Sissons, J.B. (1963). The glacial drainage system around Carlops, Peebleshire. *Trans. Inst. Br. Geogr.*, 32, 209-260.
Straw, A. (1963). Quaternary evolution of the lower and middle Trent. *East Midland Geographer*, 3, 171-189.
Watts, W.W. (1903). A buried Triassic landscape. *Geogrl. J.*, 21, 623-636.
Watts, W.W. (1905). Buried landscape of Charnwood Forest. *Trans. Leicester lit. phil. Soc.*, 9, 20-25.
Watts, W.W. (1947). *Geology of the Ancient Rocks of Charnwood Forest*. Leicester.

J.F.D. Bridger, Esq.,
Faculty of Combined Studies,
Hull College of Higher Education,
Inglemire Avenue,
Hull HU6 7LJ.

9. Quaternary History of the Southern Part of the Vale of York

G. D. Gaunt*

INTRODUCTION

During the last two decades the Quaternary deposits in much of the southern part of the Vale of York have been re-surveyed by the I.G.S., and thousands of borehole records have provided a three-dimensional assessment. The deposits are shown on the I.G.S. Goole (79) and Doncaster (88) sheets, already published, and on the forthcoming Kingston upon Hull (80) sheet, and they will be described in detail in the appropriate memoirs, now in preparation. Much of the information is available in a thesis (Gaunt, 1976a) and some aspects, mainly resulting from palaeontological work and radiocarbon dates, are already published. Only summaries of the factual data are given here, therefore, the main aim being to construct a condensed history of the region, shown diagrammatically on Fig. 1, for that part of the Quaternary represented by deposits.

There is sufficient evidence to recognise the principal episodes of the last three British Quaternary stages, the Ipswichian, Devensian and Flandrian. Older glacial deposits are present, but it is uncertain whether they result from one, or more than one, glaciation, and correlation with the Anglian and/or Wolstonian stages is not possible. In this account only one such pre-Ipswichian stage, referred to as the Older Glacial Stage, is assumed.

OLDER GLACIAL STAGE

Prelude

Most of the pre-Ipswichian glacial deposits in the southern part of the vale lie at low elevations relative to the lowest Pennine slopes to the west and Jurassic scarps to the east, showing that the region was already low lying before this glaciation, and till east of Wadworth (SK 570970) and on Misterton Carr (SK 740935) implies the existence of the Torne and Idle valleys by this time. Doubtless erosion took place in much of the earlier Quaternary, but by analogy with events during the Devensian (described below) it is possible that, as a consequence of a glacially lowered sea level, substantial denudation and river incision occurred in the Older Glacial Stage

---

* Published by permission of the Director, Institute of Geological Sciences

before the ice arrived. This suggestion is enhanced by Zagwijn's (1973) demonstration in Holland that in the Saalian Stage (probably equivalent to the Older Glacial Stage) a long sequence of ice-free stadials and interstadials preceded the glacial episode.

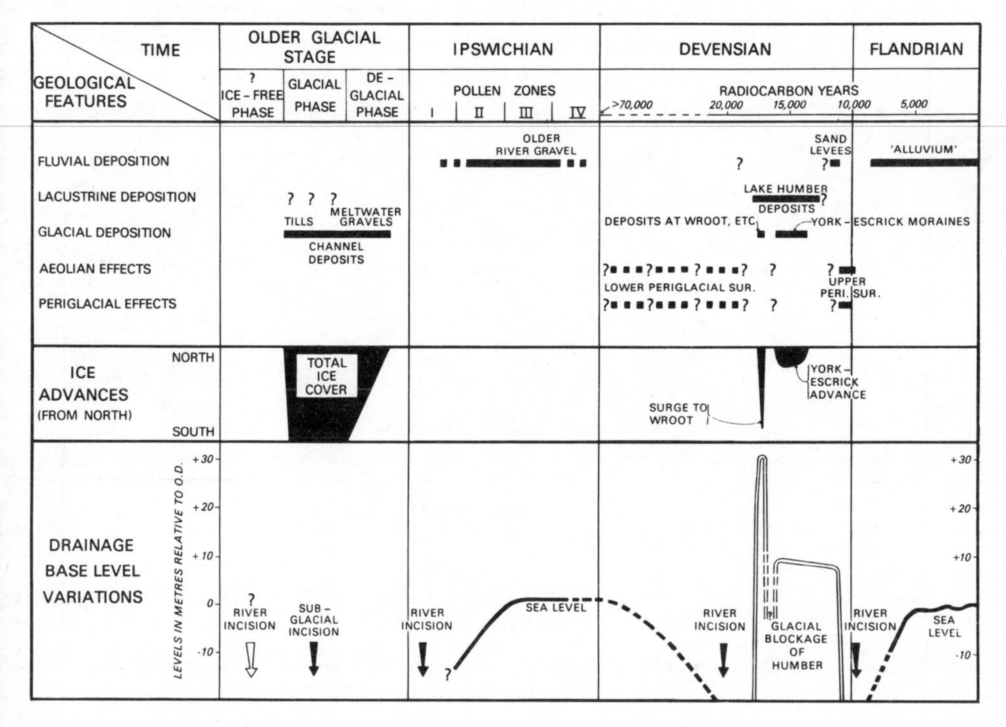

Fig. 1. Diagrammatic summary of Quaternary history

Initial Ice Advance

It has been realised for over a century (e.g. Green and others, 1878, p. 782) that, during a glaciation earlier than the last one, the entire region was covered by thick ice containing Lake District and Pennine erratics. It is logical to assume that the initial advance of this ice was by one of the lower lying trans-Pennine routes, either Stainmore (and southwards down the Vale of York) or the Aire gap. Both routes have had protagonists, Carter (1905) and Harmer (1928) favouring Stainmore and Melmore (1935) the more southerly route. The Pennine erratics are not diagnostic in this respect. Fig. 2 illustrates three lines of evidence which together possibly resolve the question.

Stone orientations in tills on the summit of Brayton Barff (SE 585305), near Selby, suggest ice movement from only slightly west of true north, in effect down the Vale of York, but orientations from till at Balby (SE 562004) suggest movement from between NNW and west (Gaunt, 1976a). The Balby till was emplaced in a SW-NE valley (coincident with a similarly aligned graben) and from the location of the orientation-sampling sites it is deduced that the ice approached from the north-west, was deflected eastwards across the valley and then resumed its former course.

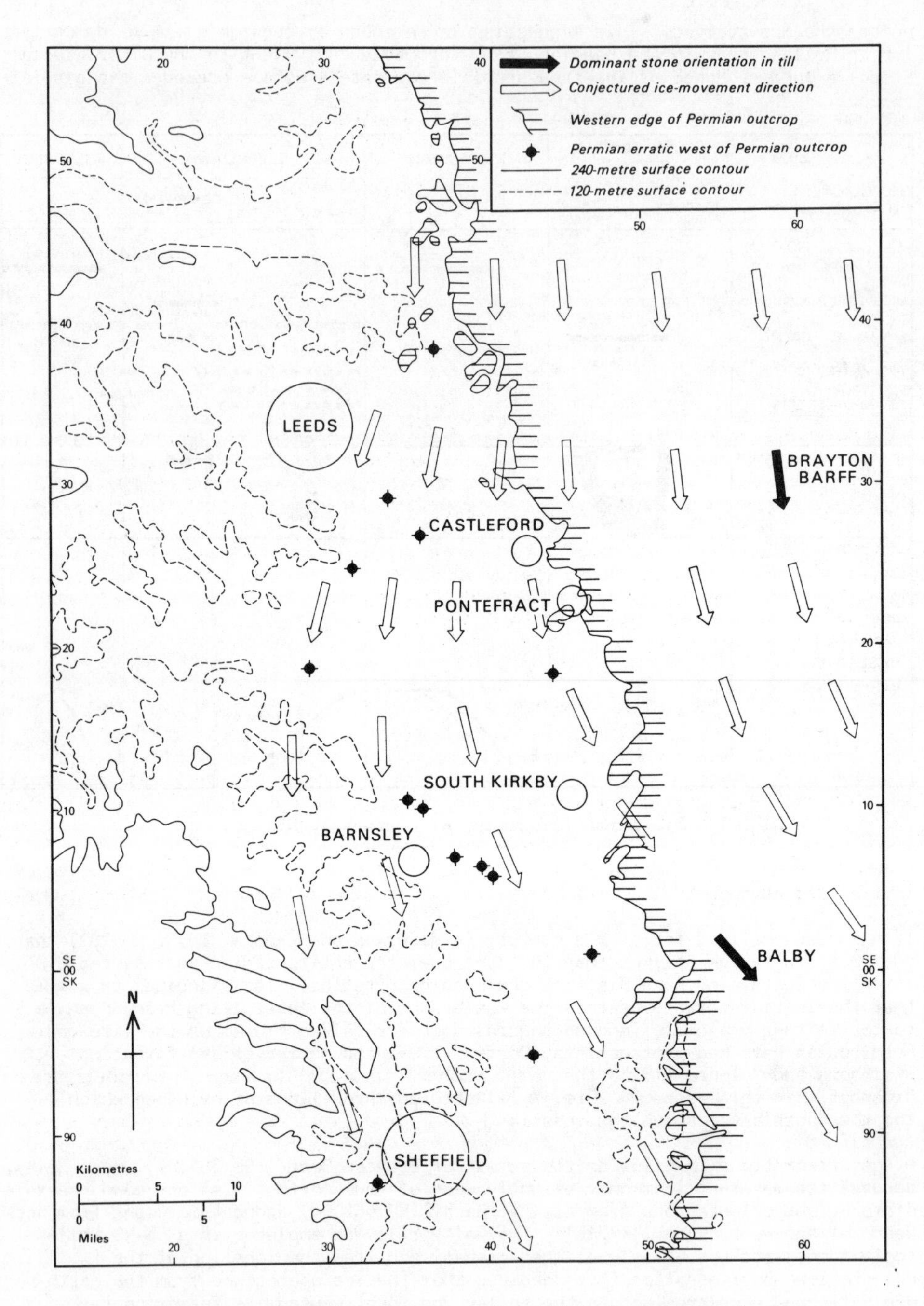

Fig. 2. Evidence of direction of initial ice advance in the Older Glacial Stage.

The presence of coal erratics in the Balby till implies a westerly derivation. Reflectance measurements and spore analyses (to give the precise rank and approximate age, respectively) on some of these erratics by Dr A.H.V. Smith* suggest sources in the Castleford-Pontefract-South Kirkby area (pers. comm.), and therefore ice movement from between NNW and NW.

A search of the literature and of I.G.S. field maps shows that Permian limestone erratics are not uncommon in the pre-Ipswichian glacial deposits lying west of the Permian outcrops in south-western Yorkshire. The only conceivable source to the north-west is the thin Belah Dolomite in the Vale of Eden, but its outcrops are so minute that it can be discounted, and the inescapable conclusion is that the Permian erratics in south-western Yorkshire come from the Yorkshire Permian outcrops, implying ice movement from between north and NE.

This apparently conflicting evidence can be reconciled if the initial ice advance was via Stainmore and the Vale of York. Ice in the middle of the vale would move southwards, as probably reflected by the stone orientations at Brayton Barff. Ice moving down the western edge of the vale would, even by merely maintaining a southward movement, cross the Permian outcrops and transgress on to Coal Measures to the west. Once this western ice had outflanked the high ground north of Leeds, however, it would probably spread out south-westwards over the lower ground to the south, carrying Permian erratics even farther west. Ultimately this ice would be deflected southwards and south-eastwards as it encroached on to higher ground west of Barnsley and Sheffield, giving the stone orientations and coal erratics at Balby. It is conceivable that contemporaneous but 'weak' ice from the Aire gap may have helped to deflect the main ice to the south-east, but ice from this source alone cannot account for all the evidence.

Sub-glacial Drainage

In the Doncaster area numerous narrow steep-sided channels, closed at both ends, have been deeply incised into bedrock, mainly Sherwood Sandstone (Fig. 3). The maximum depth known is -58 m O.D., more than twice that of the rockhead exit from the vale under the Humber. Moreover, the channels are all aligned roughly NW-SE, a direction unrelated to any known pre-existing river system in the area. The western end of each of the channels is aligned with one or other of the gaps through the Permian limestone scarps (e.g. the Went Gorge) and all the eastern ends are directed towards the Haxey gap at the southern end of the Isle of Axholme. Similarly aligned closed channels are known near Kellington (SE 550250) (where some are cut into glacial deposits), Eggborough (SE 565233) and Wrangbrook (SE 494133).

The deposits filling the channels consist largely of laminated virtually stoneless over-consolidated clay. Some stony clay till is present, but rare except in the upper parts. Sand, with and without gravel, is present, being more common in the lower parts of the channels; it also fills the eastern ends of some of them. The erratics (some up to boulder size) and pebbles in these deposits consist mainly of Carboniferous sandstone and limestone (the latter commonly grooved) and Permian limestone, and detrital coal is present on the clay laminae and in the sand, these constituents denoting derivation from the north-west. The complex arrangement of clays, sands and gravels in a temporarily exposed channel near Eggborough indicates several cycles of incision and deposition.

Since at least one channel cuts into pre-Ipswichian glacial deposits but others contain or are overlain by till the channels (and their deposits) are presumably of

---

* National Coal Board, Yorkshire Regional Laboratory, Wath upon Dearne, Rotherham

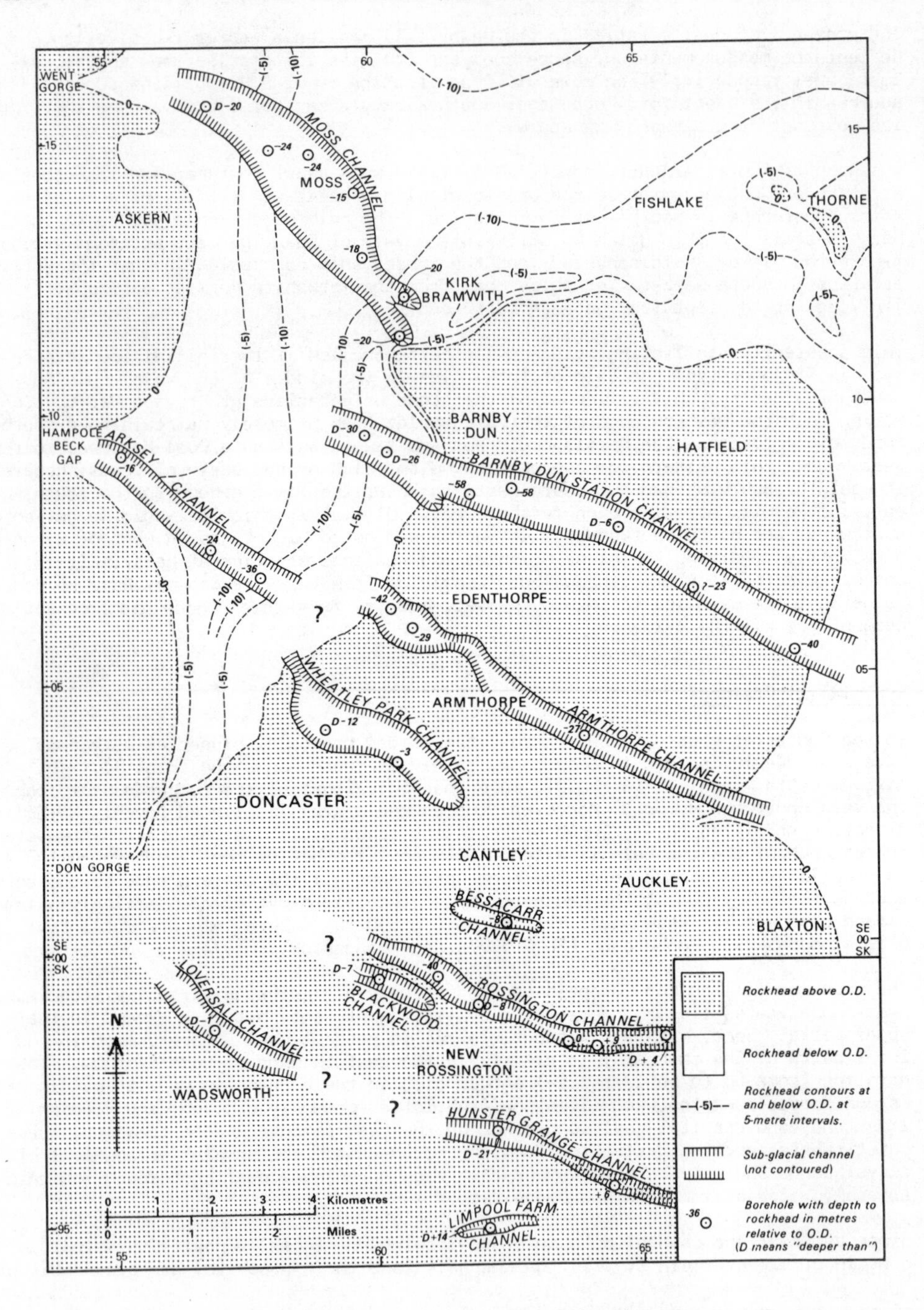

Fig. 3. Sub-glacial channels in the Doncaster area

glacial origin, and as their depth and closed nature preclude sub-aerial incision they must have been cut sub-glacially. These channels are similar to the 'tunnel-dale' of Denmark, the 'rinnentaler' of Germany and the tunnel-valleys of Britain e.g. East Anglia (Woodland, 1970), all of which are believed to have formed by sub-glacial drainage under enormous hydrostatic pressure. The channel alignments and constituents imply that here the pressure was imposed from high in the Pennines to the north-west, and there seems a distinct possibility that the gaps in the Permian scarps and the Haxey gap are genetically related to the channels.

The fact that the channel deposits do not appear to extend beyond the channel sides, the evidence of repeated cycles of cut and fill, the over-consolidation of the clays, the piling up of sand and gravel at the downstream end of some channels (suggesting that some flows may have 'choked' their own channels) and the extension of till across the top of some channel deposits all point to deposition, as well as incision, having been sub-glacial.

Conjectured Lakes

Although the ice reached over 200 m O.D. near Sheffield (Eden, Stevenson and Edwards, 1957, 158, fig. 29) and over 300 m O.D. near Chesterfield (Smith, Rhys and Eden, 1967, 223-225, fig. 36) there is no evidence that it surmounted the 480 m-high cuesta around the northern and eastern flanks of the Peak District. If the Pennine uplands were ice free, glacial lakes may have existed and provided the hydrostatic head responsible for the sub-glacial channels. Laminated clay is intimately associated with glacial deposits near Barnsley (Green and others, 1878, 776-777, fig. 26), near Rothwell (SE 340285) (Gilligan, 1918) and on top of Brayton Barff, and it is tempting to invoke pro-glacial, supra-glacial and marginal lacustrine origins in a more extensive precursor of Devensian Lake Humber.

Meltwater Drainage

The sand and gravel on the Triassic ridge running northwards past Rossington (SK 630985) possess constituents quite different from those of the locally underlying tills and channel deposits, and cannot have been derived from the ice which occupied the area. The sand is red-brown and devoid of detrital coal and clayey layers, and most of the pebbles come from the Bunter Pebble Beds of the Midlands; the size of some pebbles denotes sources in the upper Trent area. Since there are no Jurassic components the small proportion of flint pebbles present is unlikely to have been transported directly from the east and so must also have come from the south, from the 'chalky' glacial deposits (or the parent ice) in the middle Trent Valley. Northerly cross-bedding dips also imply transport from the south. Similar deposits occur on relatively higher ground farther south, reaching over 180 m O.D. near Kirkby in Ashfield (Smith, Rhys and Eden, 1967, 224-227, fig. 36).

The flints and larger Bunter pebbles must have crossed the watershed on the northern side of the middle Trent Valley, a circumstance requiring a glacial environment to mask the rockhead configuration, and all these southerly-derived deposits are believed to result from meltwater escaping northwards, in partly ice-walled channels, from the midlands. Beetles from these deposits near Scrooby (SK 652908) indicate very cold continental conditions (Maureen Girling, pers. comm.). This northerly meltwater flow presumably pre-dated the cutting of the Trent 'trench' between Nottingham and Newark and deposition of the Hilton gravels between Newark and Lincoln by easterly directed meltwater from the middle and upper Trent which, according to Straw (1963), may also have flowed in partly ice-walled channels.

The sand and gravel on the Triassic ridges running north-eastwards under Doncaster and eastwards past Kellington and Snaith (SE 643222) are similar to those on the Rossington ridge in stratigraphic context, sedimentary structures, elevation and topographic expression, but differ markedly in composition and direction of cross-bedding dips. The sand is yellow-brown, commonly clayey, with distinct clay beds and appreciable detrital coal; most of the pebbles are of Carboniferous sandstone, the rest being mainly of siltstone, ironstone and even coal. Although superficially similar in constitution to the underlying glacial deposits, the presence of coal and absence of Lake District, Carboniferous limestone and chert pebbles suggest a westerly rather than a north-westerly source, a conclusion supported by easterly cross-bedding dips. By analogy with the otherwise identical deposits on the Rossington ridge, those on the Doncaster and Snaith ridges are believed to result from transport by meltwater from the Coal Measures outcrops to the west. The absence of Permian limestone pebbles may be due to the flows having utilised ice-bound channels through the Aire and Don gaps in the limestone scarps.

The most likely reason for meltwater having entered the region from the south and west was enormous isostatic depression to the north-east. At present the only clue to correlation of the Older Glacial Stage deposits with pre-Ipswichian deposits elsewhere is the presence of flints in the Rossington ridge deposits, which shows that de-glaciation of 'Pennine' ice in the region and of 'chalky' ice in the middle Trent Valley was contemporaneous.

IPSWICHIAN STAGE

River Incision

Ipswichian Zone IIb deposits occurring as low as -12 m O.D. under Langham (SE 682213) (Gaunt, Bartley and Harland, 1974) show that earlier in the interglacial (possibly even starting late in the Older Glacial Stage) rivers had incised courses to considerable depths in response to a sea level still eustatically depressed from the preceding glaciation. Unfortunately it is impossible to trace these incised courses because two subsequent deeper incisions have destroyed most of the buried early Ipswichian topography.

River Deposition

Pollen and dinoflagellate cysts in clays, sands and gravels at between -12 and -6 m O.D. under Langham show that as sea level rose in Ipswichian Zone IIb times some estuarine sediments were forming in an area containing pine and oak woodland (Gaunt, Bartley and Harland, 1974). Insects, pollen and other floral remains in an organic silt near Austerfield (SK 662947) show that in Zone III times a marshbound lake surrounded by thick forest containing alder, pine and hornbeam existed there in a climate at least as warm as at present (Gaunt and others, 1972). The presence of a lake at about 4 m O.D. on a contemporaneous floodplain overlying fully permeable deposits and bedrock in a non-glacial environment implies that drainage base level, in effect sea level, was close to O.D. Pollen, other plant debris and dinoflagellate cysts from a thin clay near Westfield Farm (SE 63340360), Armthorpe (Gaunt, Bartley and Harland, 1974), show that at the time of the zones III/IV junction estuarine sediments were forming, with sea level at or slightly above O.D., in an area containing pine-rich woodland.

The organic deposits near Austerfield and Westfield Farm (and others from near Finningley (SK 673993) yielding an infinite radiocarbon date on abundant wood fragments) occur within the Older River Gravel of the Doncaster area. The composition of these sands with fine gravels changes markedly from north to south. In the north the sand is yellow-brown, clayey, with distinct clay beds and detrital coal, and the pebbles are mainly of Carboniferous sandstone and other durable Coal Measures rocks. South of a 1-2 km-wide transition zone running from Cantley (SE 627022) through Auckley (SE 650012) and then north-eastwards, the sand is red-brown and devoid of clay and coal, and the pebbles are mainly of 'Bunter' type with a few flints. The westerly and southerly derivation of these facies is obvious, as is their distribution where the Don and Idle valleys open into the Vale of York , and they are believed to have been deposited as these rivers filled up their incised courses and formed floodplains over the adjacent lowland. Concealed deposits apparently comparable to the Older River Gravel are present where the Aire Valley opens out into the vale near Knottingley.

DEVENSIAN STAGE

Denudation and River Incision

Contours drawn on the base of the Devensian deposits (mainly the 25-Foot Drift) show a landscape (Fig. 4) in which rivers have incised courses down to -19 m O.D. towards

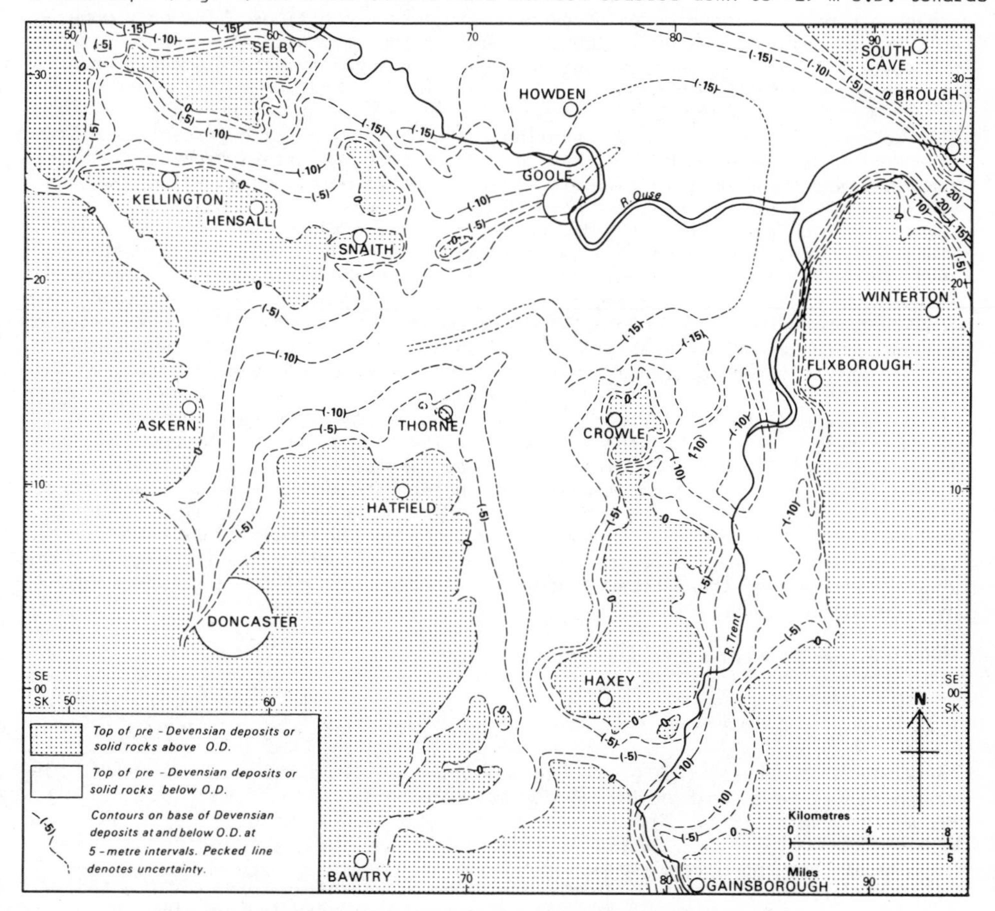

Fig. 4. Contours on the base of the Devensian deposits.

the Humber gap in wide and mature valleys, clearly the product of prolonged denudation coincident with greatly lowered sea level. The magnitude of this lowering implies a major eustatic response to glaciation, so the long phase of denudation and incision is ascribed to the Devensian.

Aeolian and Periglacial Conditions

Besides blown sand the region contains vast numbers of ventifacts and appreciable areas of 'desert pavement' as testimony to aeolian activity. Ventifacts (or dreikanter) are stones which have been sculptured by sand blasting. Desert pavement is an ill-sorted gravel layer, generally deficient in a sand matrix and depositional structures, which 'follows' the present (or former) ground surface regardless of the attitude of any underlying bedding. It is a remanié pebble assemblage resulting from winnowing out of sand from sand and gravel. In the Low Countries desert pavements are used as stratigraphical markers (Van der Hammen and others, 1967; Paepe and Vanhoorne, 1967).

In addition to head (solifluction) deposits the region contains an abundance of frozen-ground structures as evidence of periglacial conditions. The most numerous of these structures are involutions, commonly associated with clay-enriched masses and vertically aligned pebbles, but ice-wedge casts occur locally and traces of stone stripes have been noted.

All frozen-ground structures and aeolian features in the region (except rare traces near Burton Hall and South Cave, see below) occur at two stratigraphical levels, the lower and upper periglacial surfaces. The lower periglacial surface is coincident with the top of the Older River Gravel, and where that deposit is absent it transgresses on to the top of older deposits or solid rocks, but where Devensian glacial or lacustrine deposits occur it dips beneath them (see Gaunt and others, 1972, fig. 2). The associated frozen-ground structures indicate continuous permafrost. To the north of the region the lower periglacial surface is represented by ventifacts beneath Devensian glacial deposits at York (Gaunt, 1974), near Allerton Mauleverer (SE 416580) (Gaunt, 1976b) and at Aldborough (SE 406663) (Gaunt, 1970a); to the west it is recognized by ventifacts beneath Aire and Calder terrace deposits (Edwards, 1936; Bisat, 1946) and by deeply penetrating ice-wedge casts (not found in the upper periglacial surface) e.g. into Coal Measures beneath Lake Humber deposits near Darfield (SE 415045) (Gaunt, 1976a). Demonstrably Devensian, the lower periglacial surface results from one or more severe periglacial episodes during or following the long denudational phase; its widespread, in places almost continuous, presence suggests that its final development was late with respect to the denudation. West Moor (SE 650065) and other smaller depressions date from this time, and are believed to be thermokarst features, possibly alases, which can form where frozen ground thaws because of localized reduction of the insulation cover over permafrost (Gaunt, 1976a).

The upper periglacial surface is described on p. 93.

Lake Humber

Three types of deposit (high-level sand and gravel, low-level sand, laminated clay) are attributable to Lake Humber, a glacial lake in the Vale of York originally deduced by Lewis (1887a, b).

High-level sand and gravel contain constituents derived by re-working of adjacent older deposits and/or rocks. They occur along slopes at up to 33 m O.D., their elevation generally increasing to the north, and they overlie the lower periglacial surface. They have been given different names in different areas, and include the 100-Foot Strandline deposits of Edwards (1937), 'perched' deltas labelled as various terrace and glacial deposits in the Aire, Calder, Dearne and Don valleys, some of the 'glacial' sand and gravel on the slopes of Brayton Barff and on the Snaith and Rossington ridges and the 'Older Littoral Sand and Gravel' near Holme on Spalding Moor (SE 810385) and Pocklington. They are traceable farther south to the Humber, and across it into Lincolnshire.

Low-level sand (with rare gravel) includes that part of the '25-Foot Drift Sand' which is banked against containing slopes of older deposits or solid rocks (with the lower periglacial surface intervening) and which passes laterally in the opposite direction into laminated clay. Its narrow outcrops, e.g. around the base of Brayton Barff and the Snaith ridge, rarely rise above 9 m O.D. south of Selby but reach 14 m O.D. along the Escrick Moraine. In valleys to the west and south (e.g. the Idle) it is represented by low terraces.

Laminated clay ('25-Foot Drift Silt and Clay') is locally more than 20 m thick, overlies the lower periglacial surface and completely buries the deeply denuded landscape, resting at depth on thin sand (and in places gravel) which may be partly fluviatile, partly aeolian and partly lacustrine. Its surface is generally a metre or so below that of the adjacent and laterally continuous low-level sand.

The lake formed when the Humber gap was blocked, at first presumably by ice but subsequently, since it survived later than the ice, by the plug of glacial deposits in the gap and farther east. Drainage, until late in its history, was up the Trent Valley and through the Lincoln gap. Transgression to its maximum level, when the high-level sand and gravel formed round it, probably occurred soon after initiation. The date of 21,835 ± 1600 radiocarbon years from a bone fragment in or at the base of these deposits near Brantingham (SE 940295) (Gaunt, 1974) provides a rough maximum age for this phase, but a more realistic age is derived from the Dimlington Silts (Penny, Coope and Catt, 1969) which show that ice had not reached Holderness just over 18,000 radiocarbon years ago. The northerly increase in height of the high-level sand and gravel implies, however, that isostatic depression in this direction already existed. Since these 'near-shore' deposits are generally thin and patchily distributed, and equivalent 'offshore' sediments at anything like comparable heights are absent (although they probably occur as the lowest, i.e. deepest, laminated clay), the high-level lake was probably short lived. After regression, possibly when the ice ceased to provide a water-proof dam, a low-level lake was established, and survived for a considerable time in view of the thick laminated clays formed in it. Continuing northerly isostatic depression is intimated by the northerly increase in height of this clay and the low-level sand. Near Burton Hall (SE 585293) and South Cave (SE 920310) there is evidence (ventifacts and ? frozen-ground structures) that between the high and low-level lake phases the water level was lowered at least to 4 m O.D., and temporary lowering of the lake to -4 m O.D. or below, probably during the same regressive episode, is indicated by a brecciated and desiccated or ground ice-cracked layer in laminated clay from a recent borehole near Carlton. A buried soil formed on the laminated clay at West Moor and dated to 11,100 ± 200 radiocarbon years (Gaunt, Jarvis and Matthews, 1971) provides a minimum age for the final disappearance of the lake. The deposition of laminated clay to almost the same height as adjacent low-level sand (probably to the same height before compaction), the absence of regression shorelines and the upward passages from laminated clay into river levée deposits (demonstrating that the immediate post-lacustrine rivers did not incise their courses - see below) all indicate that Lake Humber finally disappeared because it silted up, not because it drained by breaching of its morainic plug.

Ice Advances

In the Humber, North Sea ice reached almost to Brough (SE 940265), Skipsea-type till (with stone orientations suggesting ice movement directed just north of west) being exposed in Elloughton Beck; south of the estuary morainic ridges mark its limit east of Winteringham (SE 930220) and Winterton, and across the Ancholme Valley at Winterton Holmes (SE 955181). The ice front then wasted eastwards (some of its deposits being overlain by lacustrine high-level sand and gravel) and probably stabilized somewhere in the North and South Ferriby area, where interbedded till and laminated clay occur on both sides of the estuary.

The initial Stainmore-Vale of York ice advance apparently surged southwards as far as Wroot (SE 715030) into the high-level waters of Lake Humber (which probably lubricated the surge), depositing sand and gravel rich in Carboniferous and Permian constituents along its western and southern sides where they rest on the lower periglacial surface and in at least one place overlie Older River Gravel (Gaunt, 1976c). The absence of other evidence of this ice suggests that it quickly melted, probably before the end of the high-level lacustrine phase, and an ice front eventually stabilized much farther north where it produced the Escrick and York moraines. These morainic deposits are too complex to be considered adequately here, but their relationship to the lacustrine deposits is significant in terms of regional history. Evidence that lacustrine laminated clay and low-level sand overlap up the morainic slopes is well documented (e.g. Dakyns, Fox-Strangways and Cameron, 1886, 36, fig. 3; Gaunt, 1970b). A study of the numerous borehole records now available from the morainic area shows, however, that at depth the lower part of the laminated clay continues northwards in places under the morainic deposits, which thin out southwards within the lacustrine clay. This enhances the long-held belief (based largely on geomorphological evidence) that the moraines were deposited into a lake, and implies that this glacial episode occurred entirely within the low-level lacustrine phase.

From these conclusions the maximum and minimum dates given earlier for Lake Humber can also be applied to the related ice advances *in toto*. One further refinement can be suggested. Pollen from deposits in a hollow on the Escrick Moraine near Tadcaster apparently indicates the Bölling climatic amelioration (Bartley, 1962), which is now included in the early part of the Windermere Interstadial, dated on botanical evidence to about 13,000 radiocarbon years ago (Pennington, 1977). Presumably, therefore, by this time ice had disappeared from the Escrick Moraine. A comparable date (13,045 ± 270 radiocarbon years) from organic deposits in The Bog (TA 274289), Roos, shows that ice had also cleared southern Holderness by this time (Madgett and Catt, 1978).

River Deposition

Much of the sand overlying the lacustrine clay forms lines of impersistent ridges and mounds with trends directed towards the Humber; it is fine-grained, increasingly silty and clayey towards the edges of the ridges and mounds, and contains thin clay beds and lenses, notably towards the base where an upward passage from the laminated clay has been seen in places. This sand is believed to be river deposits (Palmer, 1966), in effect old levées formed when rivers initiated courses towards the Humber immediately after Lake Humber disappeared. Precursors of all the present rivers, even minor ones like the Foulness and Went, can be traced (with one exception, the lower Trent, valley-bound contemporaneous deposits of which now form its first terrace). Some, such as the levée of the Derwent on the eastern side of the present river from Wheldrake (SE 680450) to Wressle (SE 710315), are simple. Others are more complex, such as those of the Aire. An early course or divergent branch of this river running south of the Snaith ridge formed narrow levées running eastwards through Highgate (SE 595195) and north of Sykehouse (SE 630170). The main or later course is traceable from West Haddlesey (SE 565267) to Carlton (SE 647240). Initially this course crossed Langham, Hales (SE 700200) and Goolefields (SE 750200), its most easterly surface evidence being at Sand House (SE 809185), but subsequently it changed to a course passing just south of Airmyn (SE 725254), east of which it joined the Ouse (represented by levée sands running from Hemingborough through Howden to Metham Hall (SE 809248).

These sand levées show that for some time after Lake Humber disappeared drainage base level remained roughly at the same height as the emergent lacustrine clay plain. Although sea level was still eustatically very low, it must have taken a considerable period for the nick point of the post-glacial 'River Humber' to erode westwards from the contemporaneous coast (well to the east of the present coast) back into the Vale of York, a period long enough for the levée deposits to form. The date from West Moor (see above) gives indirectly a minimum age for initiation of these post-lacustrine rivers, and some idea of the duration of levée deposition is provided by peat dated to 10,469 ± 60 radiocarbon years which rests on levée sand and is covered by blown sand near Cawood (SE 575377) (Jones and Gaunt, 1976).

Aeolian and Periglacial Conditions

Most of the blown sand is demonstrably post-Lake Humber, and aeolian re-distribution has affected much of the levée sand, converting some of it entirely into blown sand e.g. on Hatfield Chase and Misterton Carr. Radiocarbon dates have been obtained from organic deposits at the base of, and within, the blown sand. Those from the base include c. 11,100 at West Moor (Gaunt, Jarvis and Matthews, 1971), c. 10,470 near Cawood (Jones and Gaunt, 1976) and c. 10,280 near Messingham (SE 893045) (P.C. Buckland pers. comm.); those from within include c. 10.700 and c. 9,950 near Sutton on the Forest (SE 585647) (Matthews, 1970) and c. 10,550 near Messingham (Buckland and Dolby, 1973). These dates imply that the aeolian phase was more or less coincident with the last millenium of the Devensian, a time of climatic deterioration sometimes called the 'Younger Dryas'. The prevailing wind direction is uncertain. Sandstone and sandy deposits in the west (where abundant ventifacts associated with the upper periglacial surface also occur - see below) are the most obvious source, and the concentration of blown sand on west-facing slopes also suggests westerly winds. Most of the crescentic dunes, however, are convex to the east, implying easterly winds, and the climatic reconstructions of Lamb and Woodroofe (1970) for this period suggest anticyclonic easterly winds. Perhaps westerlies originally mobilized the sand but the final effective winds, before vegetation stabilized the surface, were easterly.

The upper periglacial surface consists of numerous ventifacts, patches of desert pavement and scattered frozen ground structures which are generally weakly developed and suggestive of seasonally frozen ground with localized discontinuous permafrost. It coincides with the top of the lacustrine high-level sand and gravel and the top of the sand and gravel formed at the edge of the Vale of York ice when it surged southwards to Wroot. Farther north it is represented by ventifacts on the Escrick Moraine at High Catton (SE 718536) (Kendall and Wroot, 1924, p. 527) and at Newton on Derwent (SE 720495) (Gaunt, 1976a), on the flanks of the York Moraine at Poppleton (SE 557542) (Edwards, 1936) and on Devensian glacial deposits near Whixley (SE 443580) (Gaunt, 1976a) and near Sutton on the Forest (Matthews, 1970); at the last locality Matthews also noted ice-wedge casts intruded into till. The upper periglacial surface has not been found either within or above the low-level lacustrine deposits, however. Absence of stones would preclude ventifact and desert-pavement formation, and frozen-ground structures do not form readily in clay, but the lack of these structures in the low-level sand is puzzling. Either they did not form because a sudden marked drop in drainage base level (see below) rendered the sand too dry, or aeolian re-distribution has destroyed them. Despite this uncertainty, the upper periglacial surface developed largely, if not entirely, after both the ice and Lake Humber had disappeared, probably as a result of the terminal Devensian climatic deterioration which facilitated blown sand formation.

FLANDRIAN STAGE

River Incision

Contours drawn on the base of the Flandrian deposits (alluvium and peat) show a landscape (Fig. 5) in which the rivers have incised narrow steep-sided and (except the smallest rivers) fairly straight courses down to nearly -20 m O.D. towards the Humber, with steeper long profiles than those of the present rivers. Even minor rivers like the Went and streams like the one south of Barlow (SE 645288) incised their channels, but interfluvial areas show no evidence of contemporaneous denudation. This vigorous incision phase clearly post-dates river levée formation

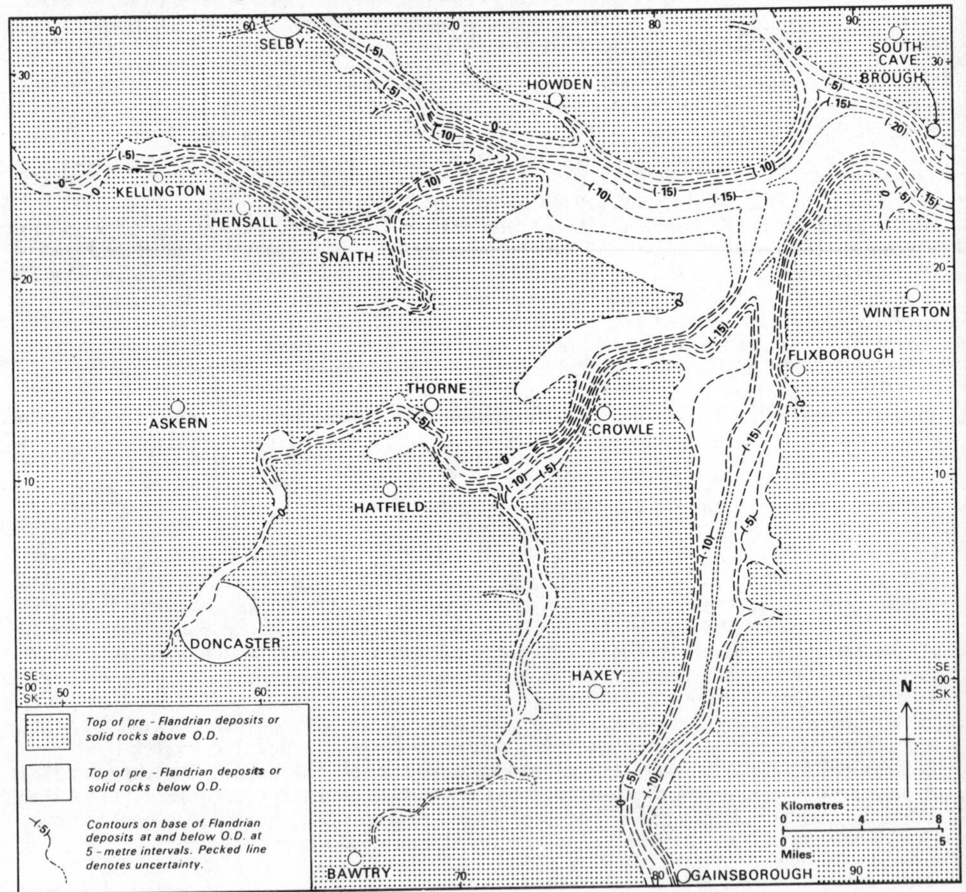

Fig. 5. Contours on the base of the Flandrian deposits.

and was of short duration, the result of a rapid marked drop of drainage base level down to the contemporaneous sea level. The onset of incision presumably occurred when the nick point of the 'River Humber' finally eroded westwards through the glacial deposits in the Humber gap and reached the soft waterlogged uncompacted lacustrine deposits in the Vale of York. The incision cannot yet be dated accurately. It may well have started before the end of the Devensian, when a suddenly lowered groundwater table would have helped to mobilize blown sand. It had probably finished by early 'Boreal' times (about 8,500 radiocarbon years ago) when shell marl dating from this time formed in a minor incised channel at Burton Salmon (SE 492274) (Norris, Bartley and Gaunt, 1971) and had certainly terminated well before 7,000 radiocarbon years ago when sea level in the Humber estuary had risen to about -9 m O.D. (Gaunt and Tooley, 1974).

River deposition

As sea level rose during the Flandrian (Gaunt and Tooley, 1974) considerable river and estuarine deposition occurred. Between 7,000 and 6,000 radiocarbon years ago, when sea level rose rapidly from about -9 m to less than -5 m O.D., thick alluviation must have virtually filled the incised channels, and in the last 3,500 radiocarbon years, as sea level has oscillated within a metre or two of O.D., alluvial deposits along the lower river courses have spread widely beyond the channel confines. The rise of the groundwater table and onset of a more oceanic 'Atlantic' climate were probably the main reasons for the spread of peat, which in areas such as Thorne Moors and Hatfield Moors ultimately developed into raised bogs (Smith, 1958).

At some time before 1410 A.D. an artificial divergent channel of the Don was cut northwards from Thorne to the Aire (Gaunt, 1975), probably as a Roman inland waterway or a Mediaeval attempt to improve drainage. Other artificial channels (e.g. the Bykers and Hek dikes) referable to one or other of these origins are known, and between 1267 and 1577 A.D. the Derwent was diverted from its old course past Howden. The largest man-made river diversions were those of Vermuyden and the Participants between 1626 and 1635 A.D. to improve drainage: these channelled parts of the Don, Torne and Idle into new or pre-existing artificial courses. In the last three centuries widespread flood-warping has been carried out to improve the agricultural value of land. This process consists of artifically inducing numerous floods of silt-laden river water into embanked areas, where the silt is allowed to settle. The earliest documented flood-warping was around Rawcliffe (SE 685230) in the 1730's; the most recent occurrence was near Yokefleet (SE 820242) in 1947.

CONCLUDING COMMENTS

The sequence of events outlined above shows that the region is particularly well endowed with evidence of its later Quaternary history. Although some events are traced from negative (i.e. non-depositional) evidence such as incised and denuded surfaces, even these features have value, e.g. in showing that erosive phases occurred mainly in the early parts, and at the ends, of glacial stages, not in the middle of interglacials as has often been assumed. Some aspects of the history are repetitive, such as drainage base-level variations, the source and route of encroaching ice and possibly even glacial lakes. Because of this cyclical tendency there is a danger that the story is over-simplified, that two or more parts of different cycles have been assigned to the same cycle, although there is virtually no possibility of this with respect to the Flandrian and Devensian. Because of recent suggestions (see review by Cox 1981, this volume) that between the Devensian and the previous glaciation in England there may have been more than one temperate phase separated by cold but ice-free intervals it is possible that the 'Ipswichian' history may have been more complex. The greatest danger of over-simplification lies in the 'Older Glacial Stage'. Although there is no indication of more than one pre-Devensian glacial episode, the evidence is so sparse that multiple glaciation, either within a single stage or in more than one stage, may have taken place and produced evidence so similar (e.g. till lithology) that its denuded remnants cannot be distinguished. Finally, there remain other problems posed by evidence elsewhere of events which must have affected the region but are not traceable in it, the most obvious examples being the interglacial sea levels of 20-30 m O.D. indicated by estuarine deposits on Speeton cliff and at Kirmington.

ACKNOWLEDGEMENTS

The paper incorporates work done by several of my I.G.S. colleagues at Leeds, namely Mr E.G. Smith, Mr J.G.O. Smart, Mr G.H. Rhys, Dr T.P. Fletcher and Mr R.J. Bull, and permission to use their results is here gratefully acknowledged.

REFERENCES

Bartley, D.D. (1962). The stratigraphy and pollen analysis of lake deposits near Tadcaster, Yorkshire. *New Phytol.*, 61, 277-287.

Bisat, W.S. (1946). Fluvio-glacial gravels at Stanley Ferry, near Wakefield. *Trans. Leeds geol. Ass.*, 6, 31-36.

Buckland, P.C. and M.J. Dolby (1973). Mesolithic and later material from Misterton Carr, Notts. An interim report. *Trans. Thoroton Soc. Notts.*, 77, 5-33.

Carter, W.L. (1905). The glaciation of the Don and Dearne valleys. *Proc. Yorks. geol. Soc.*, 15, 411-436.

Cox, F.C. (1981). The 'Gipping Till' revisited. In J.W. Neale and J.R. Flenley (Eds), *The Quaternary of Britain*. Pergamon Press, Oxford, pp. 34-42.

Dakyns, J.R., C. Fox-Strangways and A.G. Cameron (1886). *The geology of the country between York and Hull.* Mem. geol. Surv. U.K.. vi + 54.

Eden, R.A., I.P. Stevenson and W. Edwards (1957). *Geology of the country around Sheffield.* Mem. geol. Surv. U.K.. ix + 238.

Edwards, W. (1936). Pleistocene dreikanter in the Vale of York. *Mem. geol. Surv. Summ. Prog. for 1934*, Pt. 2, 8-18.

Edwards, W. (1937). A Pleistocene strand line in the Vale of York. *Proc. Yorks. geol. Soc.*, 23, 103-118.

Gaunt, G.D. (1970a). An occurrence of Pleistocene ventifacts at Aldborough, near Boroughbridge, West Yorkshire. *J. Earth Sci., Leeds*, 8, 159-161.

Gaunt, G.D. (1970b). A temporary section across the Escrick Moraine at Wheldrake, East Yorkshire. *J. Earth Sci., Leeds*, 8, 163-170.

Gaunt, G.D. (1974). A radiocarbon date relating to Lake Humber. *Proc. Yorks. geol. Soc.*, 40, 195-197.

Gaunt, G.D. (1975). The artificial nature of the River Don north of Thorne, Yorkshire. *Yorks. archaeol. J.*, 47, 15-21.

Gaunt, G.D. (1976a). *The Quaternary geology of the southern part of the Vale of York.* Unpublished Ph.D. thesis, University of Leeds.

Gaunt, G.D. (1976b). In *Annual Report for 1975.* Institute of Geological Sciences, London. 239 pp.

Gaunt, G.D. (1976c). The Devensian maximum ice limit in the Vale of York. *Proc. Yorks. geol. Soc.*, 40, 631-637.

Gaunt, G.D., D.D. Bartley and R. Harland (1974). Two interglacial deposits proved in boreholes in the southern part of the Vale of York and their bearing on contemporaneous sea levels. *Bull. geol. Surv. Gt Br.*, 48, 1-23.

Gaunt, G.D., G.R. Coope, P.J. Osborne and J.W. Franks (1972). An interglacial deposit near Austerfield, southern Yorkshire. Rep. No. 72/4, *Inst. geol. Sci.*, 13 pp.

Gaunt, G.D., R.A. Jarvis and B. Matthews (1971). The late Weichselian sequence in the Vale of York. *Proc. Yorks. geol. Soc.*, 38, 281-284.

Gaunt, G.D., and M.J. Tooley (1974). Evidence for Flandrian sea-level changes in the Humber estuary and adjacent areas. *Bull. geol. Surv. Gt Br.*, 48, 25-41.
Gilligan, A. (1918). Alluvial deposits at Woodlesford and Rothwell Haigh, near Leeds. *Proc. Yorks. geol. Soc.*, 19, 255-271.
Green, A.H., R. Russell, J.R. Dakyns, J.C. Ward, C. Fox-Strangways, W.H. Dalton and T.V. Holmes (1878). *The geology of the Yorkshire Coalfield.* Mem. geol. Surv. U.K.. xiii + 823.
Harmer, F.W. (1928). The distribution of erratics and drift. With a contoured map. *Proc. Yorks. geol. Soc.*, 21, 79-150.
Jones, R.L. and G.D. Gaunt (1976). A dated late Devensian organic deposit at Cawood, near Selby. *Naturalist, Hull*, 101, No. 939, 121-123.
Kendall, P.F. and H.E. Wroot (1924). *Geology of Yorkshire.* Vienna. xxii + 995.
Lamb, H.H. and A. Woodroffe (1970). Atmospheric circulation during the last ice age. *Quat. Res.*, 1, 29-58.
Lewis, H.C. (1887a). The terminal moraines of the great glaciers of England. *Rep. Br. Ass. Advmt Sci.*, 691-692.
Lewis, H.C. (1887b). The terminal moraines of the great glaciers of England. *Nature, Lond.*, 36, 573.
Madgett, P.A. and J.A. Catt (1978). Petrography, stratigraphy and weathering of Late Pleistocene tills in East Yorkshire, Lincolnshire and North Norfolk. *Proc. Yorks. geol. Soc.*, 42, 55-108.
Matthews, B. (1970). Age and origin of aeolian sand in the Vale of York. *Nature, Lond.*, 227, 1234-1236.
Melmore, S. (1935). *The glacial geology of Holderness and the Vale of York.* Arbroath. 96 pp.
Norris, A., D.D. Bartley and G.D. Gaunt (1971). An account of the deposit of shell marl at Burton Salmon, West Yorkshire. *Naturalist, Hull*, No. 917, 57-63.
Paepe, R. and R. Vanhoorne (1967). The stratigraphy and palaeontology of the late Pleistocene in Belgium. *Mém. Expl. Cartes Géol. Min. Belgique*, No. 8, 96 pp.
Palmer, J. (1966). Landforms, drainage and settlement in the Vale of York. In S.R. Eyre and G.R.J. Jones (Eds), *Geography as human ecology: methodology by example.* Alva. pp. 91-121.
Pennington, W. (1977). The Late Devensian flora and vegetation of Britain. In G.F. Mitchell and R.G. West (Eds), The changing environmental conditions in Great Britain and Ireland during the Devensian (last) cold Stage. *Phil. Trans. R. Soc.*, B 280, 247-271.
Penny, L.F., G.R. Coope and J.A. Catt (1969). Age and insect fauna of the Dimlington Silts, east Yorkshire. *Nature, Lond.*, 224, 65-67.
Smith, A.G. (1958). Post-glacial deposits in south Yorkshire and north Lincolnshire. *New Phytol.*, 57, 19-49.
Smith, E.G., G.H. Rhys and R.A. Eden (1967). *Geology of the country around Chesterfield, Matlock and Mansfield.* Mem. geol. Surv. Gt Br.. viii + 430.
Straw, A. (1963). The Quaternary evolution of the lower and middle Trent. *East Midl. Geogr.*, 3, 171-189.
Van der Hammen, T., G.C. Maarleveld, J.C. Vogel and W.H. Zagwijn (1967). Stratigraphy, climatic succession and radiocarbon dating of the last glacial in the Netherlands. *Geologie Mijnb.*, 46, 79-95.
Woodland, A.W. (1970). The buried tunnel-valleys of East Anglia. *Proc. Yorks. geol. Soc.*, 37, 521-578.
Zagwijn, W.H. (1973). Pollenanalitic studies of Holsteinian and Saalian beds in the northern Netherlands. *Meded. Rijks. geol. Dienst*, N.S. 24, 139-156.

Dr G.D. Gaunt,
Institute of Geological Sciences,
Ring Road, Halton,
Leeds. LS15 8TQ.

10. Patterned Ground at Wharram Percy, North Yorkshire: Its Origin and Palaeoenvironmental Implications

Stephen Ellis

INTRODUCTION

It is generally considered that, at its maximum limit, Devensian ice did not extend beyond the footslopes of the Yorkshire Wolds escarpment (Fig. 1), although precise dating of this limit is open to dispute; much of the escarpment has thus remained unglaciated since dissipation of the previous Wolstonian ice sheet, or perhaps even longer (Penny, 1974; Catt, 1979, 1980; Straw, 1979a, b). It may be argued, therefore, that during cold periods of the Devensian stage the Yorkshire Wolds experienced a periglacial climatic regime. To date, evidence for the possible existence of former periglacial conditions in this area has been reported from studies of dry valley development (Lewin, 1969), aeolian deposits (Catt, Weir and Madgett, 1974) and large scale stripe patterns (Evans, 1976). Additional cursory observations have been made on ice-wedge pseudomorphs (Matthews, 1975), angular, involuted chalk rubble (Manby, 1976) and cambering and Head (Straw, 1979b). However, as a result of recent archaeological excavations at the deserted medieval village of Wharram Percy, an area of small scale patterned ground has been discovered which provides further evidence for the occurrence and nature of past periglacial conditions on the Yorkshire Wolds.

THE STUDY SITE

The deserted medieval village of Wharram Percy is situated on the dip slope of the Yorkshire Wolds escarpment close to its northwest margin (Fig. 1). It lies on the western slopes of the valley of the Settrington Beck headwaters, which have cut through the Lower Chalk, or Ferriby Chalk Formation according to recent lithostratigraphical classification (Wood and Smith, 1978), to expose underlying Upper Jurassic clays. The study site was exposed in July 1979, forming part of a long term systematic excavation of the village by the Medieval Village Research Group (see, for example, Beresford and Hurst, 1979). It is located immediately outside the northern boundary bank at 152 m O.D. (SE 85966461) and forms Site 53 of the archaeological programme.

From a 9x9 m plot, the soil cover was removed to a depth of around 0.3 m, exposing chalk rubble and silt-rich material arranged in a series of sub-parallel bands when viewed in plan (Fig. 2). The maximum slope angle of the plot was 3° in a direction 102°; the bands ran at a slightly oblique angle to the slope in a direction of around 130°. Sectioning the bands in the southwest corner of the site showed the

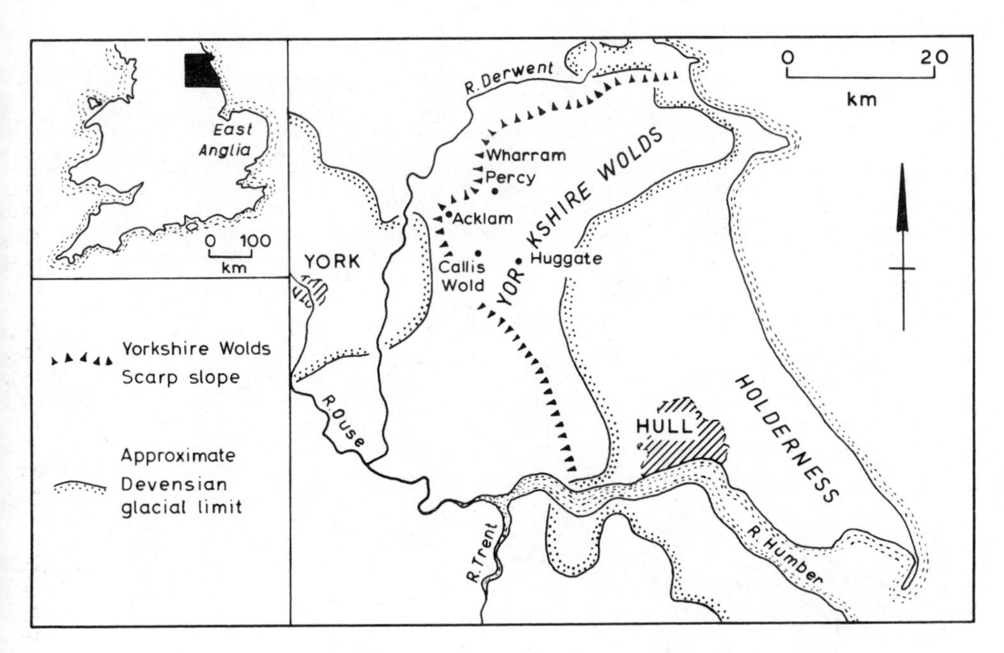

Fig. 1. Location Map

chalk rubble to be arranged in an elongated ridge and trough form, the troughs being infilled by the silt-rich material (Fig. 3). At the margin of the excavated plot, the soil layer, while also comprising predominantly silt, was seen to contain in addition a continuous horizontal concentration of chalk fragments at a depth of approximately 0.2 to 0.3 m below the present day surface along with occasional randomly distributed chalk fragments (Figs. 3 and 4).

Similar elongated ridge and trough features have been reported from East Anglia (Fig. 1), although these occur beneath predominantly sandy rather than silty material and are developed on a much larger scale, distances between successive ridge crests measuring several metres (Williams, 1964; Watt, Perrin and West, 1966; Corbett, 1973; Evans, 1976). While these examples of patterned ground have been able to be identified on panchromatic aerial photographs, this has not been possible for the features at Wharram Percy, either on vertical or oblique photographs of scales up to around 1:1200. This is perhaps due to the fact that their much smaller scale subsurface microtopography is not capable of producing differences in soil moisture content sufficiently marked to be expressed on the photographs as recognizable tonal variations. It is interesting to note, however, that on the basis of aerial photographic interpretation, large scale patterned ground features apparently similar to those of East Anglia have also been reported from the southern part of the Yorkshire Wolds (Williams, 1964; Evans, 1976).

PATTERNED GROUND FORMATION

Despite its proximity to the village settlement, the patterned ground appeared to be non-anthropogenic since the excavated plot was found to be largely devoid of archaeological materials. As a result of sectioning the ridge and trough forms, it was noticed that many long axes of the chalk rubble fragments appeared to be

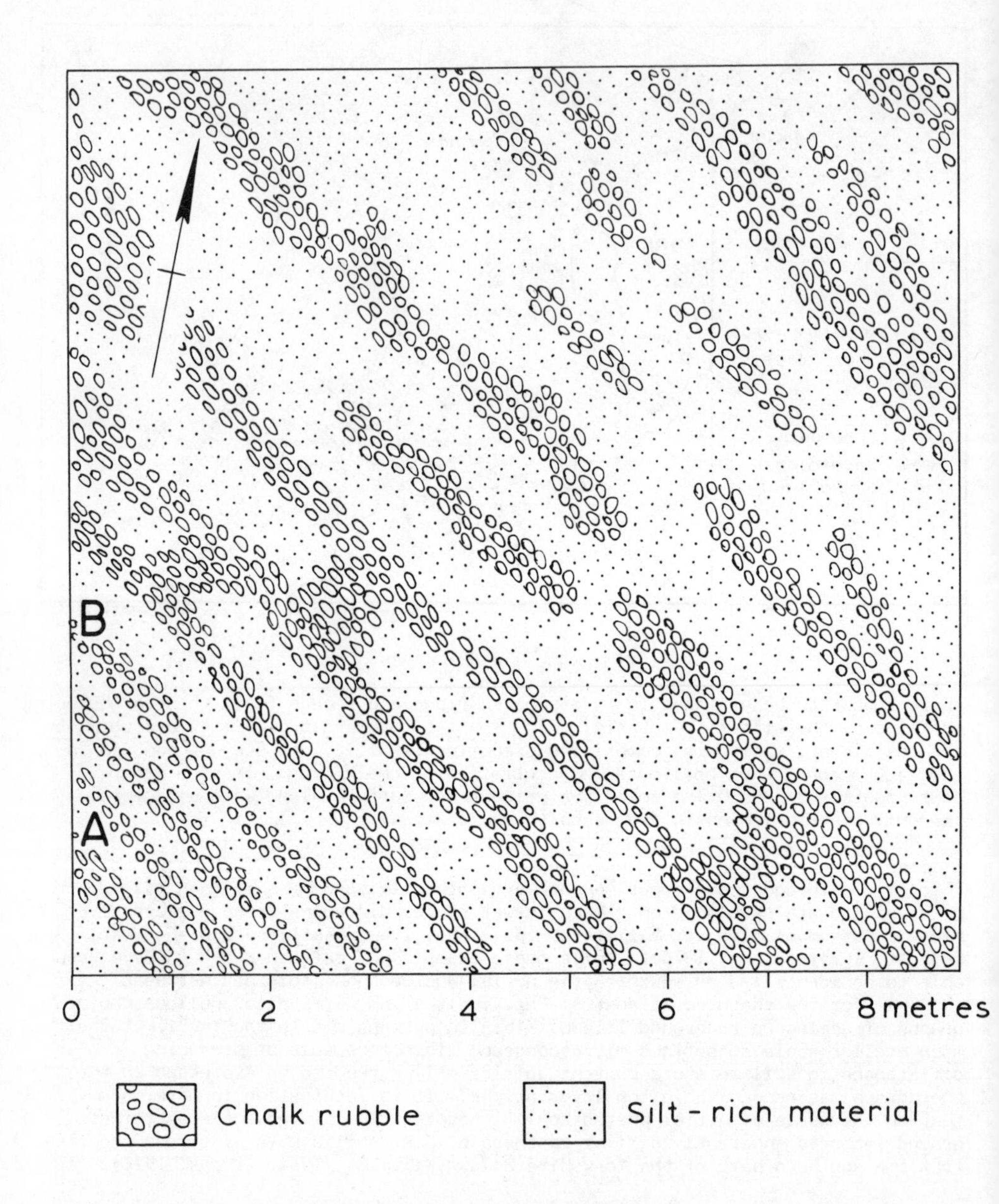

Fig. 2. Plan of the study site.

near vertically orientated. At each of two ridges (Fig. 4), azimuth and dip of 50 either rod- or blade-shaped clasts with minimum longest axis of 0.01 m were measured. Results of three dimensional vector analysis of the data (Steinmetz, 1962) are shown below.

	Ridge 1	Ridge 2
Vector strength	43.7	44.7
Resultant dip direction	306.0^{o}	321.3^{o}
Resultant dip angle	71.4^{o}	72.4^{o}

Values for vector strength at each ridge are significant at the 99% confidence level (Watson, 1956), hence both distributions are non-random. Longest axes are therefore preferentially orientated towards the vertical and have a preferred dip in approximately the opposite direction from that of the slope upon which the patterned ground is developed

Preferred dip of longest axes towards the vertical has been attributed by several authors to the process whereby clasts become progressively up-ended as a result of alternating freezing and thawing of soils or sediments (for example, Watson and Watson, 1971; FitzPatrick, 1975; French, 1976 pp. 32-3). Longest axis preferred dip in an upslope direction has been previously reported on numerous occasions, where it has been considered to be a characteristic produced by the solifluction mechanism operating in unconsolidated materials (for example, Benedict, 1970;

Fig. 3. Southwest corner of the study site showing excavated ridges and troughs (the hammer shaft is 0.3 m long).

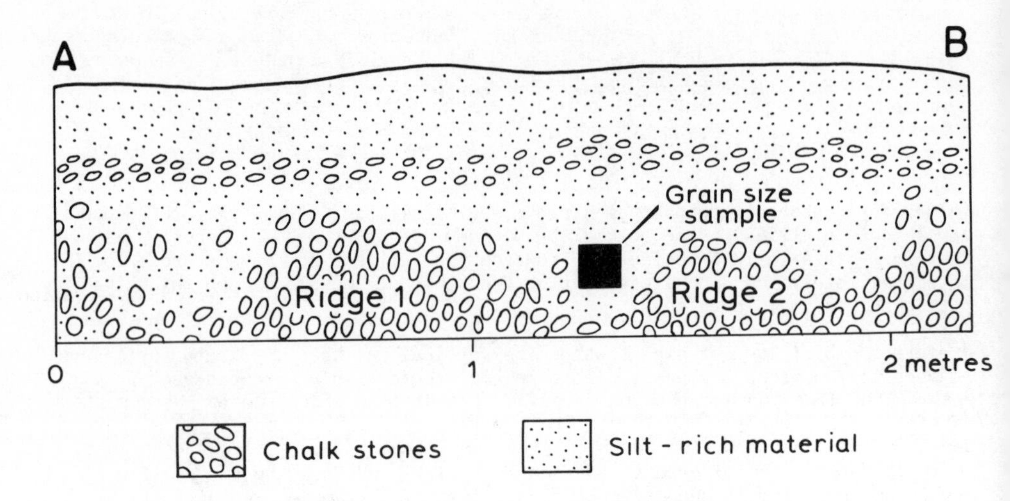

Fig. 4. Section along AB (see Fig. 2).

FitzPatrick, 1975). It would appear, therefore, that the orientation pattern of the ridge clasts might have been produced by processes of both stone up-ending and solifluction, operating during climatic conditions of a periglacial nature. However, although these mechanisms can perhaps explain clast orientation, they do not in themselves account for development of the ridge and trough forms.

Explanations of this microtopography depend largely upon whether it has been produced prior or subsequent to deposition of the silt-rich material over the chalk rubble, that is to say whether it developed as a surface or sub-surface phenomenon. If it is assumed firstly that the form is at least in part of a pre-depositional origin, then it can perhaps be considered similar to nonsorted stripes occurring in currently active periglacial environments (Washburn, 1956, 1979 pp. 151-3). Although numerous mechanisms have been proposed in order to explain the origin of stone stripes, it would appear that the precise details of formation are still not fully understood. However, Washburn (1970) has summarized the major possibilities as part of an attempt to devise a comprehensive genetic classification of patterned ground phenomena; these have been discussed more recently in greater detail (Washburn, 1979 pp. 156-70). In considering which possibilities are most feasible in the context of the Yorkshire Wolds, three observations are pertinent: (a) the features are developed in chalk rubble apparently derived from weathering of the Lower Chalk, (b) the ridges and troughs do not follow the line of maximum slope and (c) when viewed in plan some ridges and adjacent silty infills demonstrate a slightly anastomosing pattern.

Desiccation or frost cracking of the Lower Chalk bedrock along joints may have led to infilling of the resulting fissures by overlying chalk rubble. Since the local strike of the bedrock would appear to be in a direction of around 135° (Mortimer, 1885), this may explain why the ridge and trough form approximates to a direction slightly south of that of the maximum slope, assuming cracking to occur along structural weaknesses related to strike. The slightly anastomosing nature of the patterned ground suggests the possibility of surface flowing water occurring during its development (French, 1976 p. 266; Journaux 1976). This could have been of a rill network type, established subsequent to the formation of any initial topographic variations resulting from desiccation or frost cracking of the bedrock; seasonal surface flow of water may have been possible if frozen subsurface conditions existed in the chalk rubble. Once differences in ground moisture content had been established as a result of variations in microtopography, then the ridge and trough form might have been further enhanced by the operation of differential ground heaving (Nicholson, 1976). Preferred vertical orientation of chalk clasts in the ridges could well suggest the operation of frost heave at some stage during microtopographic development. Clast orientation would also appear to have become affected to some extent by solifluction, as suggested earlier.

If it is now assumed that the ridge and trough form developed entirely subsequent to deposition of the silt-rich material over the chalk rubble, then processes other than those capable of producing stone stripes should be envisaged. Such a mechanism has been proposed by Watt, Perrin and West (1966) in their consideration of the larger scale East Anglian patterned ground features. They have suggested an origin whereby variations in rates of seasonal freezing plane penetration from the surface down towards a subsurface permafrost layer result in the generation of cryostatic pressure, producing subsurface involutions in the material which remains unfrozen and also causing long axis vertical orientation of clasts. This, they proposed, may result in a series of polygonal involutions when viewed in plan, which become subsequently elongated into stripe forms by solifluction. Clast orientation suggests that the ridges at Wharram Percy have been subjected to solifluction although, as recognized above, this could equally well have occurred had the features developed at the surface prior to deposition of the silt-rich material.

Problems arise in the explanations of both surface and subsurface origins. In the former case, the extremely high angle (up to 70°) of the ridge flanks would be far in excess of the angle of rest for chalk rubble in a feature developed at the surface and would therefore appear to discount a purely surface-development hypothesis. In the case of subsurface formation the problems are three-fold: (a) the concept of cryostatic pressure has several mechanical problems associated with it, and still remains to be demonstrated convincingly (Washburn, 1979 pp. 98-100), (b) subsurface pressure of the nature envisaged might be expected to result in a much more irregular interface between the chalk rubble and the overlying material, similar to that reported by Watt, Perrin and West (1966) and (c) subsequent solifluction of polygonal forms would appear unlikely to produce such regularly spaced, narrow and yet laterally continuous ridges, especially without disturbing the marked vertical orientation of clasts. A possible explanation to help overcome this latter problem is that development of shallow rill networks on the surface of seasonally frozen silt-rich material might have led to establishment of vegetation in linear concentrations on the more freely draining inter-rill areas. This could have given rise to varying ground thermal conductivities, differential rates of freezing plane penetration and the associated involution of subsurface material remaining unfrozen. If so, then it is not necessary to envisage solifluction as an integral process in ridge and trough formation; it could have operated contemporaneously with and/or subsequent to formation, but to a limited extent only, thereby modifying the dip directions of the chalk clasts without destroying their preferred vertical orientation.

An alternative possibility, as yet unconsidered, is that the patterned ground originated from a combination of processes operating both prior and subsequent to deposition of the silt-rich material. For example, an initial, but rather less steep sided, ridge and trough form may originally have developed at the surface by the processes discussed earlier. Following subsequent burial by silt-rich material, differing moisture contents may have become established, greatest contents occurring in the material above the troughs due to higher moisture retention of silt relative to that of chalk rubble. Differences in moisture content and their associated variations in ground thermal conductivity may then have led to differential frost heave and subsequently to enhancement of the ridge and trough form which by this stage had become a subsurface feature.

In short, it is not clear whether formation of the ridges and troughs in the chalk rubble occurred entirely subsequent to deposition of the silt-rich material or whether the features are polygenetic, resulting from processes operating both prior and subsequent to deposition. However, in either event, it appears likely that development occurred under periglacial climatic conditions during which freeze-thaw and solifluction were active.

DERIVATION OF THE SILT-RICH MATERIAL

In order to examine the derivation of the silt-rich material, a grain size analysis (Bascomb, 1974) was performed on a sample of this material taken at a depth of 0.4 to 0.5 m below the present day surface (Fig. 4). Results show the material comprises predominantly silt (2-60 μm), with moderate clay (< 2 μm) and low sand (60-2000 μm) percentages (Fig. 5). This is seen to bear a close similarity to the soil grain size characteristics of samples examined by Catt, Weir and Madgett (1974) from depths of 0.4 to 0.5 m and 0.25 to 0.3 m respectively from the neighbouring localities of Huggate and Callis Wold (Figs. 1 and 5). Catt, Weir and Madgett (1974) have considered the material from these sites to be composed mainly of loess

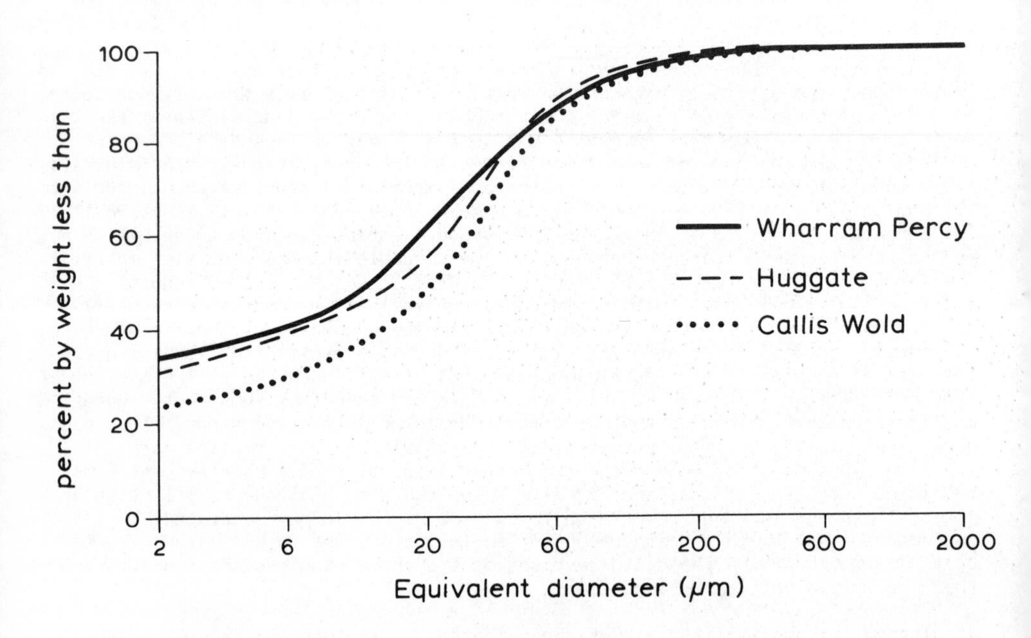

Fig. 5. Grain size cumulative frequency curves for < 2 mm material from three Yorkshire Wolds sites (Huggate and Callis Wold data from Catt, Weir and Madgett 1974).

derived from Devensian glacial debris. More recently, soils of similar grain size and origin have been reported from nearby Acklam by Matthews (1975) (Fig. 1). In view of these apparent similarities, it might therefore seem feasible to suggest that the material overlying the chalk rubble at Wharram Percy is also of an aeolian origin, having been derived from Devensian glacial debris.

Although its silt-rich nature is clearly marked, it has been noted that the material also contains a horizontal concentration of chalk fragments along with occasional randomly distributed chalk fragments (Fig. 4). These have probably been derived from the underlying chalk rubble, having been incorporated into the aeolian material both by cryoturbation and mesofaunal activity and to some extent subsequently comminuted by solution weathering; burrowing by earthworms is a widely recognized mechanism by which stones in soils become concentrated into sub-surface horizontal bands (Atkinson, 1957).

PALAEOENVIRONMENTAL IMPLICATIONS

It has been suggested that formation of aeolian deposits on the Yorkshire Wolds occurred during Late Devensian times, largely before the end of this sub-stage, since the source area of loess probably later diminished as a result of westward advance of the ice sheet across Holderness (Fig. 1) (Catt, Weir and Madgett, 1974; Catt, 1978). If so, and if the patterned ground originated entirely subsequent to deposition of the aeolian silt-rich material, then its development can probably be assigned to the latter part of the Late Devensian. According to the mechanisms of formation discussed earlier, this period would therefore have probably been one in which permafrost existed within the top metre or so of the surface and in which mean annual air temperatures were -6°C or below (Williams, 1969, 1975).

If, however, the ridges and troughs developed at least in part prior to aeolian activity, then a pre-Late Devensian age for initiation must be established. It has been argued that pre-Devensian superficial material on the Yorkshire Wolds was largely removed by solifluction, mudflows and fluvial activity during the ensuing Devensian cold periods, and that the Late Devensian loess was therefore deposited on a bare chalk surface exposed to frost shattering and disturbance (Catt, Weir and Madgett, 1974). Assuming this to be the case, initial formation of the patterned ground may be assigned tentatively to the Early or Middle Devensian. It was suggested previously that if the ridges and troughs originally developed as surface features they might have formed as a result of desiccation or frost cracking. If so, then formation may have occurred under intensely cold conditions, with mean annual air temperatures of perhaps -8°C or lower (Williams, 1975). However, the severity of the temperatures required for frost cracking suggested by Williams has been questioned (Washburn, 1979 p. 137), and it now appears that this may have been able to occur at slightly warmer temperatures.

The high degree of preservation of the patterned ground would suggest that slope erosional processes during the Flandrian have been minor in this locality. This is in accord with the observations of Williams (1964, 1968) regarding preservation of the larger scale patterned ground features in East Anglia. Although, on the basis of aerial photograph examination, the large scale type of patterned ground has been recognized extensively in East Anglia, and to a lesser degree in other parts of eastern England (Williams, 1964; Evans, 1976), the small scale patterns at Wharram Percy appear to be the first reported examples of their type in the British Isles. However, the apparent absence of these features from aerial photographs possesses implications for their more widespread occurrence; their discovery was facilitated by the geomorphologically uncommon nature of the site excavation technique, but there is no reason to suppose that they should not be encountered at similar sites elsewhere if this technique were adopted.

CONCLUSION

Patterned ground has been observed in the form of an elongated ridge and trough subsurface microtopography developed in chalk rubble. This is similar in gross morphology to that previously reported from East Anglia, although a marked difference in scale exists, the patterns at Wharram Percy being of much smaller dimensions. These appear to be the first reported examples of their type in the British Isles. The precise mechanisms responsible for their formation are uncertain, but various possibilities have been recognized, depending upon whether they developed originally as surface or subsurface phenomena. In either event, they are considered to have formed under periglacial climatic conditions during which freeze-thaw and solifluction were active. The chalk rubble in which the patterns have developed has become covered by a thin veneer of silt-rich material thought to be of aeolian origin and deposited in Late Devensian times. The age of the patterned ground is unclear, but it is considered likely to have developed at some time during the Devensian.

ACKNOWLEDGEMENTS

I am grateful to members of the Wharram Percy archaeological excavation for bringing the site to my attention and for the interest and hospitality shown during my visits. Thanks are due in particular to John Hurst, Dick Porter and David Andrews. Grain size analysis was performed, and illustrative material prepared, by members of the Geography Department technical staff at the University of Hull. For helpful criticism of the manuscript I thank Dr Peter Worsley.

REFERENCES

Atkinson, R.J.C. (1957). Worms and weathering. *Antiq.*, 31, 219-233.

Bascomb, C.L. (1974). Physical and chemical analyses of < 2 mm samples. In B.W. Avery and C.L. Bascomb (Eds), *Soil Survey Laboratory Methods*. Soil Surv. Tech. Monogr., 6., Rothamsted, Harpenden, pp. 14-41.

Benedict, J.B. (1970). Downslope soil movement in a Colorado alpine region: rates, processes, and climatic significance. *Arct. Alp. Res.*, 2, 165-226.

Beresford, M.W. and J.G. Hurst (1979). Wharram Percy: a case study in microtopography. In P.H. Sawyer (Ed.), *English Medieval Settlement*. Arnold, London, pp. 52-85.

Catt, J.A. (1978). The contribution of loess to soils in lowland Britain. In S. Limbrey and J.G. Evans (Eds), *The Effect of Man on the Landscape: the Lowland Zone*. Counc. Br. Archaeol. Res. Rep., 21, pp. 12-20.

Catt, J.A. (1979). Soils and Quaternary geology in Britain. *J. Soil. Sci.*, 30, 607-642.

Catt, J.A. (1980). Till facies associated with the Devensian glacial maximum in eastern England. *Quaternary Newsl.*, 30, 4-10.

Catt, J.A., A.H. Weir and P.A. Madgett (1974). The loess of eastern Yorkshire and Lincolnshire. *Proc. Yorks. geol. Soc.*, 40, 23-39.

Corbett, W.M. (1973). Breckland Forest Soils. *Soil Surv. Spec. Surv.*, 7., Rothamsted, Harpenden, 120 pp.

Evans, R. (1976). Observations on a stripe pattern. *Biul. Peryglac.*,25, 9-22.

FitzPatrick, E.A. (1975). Particle size distribution and stone orientation patterns in some soils of north east Scotland. In A.M.D. Gemmell (Ed.), *Quaternary Studies in North East Scotland*. Univ. Aberdeen, Dept Geogr. Aberdeen, pp. 49-60.

French, H.M. (1976). *The Periglacial Environment*. Longman Inc., New, York, 309 pp.

Journaux, A. (1976). Alternances du ruissellement et de la solifluction dans les millieux périglaciaires: exemples Canadiens et expérimentations. *Biul. Peryglac.*, 26, 269-273.

Lewin, J. (1969). The Yorkshire Wolds. A study in Geomorphology. *Univ. Hull Occas. Pap. Geogr.* 11, 89 pp.
Manby, T.G. (1976). The excavation of the Kilham long barrow, East Riding of Yorkshire. *Proc. prehist. Soc.*, 42, 111-159.
Matthews, B. (1975). Soils in North Yorkshire I. *Soil Surv. Rec.*, 23., Rothamsted, Harpenden, 218 pp.
Mortimer, J.R. (1885). On the origin of the chalk dales of Yorkshire. *Proc. Yorks. geol. Soc.*, 9, 29-42.
Nicholson, F.H. (1976). Patterned ground formation and description as suggested by low arctic and subarctic examples. *Arct. Alp. Res.*, 8, 329-342.
Penny, L.F. (1974). Quaternary. In D.H. Rayner and J.E. Hemingway (Eds), *The Geology and Mineral Resources of Yorkshire.* Yorks. geol. Soc., Leeds, pp. 245-264.
Steinmetz, R. (1962). Analysis of vectorial data. *J. sediment. Petrol.*, 32, 801-812.
Straw, A. (1979a). An Early Devensian glaciation in eastern England? *Quaternary Newsl.*, 28, 18-24.
Straw, A. (1979b). Eastern England. In A. Straw and K.M. Clayton (Eds), *Eastern and Central England.* Methuen, London, pp. 1-139.
Washburn, A.L. (1956). Classification of patterned ground and review of suggested origins. *Bull. geol. Soc. Am.*, 67, 823-865.
Washburn, A.L. (1970). An approach to a genetic classification of patterned ground. *Acta geogr. Univ. Lodz.*, 24, 437-446.
Washburn, A.L. (1979). *Geocryology.* Arnold, London, 406 pp.
Watson, E. and S. Watson (1971). Vertical stones and analogous structures. *Geogr. Annlr.*, 53A, 107-114.
Watson, G.S. (1956). A test for randomness of directions. *Mon. Not. R. Astr. Soc. geophys. Suppl.*, 7, 160-161.
Watt, A.S., R.M.S. Perrin and R.G. West (1966). Patterned ground in Breckland: structure and composition. *J. Ecol.*, 54, 239-258.
Williams, R.B.G. (1964). Fossil patterned ground in eastern England. *Biul. Peryglac.*, 14, 337-349.
Williams, R.B.G. (1968). Some estimates of periglacial erosion in southern and eastern England. *Biul.Peryglac.*, 17, 311-335.
Williams, R.B.G. (1969). Permafrost and temperature conditions in England during the last glacial period. In T.L. Péwé (Ed.), *The Periglacial Environment.* McGill-Queen's Univ. Press, Montreal, pp. 399-410.
Williams, R.B.G. (1975). The British climate during the Last Glaciation; an interpretation based on periqlacial phenomena. In A.E. Wright and F. Moseley (Eds), *Ice Ages: Ancient and Modern.* Seel House Press, Liverpool, pp. 95-120.
Wood, C.J. and E.G. Smith (1978). Lithostratigraphical classification of the Chalk in North Yorkshire, Humberside and Lincolnshire. *Proc. Yorks. geol. Soc.*, 42, 263-287.

Dr Stephen Ellis,
Department of Geography,
The University,
Hull HU6 7RX.

11. The Tills of Filey Bay

C. A. Edwards

The everchanging and short-lived exposures of till in Filey Bay have long attracted Quaternary Geologists' attention. In the early nineteenth century, John Phillips (1829, 1835) mentioned the good exposures of 'diluvial clay and pebbles' between Filey and Speeton.

It was almost fifty years later that G.W. Lamplugh began recording exposures and excavations in the Filey Bay area. He realised the value of constantly revisiting the rapidly eroding coastline, and over a period of about thirty years, he amassed a wealth of stratigraphic information (G.W. Lamplugh, 1879, 1881a, 1881b, 1889, 1891 and 1911). He was one of the first Quaternary geologists to draw conclusions solely from his own observations; it is a tribute to him that they have stood the test of time and can be relied upon to be an accurate record of what he saw.

During the nineteenth century, the till sequence of north and east Yorkshire was divided into two units separated by the 'middle sands and gravels' as portrayed by Martin Simpson (1859), Curator of Whitby Museum, in his long section of the Yorkshire coast from Whitby to Spurn Point. Harrison (1895) suggested these sands and gravels were of fluvial origin thus contradicting Wood (1871) who attributed them to a marine transgression. Harrison's views became accepted and remained in vogue until 1939 when Carruthers expressed the view that the sands and gravels had a glacial origin and that the coastal deposits of north and east Yorkshire were the product of one glaciation. This is now accepted.

Bisat (1940) examined the Filey Bay tills, introduced the term 'Drab' till and regarded the Hessle till as part of the Upper Purple till. Madgett's (1975) work in Holderness has supported the findings of Bisat and resolved a major problem in the glacial stratigraphy of east Yorkshire. Table 1 shows the various interpretations of the glacial stratigraphy of coastal north and east Yorkshire between 1879 and 1978.

The coastal glacial succession in north Yorkshire (Filey Bay) is broadly similar to that of Holderness (Catt and Penny, 1966) with which it correlates laterally. The coastal tills of Holderness are remarkably consistent, displaying a three-tier succession of Basement, Skipsea and Withernsea tills with the Dimlington silts between the Wolstonian Basement and Devensian Skipsea tills. This consistency of lithology and sequence is largely due to the topographic consistency of the underlying chalk bedrock which forms a gentle easterly dipping basement at some metres below O.D. and which provided an unobstructed path for the advance of ice from the north-east, and a uniform basement on which melting occurred producing thick sequences of lodgement till.

North of Flamborough Head, the pre-Devensian topography was much more varied ranging from the soft Mesozoic clay area of Filey Bay to the resistant chalk buttress of Flamborough Head. In contrast to Holderness, coastal north Yorkshire provided many

<table>
<tr><th>LAMPLUGH (1879) FILEY BAY</th><th>LAMPLUGH (1891) FLAMBRGH</th><th>MUFF et al (1896) ROB. H's BAY</th><th>BISAT (1940) FILEY BAY</th><th>CATT (1963) HLDRNESS</th><th>MITCHELL et al (1971) N.E. YORKS</th><th>EDWARDS (1977) N.E. YORKS</th></tr>
<tr><td>HESSLE TILL</td><td>UPPER TILL</td><td>HESSLE TILL</td><td>SANDY TILL U. PURPLE TILL</td><td>HESSLE TILL</td><td>UPPER TILL</td><td rowspan="2">UPPER TILL SERIES</td></tr>
<tr><td>BROWN TILL</td><td rowspan="4">INTER-STRATIFIED SERIES</td><td>UPPER TILL</td><td>L. PURPLE TILL</td><td rowspan="2">PURPLE TILL</td><td>UNNAMED TILL OF FILEY BAY</td></tr>
<tr><td rowspan="2">GRAVEL</td><td rowspan="2">MIDDLE SANDS</td><td rowspan="2">SAND, SILT & GRAVEL</td><td rowspan="2">GRAVEL</td><td rowspan="2">GRAVEL</td></tr>
<tr><td rowspan="2">DRAB TILL</td></tr>
<tr><td>GREENISH-PURPLE TILL</td><td>LOWER TILL</td><td>GREY, SUB & STONY DRAB TILLS</td><td>LOWER TILL OF COAST</td><td>LOWER TILL SERIES</td></tr>
<tr><td rowspan="2">BASEMENT TILL</td><td rowspan="4"></td><td rowspan="4"></td><td rowspan="4"></td><td rowspan="2">DIMLINGTON SILTS BASEMENT TILL</td><td>CHALK RUBBLE</td><td>CHALK RUBBLE</td></tr>
<tr><td>BASEMENT TILL</td><td>BASEMENT TILL</td></tr>
<tr><td rowspan="2">CHALK RUBBLE</td><td rowspan="2"></td><td></td><td>CHALK RUBBLE</td></tr>
<tr><td>SPEETON SHELL BED</td><td>SPEETON SHELL BED</td></tr>
</table>

Table 1. Comparison of nomenclature for coastal exposures

topographic barriers to the ice advance and an extremely varied pre-Devensian basement upon which melting occurred. The result is greater within-till variation than in Holderness with the additional formation of till which is largely local in origin. Because of this greater variation, the coastal tills of north Yorkshire have been named as the Lower and Upper Till Series. The Lower is correlated with the Skipsea and the Upper with the Withernsea till of Holderness.

Within the North Sea basin it is considered that northern Pennine ice was overridden by ice originating in County Durham and that the compound glacier thus formed moved southwards between higher land to the west and Scandinavian ice to the east. Pressure from the Scandinavian ice caused the compound glacier to impinge on the coastline wherever there were no barriers and subsequent melting of the ice has produced, in such areas as Filey Bay, a two-tier Devensian depositional sequence similar to that of Holderness.

Towards the south-eastern end of Filey Bay, the ice was obstructed by the uphill

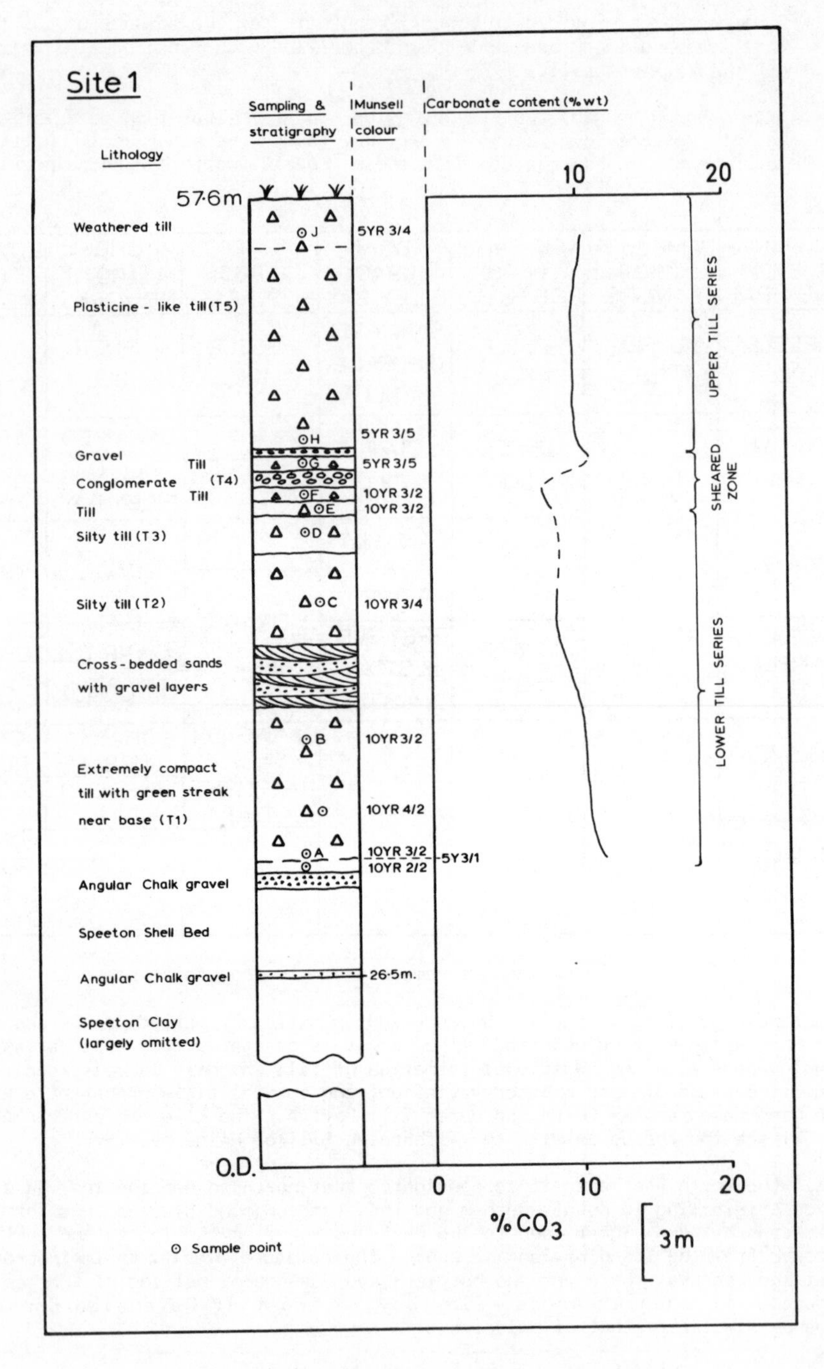

Fig. 1. Measured Section: Site 1 (Filey Bay)

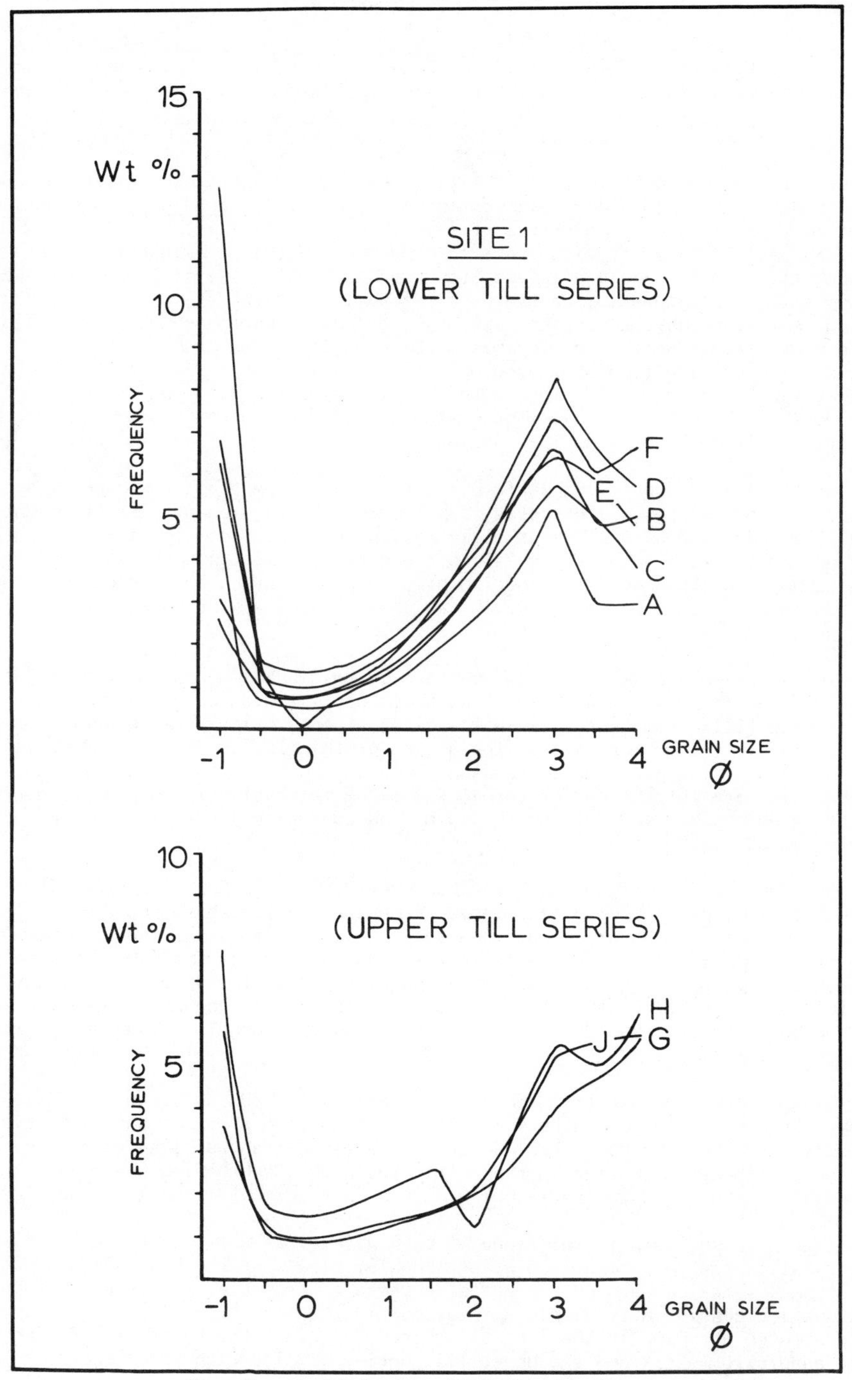

Fig. 2. Particle Size Distribution Curve

gradient of the flanks of the Flamborough headland. This resulted in much en-glacial shearing and the formation of many narrow band tills which are laterally impersistent. These have been particularly well exposed at TA 14707585 (Fig. 1).

The base of the succession consists of an extremely hard, compact till about 7 m thick resting on the angular chalk gravel which overlies the Speeton Shell Bed. Thirty five cm above the base of the till there is a 10 cm thick greenish band (5Y 3/1) contained within a clay-rich brown till (10YR 2/2) containing secondarily derived igneous erratics. Above the greenish band, the till matrix is sandy and the well-rounded igneous erratics are less numerous. Within the lowermost metre of this basal till, angular chalk erratics are numerous, making the till carbonate-rich. The basal till is interpreted, on the basis of particle size data and Munsell colour, as being equivalent to the Skipsea till of Holderness. By the same count, the greenish streak of till is most likely derived from the Wolstonian glaciation. The well-rounded igneous erratics were brought into the area by the Wolstonian ice and incorporated into the base of the Devensian till sequence with the Wolstonian till. The angular chalk fragments in the Devensian till have been derived from the soliflucted chalk overlying the Speeton Shell Bed.

The Shell Bed and underlying chalk gravel have both been folded by the same glaciotectonic event which is considered to be the Wolstonian ice advance, thus dating the lower periglacial surface as early Wolstonian following deposition of the Shell Bed during the Hoxnian interglacial as supported by microfauna, lithology, tectonic disturbance and Ipswichian weathering of the upper surface upon which lies the tectonically undisturbed early Devensian chalk gravel at the base of the till sequence.

The lowest Devensian till unit is overlain by en-glacial sands and gravels which in turn are overlain by 4 m of silty till which has lithological affinities with the underlying till. The next till unit is 1.3 m thick, compact, with a fine sand/silt matrix and infrequent erratics. The series of thin layers of till and gravel around 47 m O.D. represents the contact between the Lower and Upper Till Series. These tills are interpreted as thrust slices of melt-out till formed by en-glacial shearing which occurred between the lower and upper ice layers in the compound North Sea glacier.

Wherever shearing has occurred near the Lower/Upper Till Series contact in Filey Bay, thrust slices of Lower Till Series occasionally lie above the contact indicating that the shearing was upwards. At TA 14707585, the conglomerate within the transitional narrow band tills is taken as the base of the Upper Till Series. Though thin layers of Lower Till Series do not overlie the Upper Till Series base at this site, it is apparent that the shearing processes occurred in a zone above and below the actual boundary, which is based on the Munsell colour change and change in lithology.

At this site the topmost till (Upper Till Series) is reddish-brown (5YR 3/4) with a clay-rich matrix. It contains many north-east English erratics and is less well consolidated than the Lower Till Series. The top 2 m shows blue-faced joints and an increased fine sand content in the matrix which is a weathering phenomenon associated with the downward leaching of clay particles (Mitchell and Jarvis, 1956).

Particle size analysis in the range -1 to 4 Ø (Fig. 2) shows that the Lower and Upper Till Series are readily distinguishable. Tills of the Lower Series consistently show a maximum around 3 Ø and a decrease towards 4 Ø. Conversely, the Upper Till Series consistently displays a maximum in the clay fraction. Figure 2 demonstrates graphically the long recognised fact that the Lower Till Series has a fine sand/silt matrix and the Upper Till Series is clay-rich

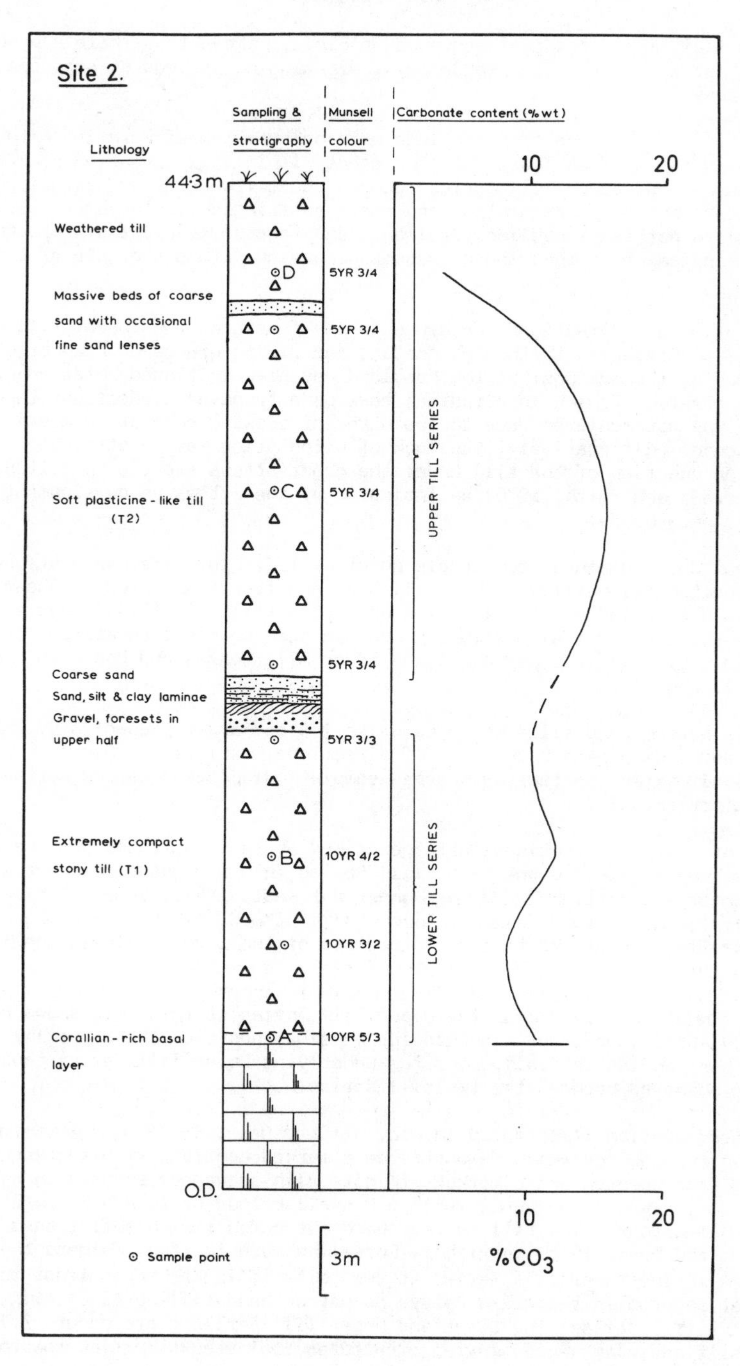

Fig. 3. Measured Section: Site 2 (Filey Bay)

Carbonate analysis of the till sequence in Fig. 1 shows that there are no weathered horizons within the sequence, indicating no periods of subaerial erosion during deposition.

Directional data obtained from the long axis orientation of erratics with an a-axis/b- axis ratio of at least 2:1 show great within-till variation at TA 14707585. At the base of the Lower Till Series rounded quartzite erratics are numerous though their lack of definite orientation and their restriction to the basal till indicates that they are not far-travelled (Dreimanis and Vagners, 1969) therefore their shape is not an indicator of shearing processes occurring within the sole of the Devensian ice.

The lack of strong directional evidence in the fabric of the lowest till unit at this site is attributed to the ice meeting the rising ground of Flamborough Head. The advance of the ice against the headland may have initiated rotation of elongated erratics from a longitudinal towards a transverse position, though the end-point was not achieved, due to ice movement ceasing soon after meeting the rising ground. Alternatively, the lack of orientation may be attributed to deformation and flow of the till under the great stress exerted upon it at this point (Andrews and Smith, 1970) as indicated by the extensive development of narrow band tills hereabouts.

Directional data obtained from sample point C (T.2, Fig. 1) shows a strong N.N.E.-S.S.W. longitudinal maximum. This till has been free from internal shearing processes and is a true lodgement till. The overlying till unit (T.3) shows a north-east to south-west longitudinal maximum, hence the bedding plane separating it from the underlying T.2. T.3 is a member of the Lower Till Series and has been produced by within-till shearing.

Two of the narrow band tills which comprise T.4 and are represented by the sample points E and F display a wide range of directional variation. Sample E displays no longitudinal maximum whatsoever whereas sample F displays a strong north-east to south-west orientation.

The lack of unified directional data described above illustrates how the shearing processes have subdivided the Lower Till Series into a sequence of narrow band tills (Donner and West, 1956; Glen, Donner and West, 1957), some of which have been internally disturbed to a greater degree than others. These narrow band tills are a direct reflection of the tectonic structure of the late-Devensian ice in this area.

Directional data obtained from the Upper Till Series at this site shows a longitudinal N.N.W.-S.S.E. maximum. In this ice marginal position the Upper Till Series may have been moving laterally over the underlying Lower Till Series thus creating the shear observed between the two till series.

The measured section illustrated in Fig. 3 (Quay Hole, TA 125816) shows a complete depositionary sequence which demonstrates a marked contrast with the sequence described from Speeton and a marked affinity with sequences exposed in Holderness. At TA 125816, Lower Till Series with a Munsell colour of 10YR 3/2 - 4/2 overlies Corallian limestone. The till is very hard and compact with a fine sand/silt matrix and the basal 15 cm is rich in Corallian material that colours the matrix (10 YR 5/3). The Lower Till Series at this site is in marked contrast to that exposed at Speeton in that it displays no narrow band till development. It is true lodgement till. Between the Lower and Upper Till Series there occurs 2.7 m of sand, gravel, silt and clay whose fluvial structures indicate water flow towards the south west. The overlying Upper Till Series is compact with a clay-rich matrix, being characteristically soft when wet and has a Munsell colour of 5YR 3/4

Particle size data can be used to distinguish between the Lower and Upper Till Series at this site. Curves A and B of Fig. 4 demonstrate that the basal till sequence is a member of the Lower Till Series with a maximum between 3 and 4 Ø followed by a sharp decline in clay content. Curves C and D show the characteristic shape of the Upper Till Series with no sharply defined maximum in the 3 or 4 Ø range and a slight increase towards the clay fraction which is in accord with the plasticine-like nature of the till. Curve A is unusual in that it displays a large percentage of sand. This is due to the presence of a large number of uncomminuted

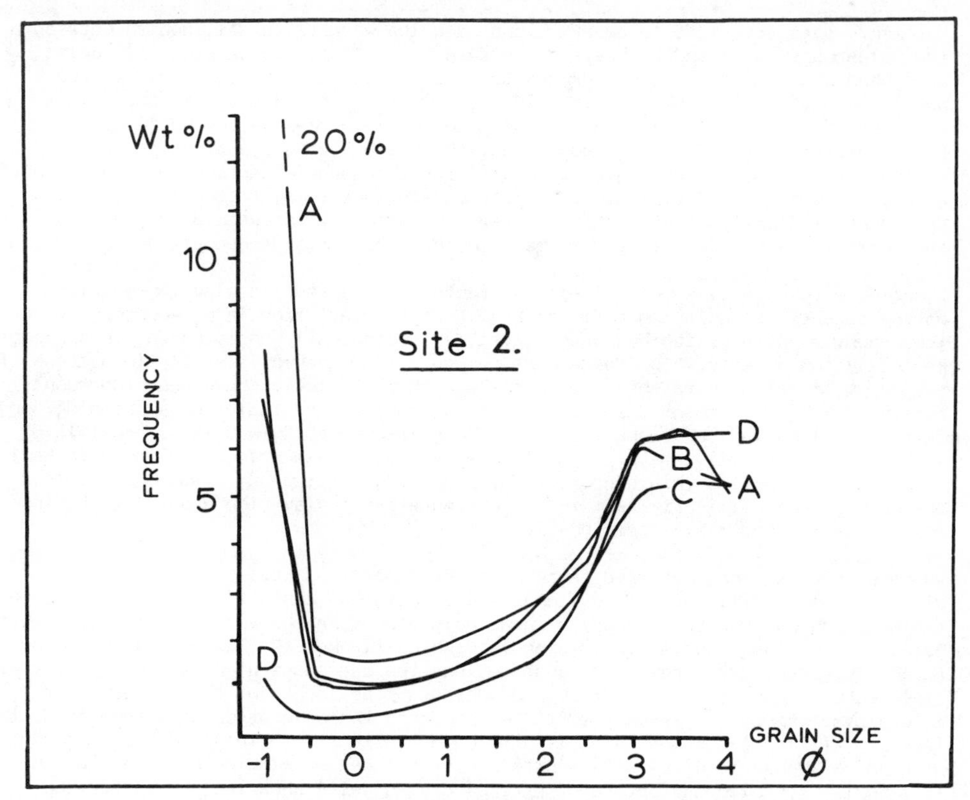

Fig. 4. Particle Size Distribution curve

limestone ooliths derived locally from the underlying limestone basement and not reduced to their terminal size mode.

Directional data at this site have been obtained from stone orientation counts and historical records. Stather (1897) observed striated limestone on the north side of Carr Naze which demonstrated a N.N.E.-S.S.W. ice flow. In 1891 Lamplugh observed striations at Quay Hole and these were visible again in 1975. Ten separate striations ranged from 22° - 70° (true) and gave a mean of 34°. A stone orientation analysis in the overlying Lower Till Series displayed a north to south longitudinal maximum with a minor transverse maximum normal to the basal striation's mean trend.

20 m south of TA 125816 there occurs a lens of overfolded gravel within which platy pebbles have their long axes orientated normal to the fold axis. The feature indicates compression from north to south suggesting that either the basal till was subject to flow after deposition or was disturbed by overriding pressure from the

north, probably by the upper ice of the Lower Till Series as there is no sign of shearing or folding between the Lower and Upper Till Series, and the Lower Till Series here is a true lodgement till. The absence of numerous pebbles orientated parallel to the fold axis in the axial planar zone indicates that very little near horizontal shear occurred.

CONCLUSIONS

For many years visual field observations have led workers to distinguish between the Holderness tills on the basis of texture and colour, the Skipsea till being sandy and drab whereas the Withernsea till is plastic and purple. This method proved adequate for fresh till in Holderness where both types are in the main undisturbed lodgement till deposited on a chalk plain some metres below present O.D. However, north of Flamborough Head, the pre-Quaternary surface presents extreme variety both lithologically and in its topographic expression. Consequently, greater textural and lithological variety occurs in the tills of Filey Bay than Holderness indicating that such parameters are strongly influenced by the lithology and form of the pre-Quaternary basement in the local area of deposition.

Shearing within the ice most directly affects texture and lithology, producing narrow band tills and being a direct result of the variation in pre-Quaternary topography. Wherever ice has been relatively tectonically undisturbed, thick lodgement till characteristic of Holderness occurs. This grades laterally into areas of excessive shearing where the Holderness-type characteristics have been thoroughly destroyed. The lodgement till is replaced by narrow band tills whose lithology and texture reflect changes brought about by localised shear. Thus it is possible within the Lower Till Series, whose lateral equivalent in Holderness is characteristically sandy, to find thin bands of clay-rich till, especially towards the base of the succession. Similarly, within the characteristic clay-rich Upper Till Series, thin bands of sandy till occasionally occur.

Wherever shearing has occurred near the Lower/Upper Till Series contact, thrust slices of Lower Till Series occasionally lie above the contact plane indicating that the shearing was upward. Shearing invariably occurs in the vicinity of pre-Quaternar basal obstructions such as the rising land which flanked the Flamborough Headland. As one traverses away from these obstructions the sequences of narrow band tills are replaced by sequences of undisturbed Holderness-type lodgement till. Much of the lateral variation in Filey Bay, which has rendered the allocation of a type locality for both till series impossible, is attributed to this shearing process. The presence of Mesozoic clay rafts and folded gravel lenses within the Filey Bay Lower Till Series are also indicators of extensive en-glacial shearing. Despite the difficulties of readily determining the Lower/Upper Till Series contact wherever shearing has occurred, it is possible to find it by Munsell colour notation and particle size parameters in the range -1 to 4 Ø.

Thus the boundary can be traced through these shear zones and linked to more obvious contacts where there is undisturbed lodgement till. At such lodgement till sites, the boundary is frequently marked by sands and gravels. It is not unusual that such sands and gravels are found wherever a two-tier ice sheet existed as they are simply an expression of meltwater flowing between the two major ice layers, being unable to penetrate to any marked extent into the tectonically undisturbed ice from which the lodgement till was produced. Conversely, within the shear zones, thick sequences of sand and gravel may occur at any horizon. This is an indication of tectonically disturbed ice allowing water to penetrate wherever weaknesses in the form of shear planes and associated joints allowed. Thus at the south-eastern end of Filey Bay, sands and gravels occur within the Lower Till Series. Due to the lateral impersistence of the shear planes, the sands and gravels are extremely local in distribution and do not constitute reliable marker horizons. The same is true of narrow band tills, some of which are characteristically coloured. Reddened narrow band tills are most usual and they may occur at any horizon within the Lower Till Series.

Narrow band tills are conspicuous by their absence in the Upper Till Series. Upper Till Series ice overlay that of the Lower Till Series in the North Sea basin and was erosionally less active. The lower ice eroded the floor of the North Sea basin and was continuously incorporating detrital material into its matrix. The material did not totally achieve its terminal particle size mode, thus the Lower Till Series has a sandy matrix. The Upper Till Series ice was not constantly incorporating new material from the terrain it traversed hence the till retains colour properties which were established in the Permo-Triassic source area of North-East England and the till matrix has largely been comminuted to its terminal clay mode.

REFERENCES

Andrews, J.T. and D.I. Smith (1970). Statistical analysis of till fabric: methodology, local and regional variability. *Q. Jl geol. Soc. Lond.*, 125, 503-542.

Bisat, W.S. (1940). Older and Newer Drift in East Yorkshire. *Proc. Yorks. geol. Soc.*, 24, 137-151.

Carruthers, R.G. (1939). On northern Glacial Drifts: some peculiarities and their significance. *Q. Jl geol. Soc. Lond.*, 95, 299-333.

Catt, J.A. and L.F. Penny (1966). The Pleistocene Deposits of Holderness, East Yorkshire. *Proc. Yorks. geol. Soc.*, 35, 375-420.

Donner, J.J. and R.G. West (1956). The Quaternary Geology of Brageneset, Nordaustlandet, Spitzbergen. *Skr. norsk Polarinst.*, 109, 5-31.

Dreimanis, A. and U.J. Vagners (1969). Lithologic relationship of till to bedrock. In J.E. Wright (Jr.) (Ed.), *Quaternary Geology and Climate*. Amer. Nat. Acad. Sci. Publn., 1701, 93-98.

Glen, J.W., J.J. Donner and R.G. West (1957). On the mechanism by which stones become orientated. *Am. J. Sci.*, 255, 194-205.

Harrison, W.J. (1895). Notes on the glacial geology of the Yorkshire coast (chiefly near Whitby). *Glacialists' Magazine*, 3, 67-89.

Lamplugh, G.W. (1879). On the divisions of the glacial beds in Filey Bay. *Proc. Yorks. geol. Soc.*, 7, 167-177.

Lamplugh, G.W. (1881a). On a shell bed at the base of the drift at Speeton, near Filey. *Geol. Mag.*, 8, 174-180.

Lamplugh, G.W. (1881b). On the Bridlington and Dimlington glacial shell beds. *Geol. Mag.*, 8, 535-546.

Lamplugh, G.W. (1889). Glacial sections near Bridlington. Part IV. *Proc. Yorks. geol. Soc.*, 11, 275-295.

Lamplugh, G.W. (1891). On the drifts of Flamborough Head. *Q. Jl geol. Soc. Lond.*, 47, 384-431.

Lamplugh, G.W. (1911). On the shelly moraine of the Sefstrom glacier and other Spitzbergen phenomena illustrative of British glacial conditions. *Proc. Yorks. geol. Soc.*, 17, 216-241.

Madgett, P.A. (1975). Re-interpretation of Devensian Till stratigraphy in eastern England. *Nature, Lond.*, 253, 105-107.

Mitchell, B.D. and R.A. Jarvis (1956). *The Soils of the Country round Kilmarnock*. Mem. geol. Surv. U.K., 234 pp.

Phillips, J. (1829). *Illustrations of the Geology of Yorkshire*. 1st Edition. York. xvi + 192 pp.

Phillips, J. (1835). *Illustrations of the Geology of Yorkshire: or a description of the strata and organic remains. Part I. The Yorkshire Coast.* 2nd Edition, London. vi + 184 pp.

Simpson, M. (1859). *A guide to the geology of the Yorkshire Coast*. 3rd Edition. Silvester Reed, Old Market Place, Whitby.

Stather, J.W. (1897). A glaciated surface at Filey. *Proc. Yorks. geol. Soc.*, 13, 346-349.

Wood, S.V. (jn.) (1871). Mr Croll's hypothesis of the formation of the Yorkshire Boulder Clay. *Geol. Mag.*, 8, 92.

Dr C.A. Edwards,
2 Hapton Way,
Loveclough,
Rossendale,
Lancs. BB4 8QG.

12. The Work of W. S. Bisat F.R.S. on the Yorkshire Coast

J. A. Catt and P. A. Madgett

INTRODUCTION

As Lewis Penny has already written (Stubblefield, 1974), W.S. Bisat's interest in the Quaternary of eastern Yorkshire was only a side-line to his professional work as an engineer and his research interests in the Carboniferous. Nevertheless, between 1932 and 1952 he wrote nine important papers on the glacial and postglacial deposits of the Holderness coast, and also led numerous field excursions in the area. In addition he published papers on Quaternary deposits in the North Ferriby and Wakefield areas.

After his death in 1973, Bisat's geological notes and correspondence passed to the Leeds office of the Institute of Geological Sciences to be stored in their achives. Items of Quaternary interest were carefully sorted and annotated by Lewis Penny, who realised that they contained a wealth of unpublished records and ideas concerning the glaciation of eastern England. Through him, we were able to borrow the notes from Dr W.H.C. Ramsbottom of I.G.S., and have already commented briefly on some of the contained items (Madgett and Catt, 1978, 93-94). The present volume seemed an appropriate means of publishing other contents of interest to Quaternary geologists, particularly those describing the coastal sections, which attracted so much of Bisat's attention.

SUMMARY OF CONTENTS OF BISAT'S NOTES

Probably the most significant aspect of Bisat's work in east Yorkshire was his meticulous recording of the continually changing coastal exposures. Details were inserted on carefully drawn sections with horizontal scales ranging from 1:240 to 1:126,720. A complete section of the Holderness coast between Easington and Sewerby (total length 59.4 km) was drawn between 1932 and 1934, and many parts of this were redrawn in subsequent years up to 1951. There are also almost complete sections of the cliffs in Filey Bay (Speeton-Filey), and shorter sections at Danes Dyke, Robin Hood's Bay, Whitby and Upgang. Written notes on all these parts of the coast in the 1930's are contained in a loose-leaf notebook, together with details of erratics and till matrix colours.

In addition to the large number of cliff sections and the related notebook, there are drafts of at least three unpublished papers and numerous other handwritten notes on coastal exposures. We cannot reproduce all of these, but have attempted instead to summarise Bisat's opinions on matters that still have some significance.

NOMENCLATURE OF THE HOLDERNESS TILLS

During the past century the tills exposed on the Holderness coast have undergone many changes in nomenclature (Catt and Penny, 1966). Table 1 summarises the succession recognised by Bisat in the 1930's and shows the probable correlations with successions published since then and with the sequence of British Quaternary stages. The names Basement and Purple had previously been used by many workers, though often with slightly different meanings. Bisat was the first to use "Drab", and probably took the name from colour charts of the West Riding woollen trade.

BISAT'S SUBDIVISION OF THE DRAB OR SKIPSEA TILL

From his fieldwork in the 1930's, Bisat (1939, 1940) subdivided his Drab Clay both at Dimlington in southern Holderness and in the Hornsea area of northern Holderness, though correlation of the two successions was not completely satisfactory. Catt and Penny (1966) failed to retrace these subdivisions, and Madgett and Catt (1978)

<table>
<tr><th>Bisat (1939, 1940)</th><th>Catt and Penny (1966)</th><th>Madgett and Catt (1978)</th><th>British Quaternary Stages, Mitchell <u>et al.</u> (1973)</th></tr>
<tr><td>Upper Purple Clays (2 beds)
Gravels
Lower Purple Clays (3 beds)</td><td>Hessle Till
Purple Till</td><td>Withernsea Till</td><td rowspan="2">Late Devensian
(18,000-13,000 B.P.
approximately)</td></tr>
<tr><td>Upper Drab Clay
Middle Drab Clay
Chalk rafts
Lower Drab Clay
Sub Drab Clay
Basement Drab Clay</td><td>Drab Till</td><td>Skipsea Till</td></tr>
<tr><td rowspan="2">Basement Clay
Sub-Basement Clay</td><td>Dimlington Interstadial Beds</td><td>Dimlington Silts</td><td>Late Devensian (18,240±250 B.P.)</td></tr>
<tr><td>Basement series</td><td>Basement Till</td><td>Wolstonian</td></tr>
</table>

Table. 1. Subdivisions of the Holderness glacial sequence recognised by W.S. Bisat, and their relation to those of later workers

renamed the sequence the Skipsea Till, finding little petrographic heterogeneity within it (Table 1). However, more detailed work in the late 1940's allowed Bisat to refine his subdivision somewhat, and revise the correlation between the Hornsea and Dimlington sections, which are separated by an 11.5 km gap (from Holmpton to Tunstall), where the Skipsea Till is below low-tide level and the cliffs expose only the overlying Withernsea Till (= Purple of Bisat).

In the northern section (Tunstall to Bridlington), centred on Hornsea, Bisat previously recognised four main divisions of the Drab (the Upper, Middle, Lower and Sub Drab Clays). The occurrence of these in the cliff sections is clearly shown by the complete Holderness coast section constructed between 1932 and 1934. Short fragments of this (at Dimlington, Hilston and Mappleton) were given by Bisat (1940), but the complete section is redrawn as Fig. 1, A-U; the original is 8.2 m long and has a horizontal scale of 1:7200. The uppermost division (the Upper Drab) rises into the cliff foot near Tunstall (Fig. 1, G) and reaches the cliff top north of Hornsea (Fig. 1, N); Bisat described it as a homogeneous dark grey clay full of hard white chalk erratics, and also containing much Liassic shale and limestone, Magnesian Limestone, Carboniferous Limestone, Bunter Sandstone, grey flints, green greywackes, Cheviot porphyrites, and a few black flints and pieces of pink chalk. Sand separated from this bed had a salmon pink colour.

The underlying Middle Drab comes into the cliff base at Aldbrough (fig. 1, J), and forms the lower part of the cliff northwards to Hornsea (Fig. 1, M), a section in which it is separated from the Upper Drab by a prominent band of red clay (the "Red Band" of Reid, 1885). Bisat wrote that the Middle Drab generally has less chalk than the Upper Drab, though it is fairly common towards the base; it contains more black flints, and occasional granites and pieces of rhomb porphyry (but no other Norwegian rocks, such as larvikite). In addition to the Red Band, Bisat recognised four minor subdivisions within the Middle Drab: a reddish layer immediately below the Red Band, 0-4 m thick, with some Magnesian Limestone and a few black and grey flints, greywackes and Cheviot porphyrites; a browner layer beneath, again with Magnesian Limestone, greywackes and Cheviot porphyrites, but also some Carboniferous Limestone and rhomb porphyry; the more chalky layer with grey flints and greywackes; and at the base a leaden-coloured and often laminated bed with few stones.

The next lower division (the Lower Drab) is also composite, though Bisat never recognised any laterally persistent subdivisions. It rises into the cliff base immediately north of Hornsea (Fig. 1, N), and has an irregular, hummocky surface, often marked by streaks of chalk up to 30 cm thick and containing *Belemnitella mucronata*. The maximum thickness of this division is approximately 7 m. Near the top black and grey flints are abundant, and Carboniferous Limestone is more common than in the Middle Drab. The upper parts also contain a large suite of igneous rocks, including Scottish granites, larvikite, rhomb porphyry, quartz porphyry and vesicular and amygdaloidal basalts. Northern English erratics (Liassic shale and limestone, Bunter Sandstone, Magnesian Limestone, Cheviot porphyrites) and greywackes occur only sporadically. Fragments of Pleistocene marine shells (*Tellina balthica, Arctica islandica* and *Dentalium* sp.) are also fairly common in the upper part. Lower parts of the Lower Drab include laminated till with interleaved streaks of varve-like silts, the bedding planes of which are "marbled" and very irregular. Where the upper surface of the Lower Drab descends into hollows, the upper 20-50 cm are often chalk-free and paler in colour, with thin red, green and blue streaks containing small erratics of greywacke, Carboniferous Limestone and yellow sandstone. Lower Drab was not recognised south of Hornsea, but streaks of rubbly chalk seen on the foreshore at low tide beyond the southern end of Hornsea sea wall may mark the top of the division, as they do further north.

The lowermost division of the Drab in northern Holderness is the Sub-Drab. This occurs in the core of an asymmetrical anticline at Beacon Hill, Skirlington (Fig. 1, O), and possibly also near Skipsea (Fig. 1, P) and between Fraisthorpe and Bridlington (Fig. 1, R & S). It is a dull brown till with very little chalk and only occasional pieces of greywacke, Carboniferous Limestone, Magnesian Limestone, Bunter Sandstone, vesicular basalt, Liassic shale, black flint and Cheviot porphyrite, and a few white quartz pebbles. Pockets of greenish shelly Basement Till occurred within the sub-Drab on the foreshore at Skirlington.

In southern Holderness, the cliffs between Kilnsea and Holmpton, on either side of Dimlington High Land (at 30 m + O.D., the highest part of Holderness), expose a rather different succession in the Skipsea or Drab Till. Originally Bisat (1939) recognised two main divisions, the upper of which he correlated with the Sub-Drab of northern Holderness, and the lower he named the Basement Drab; a 2 m bed with abundant chalk and black flint above the Sub-Drab at Holmpton was tentatively correlated with the basal part of the Lower Drab. Postwar refinement of the Dimlington succession led to the recognition of five divisions, with the following characteristics:

Division 1 consists of two beds, a lower leaden coloured till with little chalk, and an upper very full of chalk. Both reach a maximum thickness of about 4 m near Dimlington Farm (Fig. 1, A), and die out to the north near Cliff Farm (Fig. 1, B). South of Dimlington Farm, chalk streaks with black flints occur between the two beds, and in one place a raft of transported Basement Till occurred at this horizon. At the top of the upper bed there are red streaks containing Palaeozoic rocks, chalky gravels and sometimes a greenish band with greywackes and sandstones. Far-travelled igneous rocks occur sporadically in both beds, and the lower bed also contains Magnesian Limestone and coal.

Division 2 consists of three thin beds, the main one being a dark grey till of irregular thickness, which is seen at Kilnsea and Easington (Fig. 1, A), but thins northwards. This part contains many granites, other igneous rocks, chalk and Carboniferous Limestone, and a few Pleistocene shells. It is locally underlain by a reddish-brown layer, and is succeeded northwards, towards Holmpton (Fig. 1, C), by a dark reddish-brown bed with porphyrites but little chalk. Divisions 1 and 2 are together equivalent to the Basement Drab.

Division 3 contains abundant small chalk erratics, and its upper part has contorted streaks of red (Triassic) material and chalk with *Belemnitella mucronata*. It is 60-150 cm thick, and at Dimlington can be divided into a pale upper part and less pale lower part. A persistent band of blue clay containing many Carboniferous Limestone, greywacke and sandstone erratics occurs beneath the red and white streaks in the northern part of Dimlington cliff.

Division 4, in the central and northern part of Dimlington cliff (Fig. 1, B), consists of thick sands and silts, underlain by a thin gravel and capped by varved clays interleaved with gravels. Elsewhere between Kilnsea and Holmpton, this horizon is represented only by occasional shallow basins filled with silts and gravels.

Division 5 consists of two dark reddish-brown layers, the lower of which is up to 50 cm thick and contains much Liassic material, and the upper is 1.5-3 m thick and contains abundant chalk, and many black flints and igneous rocks including pieces of larvikite, as well as some Liassic and Triassic erratics. The less chalky lower layer is irregular in thickness, and best seen at Kilnsea, north of Easington lane end (Fig. 1, A) and in the southern part of Dimlington Cliff. The upper bed is exposed at Holmpton lane end, immediately beneath the Purple (=Withernsea) Till (Fig. 1, C) and forms the cliff top at Easington and Kilnsea, where it is sometimes separated from the lower layer by irregular pockets of gravel with a dark reddish marly matrix.

In his later notes, Bisat referred to divisions 3-5 as the Higher Drab, probably to avoid giving the impression of a clear-cut correlation with any part of the succession in northern Holderness. He considered various possible correlations based on the vertical distribution of erratics, and another on the occurrence of stratified waterlain sediments and other features of till lithology. The erratics suggested that the Basement Drab of Dimlington includes both Lower Drab and Sub Drab further north, and that the Higher Drab is roughly equivalent to the Upper Drab of the Hornsea area. In both areas Triassic material increases upwards, becoming common in the Higher Drab of Dimlington and the Upper Drab of Aldbrough (Fig. 1, J). Larvikite persists into the highest bed at Dimlington, but at Hornsea it almost disappears at the top of the Lower Drab. Amygdaloidal porphyrite occurs in the uppermost Basement Drab at Holmpton (Fig. 1, C), but is absent above the Lower Drab at Hornsea. The Lower Drab contains a very varied suite of igneous rocks, which is to some extent matched by the assemblage in the main bed of division 2 at Dimlington, and the same horizons yield Pleistocene shells in both areas. The lowest bed of the Basement Drab resembles the Sub Drab at Skirlington (Fig. 1, O) in matrix colour, paucity of chalk, and the occurrence of other erratics, including masses of Basement Till. Bisat concluded that, if these correlations are correct, black flints and chalk must become more abundant southwards, and Scandinavian erratics must persist higher in the succession at Dimlington than further north. The first was in fact confirmed for the Skipsea Till as a whole by Madgett and Catt (1978).

It is possible that division 3 at Dimlington should be correlated with the Lower Drab (and therefore more appropriately included with the Basement Drab), as the pale bed in this division and the upper part of the Lower Drab both contain chalk streaks with *Belemnitella mucronata* and also red and blue streaks. If correct, this suggests that the Middle Drab is either absent at Dimlington or represented by the sands and silts of division 4.

Another possible correlation based on erratics equates the Higher Drab of Dimlington with the Lower Drab of Hornsea, on the evidence that both contain larvikite and black flints; higher parts of the Hornsea succession contain grey flints and no larvikite. This is supported by the idea of a progressive upward increase in Liassic and Triassic erratics through the Drab succession, the Basement Drab containing approximately the same amounts as the Sub Drab, the Higher Drab the same as the Lower Drab, and all containing less than the Middle and Upper Drab of northern Holderness.

Yet another possible correlation arose from the idea that the sands and silts of division 4 at Dimlington are equivalent to the bed of sand and gravel above the Lower Drab between Atwick and Skipsea (Fig. 1, O-P), and also to thick sands, silts and gravels further north at Barmston (Fig. 1, Q & R) and Sewerby (Fig. 1, U). This is supported by the lithological similarities (already referred to) between the pale upper part of Division 3 at Dimlington and the Lower Drab of northern Holderness. Greenish streaks with greywacke and yellow sandstone fragments, occurring just below the gravels at Hilderthorpe (immediately south of Bridlington), resemble the streaks near the top of the Lower Drab north of Hornsea and also the blue band with similar erratics near the top of Division 3 at Dimlington. At Sewerby the Lower Drab overlies a less chalky layer (probably Sub Drab), and this in turn rests on a grey bed, which could be equivalent to the dark grey till of Division 2 at Dimlington or the leaden-coloured lower layer of Division 1 (Fig. 1, U). This suggests that the Sub Drab is equivalent to (a) the less pale lower part of Division 3, or (b) the dark reddish brown bed with porphyrites at the top of Division 2 at Holmpton, or (c) the chalky upper part of Division 1.

There is no indication that Bisat ever decided which of these possible correlations between northern and southern Holderness is correct. However, all of them equate

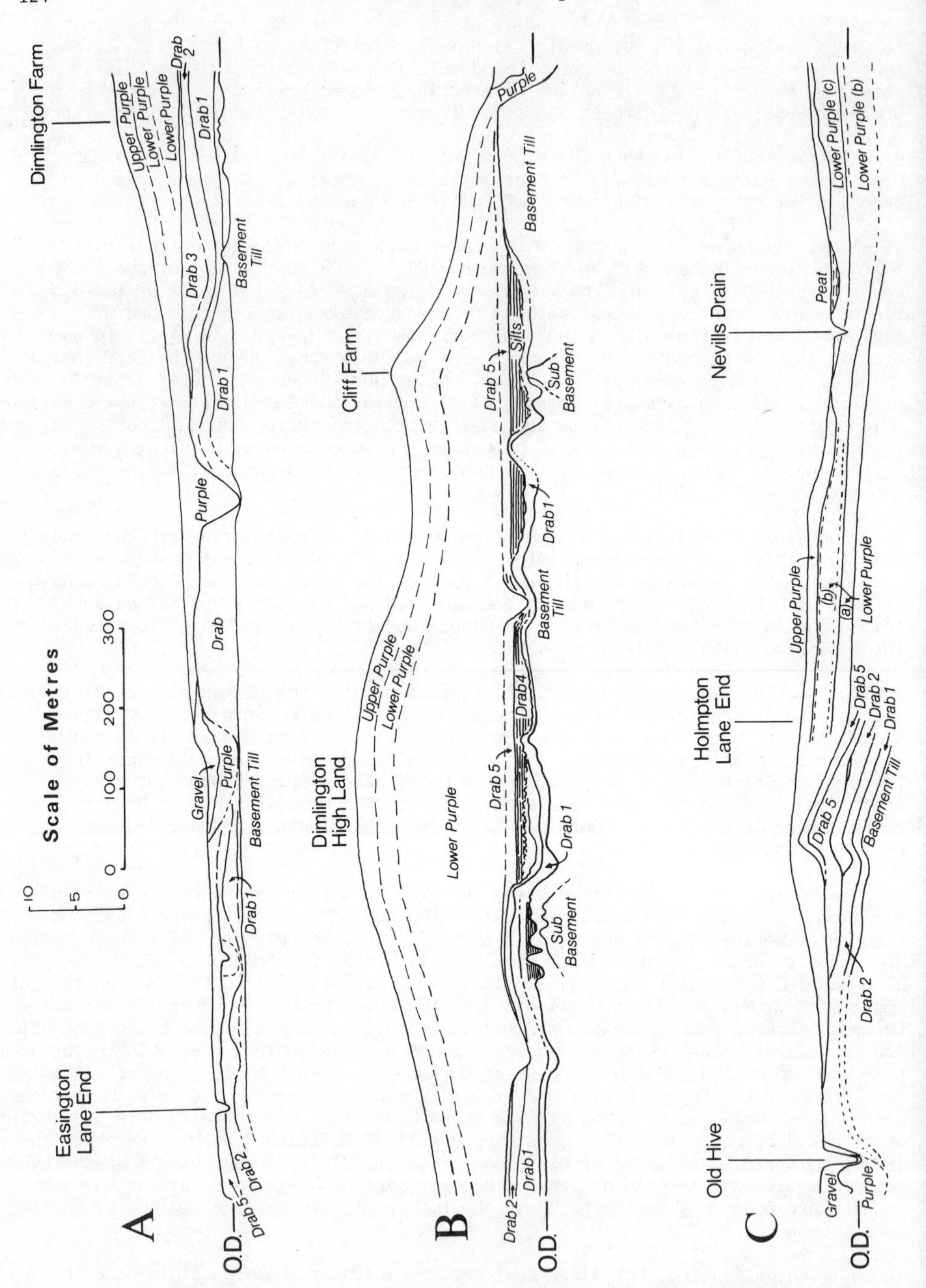

Fig. 1. A-U. Bisat's section of the Holderness coast between Easington and Sewerby, largely drawn in the 1930's. A-C. Easington Lane End to Nevills Drain.

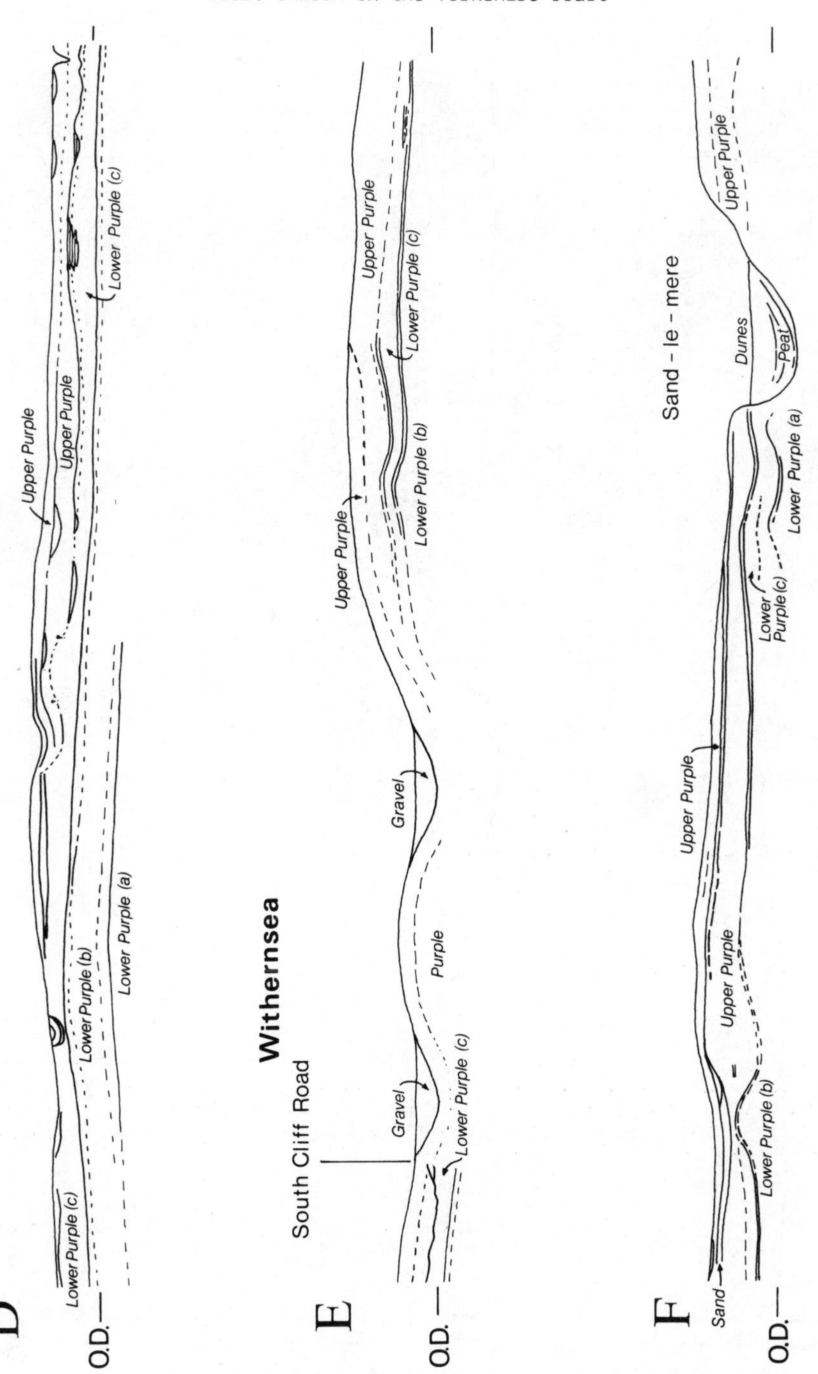

Fig. 1. D-F. Withernsea to Sand-le-mere.

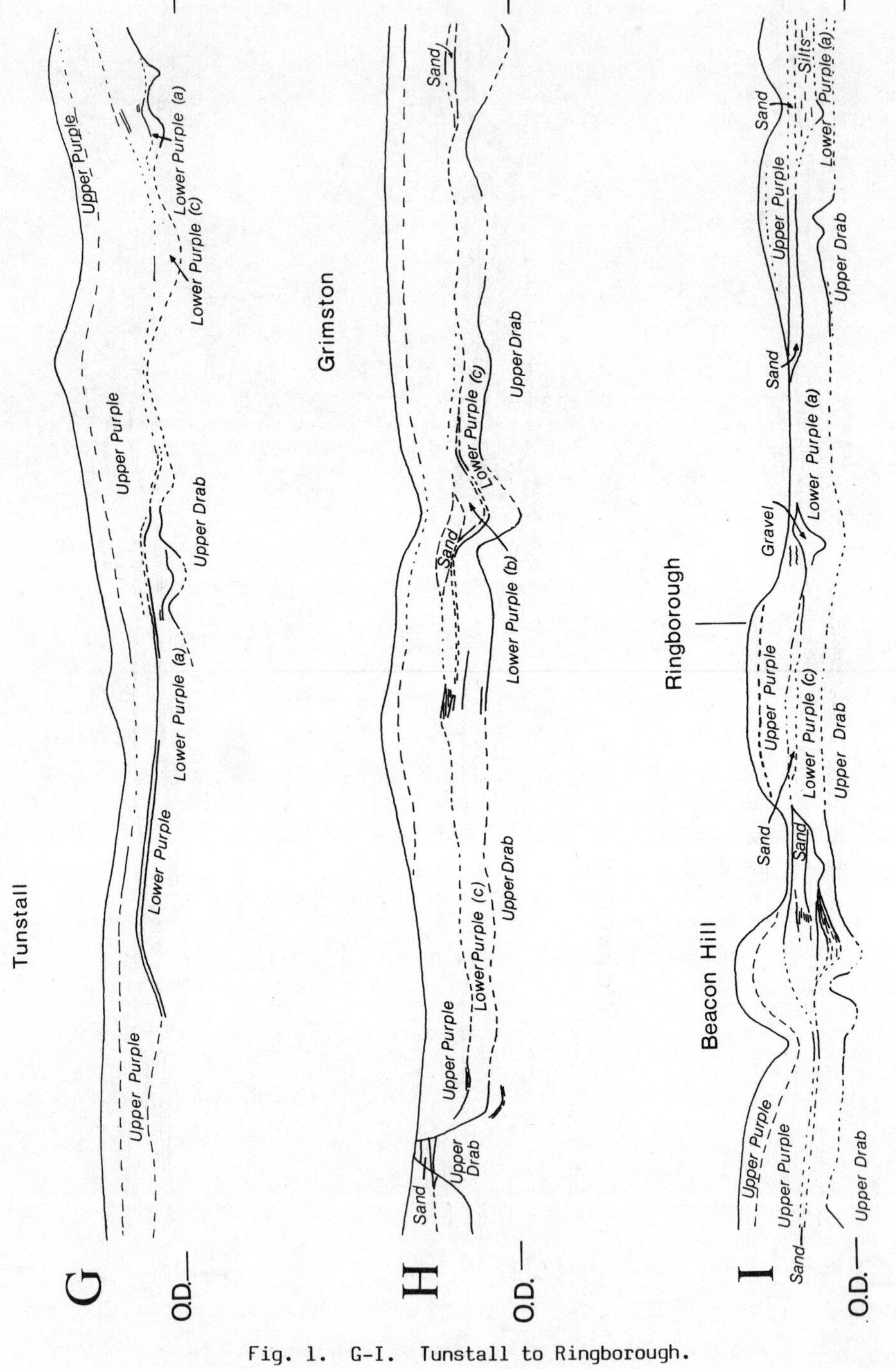

Fig. 1. G-I. Tunstall to Ringborough.

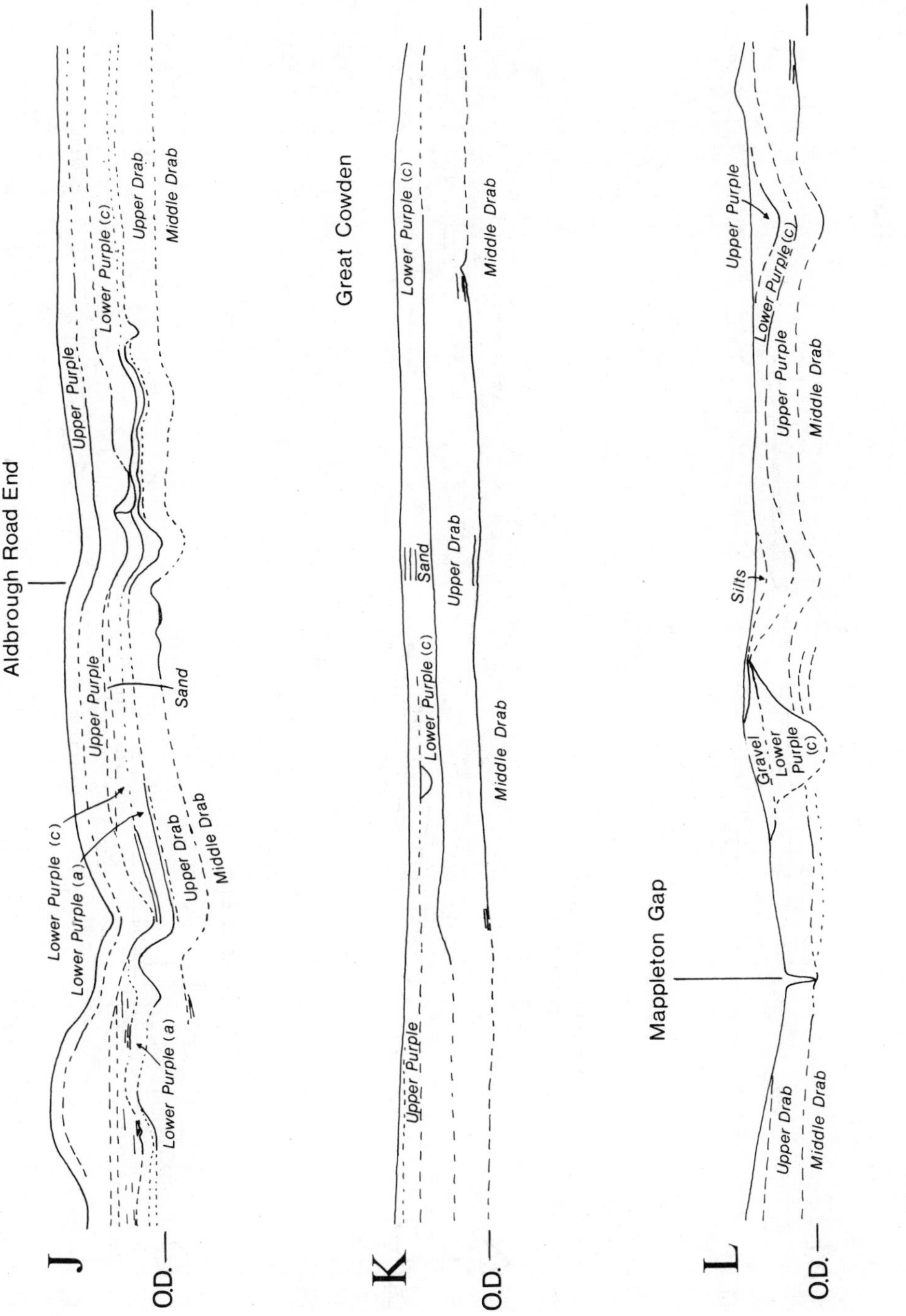

Fig. 1. J-L. Aldbrough Road End to Mappleton Gap.

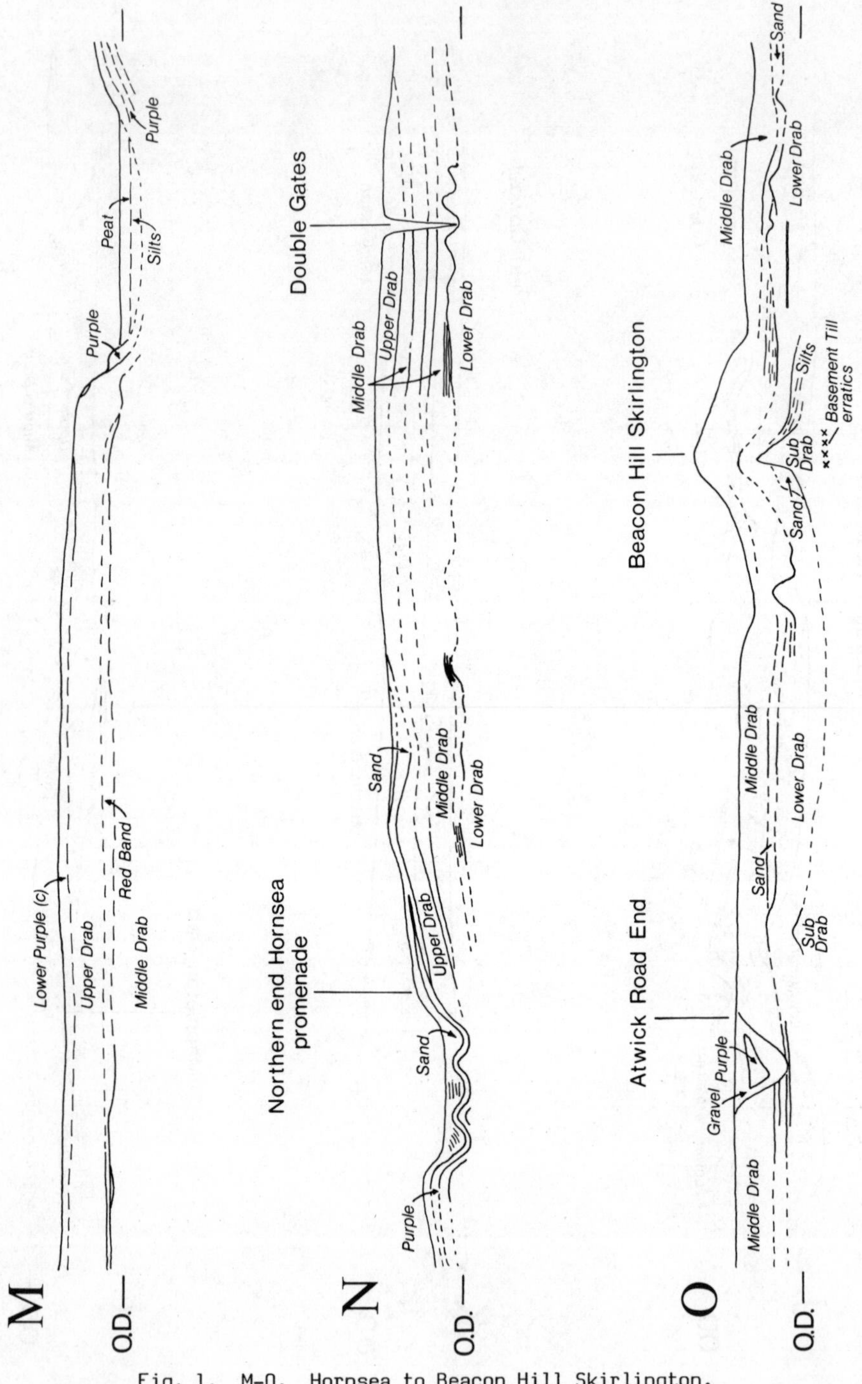

Fig. 1. M-O. Hornsea to Beacon Hill Skirlington.

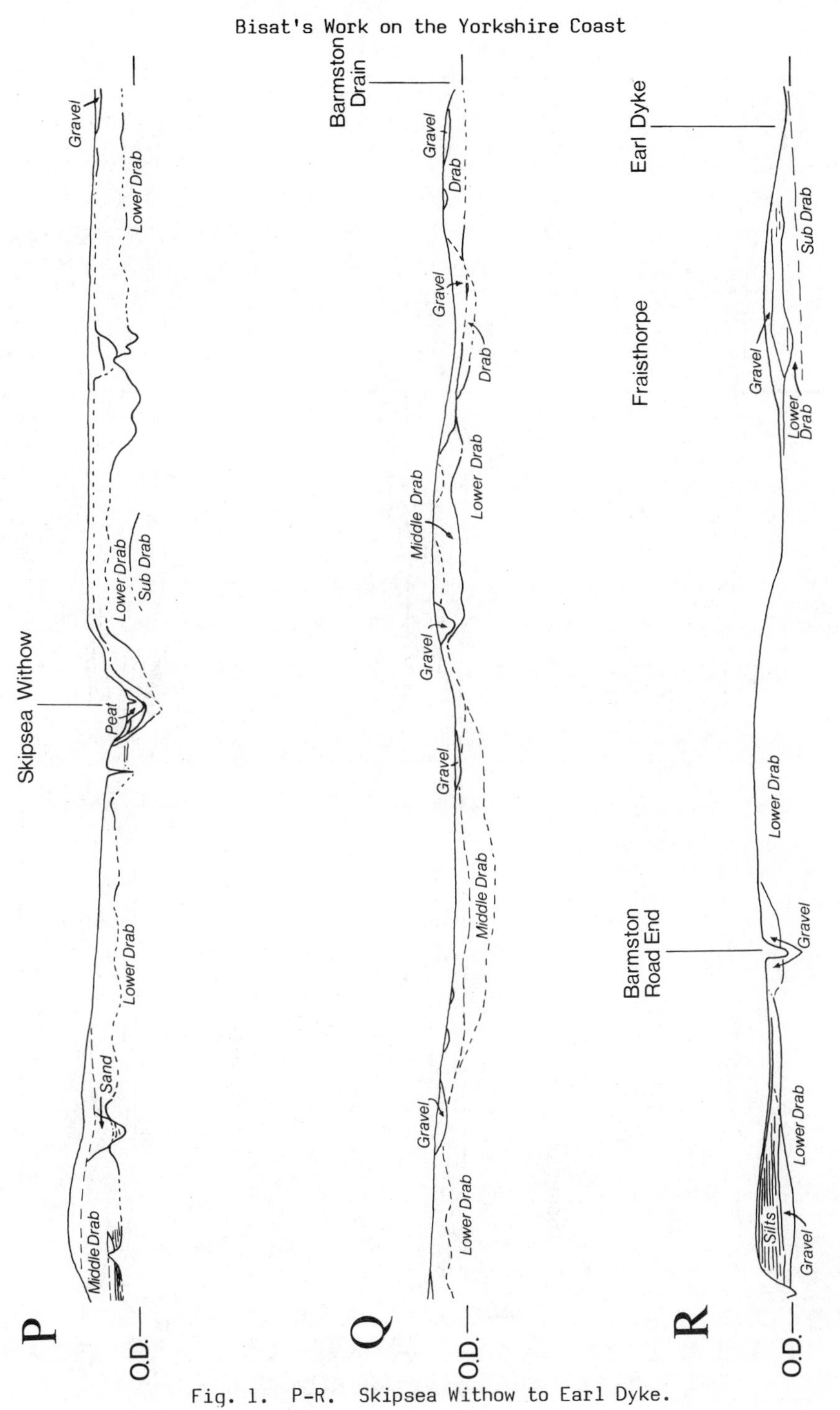

Fig. 1. P-R. Skipsea Withow to Earl Dyke.

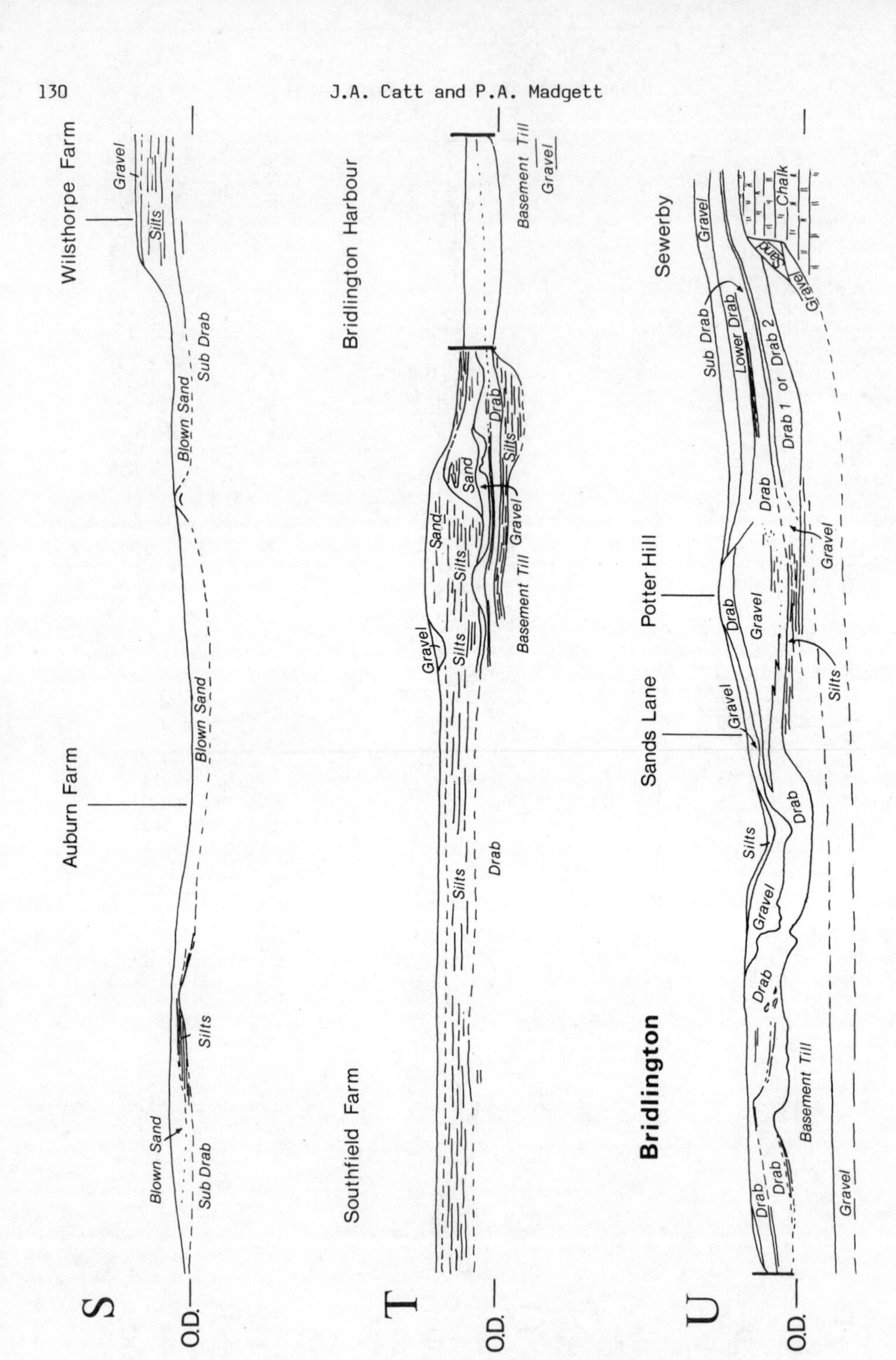

Fig. 1. S-U. Auburn Farm to Sewerby.

the Sub-Drab with lower parts of the Dimlington succession than his early (pre-war) work, and thus suggest that the Basement Till should occur at no great depth below the cliff-base and foreshore at Beacon Hill, Skirlington (Fig. 1, O), and possibly some other points along the north Holderness coast. This could easily be verified by augering. Bisat clearly felt that further observation of the coastal sections would ultimately clarify the successions on either side of the "Purple basin" centred on Withernsea, and the correlation between them. This is work which could and should be undertaken by a local geologist, able to visit the coast frequently over a number of years, and we hope that this summary of Bisat's ideas will provide a useful start for such a person. The modern Munsell Color Chart would obviously provide a more objective method of till matrix colour description than Bisat used, and grouping of erratics should be done in more detail than the broad lithological divisions used by Madgett and Catt (1978).

BISAT'S SUBDIVISION OF THE PURPLE OR WITHERNSEA TILL

Bisat's pre-war work allowed him to divide the Purple Till into Upper and Lower parts (Bisat, 1939, 1940), and the "dark Arabian brown" Lower Purple into three beds (a, b and c), the reddish-brown Upper Purple into two (Table 1). Notes made some time after publication of the 1940 paper give many details about the occurrence of these subdivisions in cliff sections both south and north of Withernsea.

At the southern end of Withernsea promenade (then at South Cliff Road, Fig. 1, E), the Upper and Lower Purple are separated by 1.2 m of silts and gravels, lying at about highwater mark of neap tides. The Upper Purple is a reddish mottled clay, 3 m thick, and containing many small pieces of chalk; the highest bed (c) of the Lower Purple is a uniform dark reddish-brown clay with some chalk, Carboniferous Limestone and other erratics. Southwards the silts and gravels are replaced by a brown clay with very little chalk, which has greenish silty streaks at the top. The surface of the Lower Purple rises southwards, and about 320 m south of South Cliff Road locally cuts out the Upper Purple. It encloses two silt basins further south, and about 1.3 km from South Cliff Road the middle bed (b) of the Lower Purple rises into the cliff. This is darker and less chalky than Lower Purple (c), and reaches about 4 m in thickness; the junction between (b) and (c) is frequently marked by streaks of greenish clay with greywackes and reddish (Triassic) clay. The brown silty base of Lower Purple (b) and the red top of (a) rise into the cliff base about 2.5 km south of the promenade (Fig. 1, D), but sink again on approaching Nevills Drain (Fig. 1, C). The Upper Purple is absent at the drain, the shelly postglacial deposits resting directly on Lower Purple (c). Further south Lower Purple (c) persists only for 50 m as a thin remnant at the cliff top, and the Upper Purple returns, resting on Lower Purple (b), with gravels, silts and green streaks separating them. As the cliff increases in height, the Upper Purple can be divided into two, the base of the upper part being marked by abundant Lias shale fragments. 100 m north of Holmpton Lane end (Fig. 1, C) the upper part is nearly 5 m thick, but the surface of the Drab (Skipsea Till) rises rapidly southwards and cuts out all but a metre or so of the Purple. About 1 km south of here the Purple again thickens rapidly towards Dimlington High Land, where it rests on a plateau of Drab that rises gently from south to north.

North of Withernsea promenade the cliffs are higher than on the south side of the town, mainly because the lower bed of the Upper Purple is thicker. The same general sequence can be traced northwards to Hornsea, though the Lower Purple is generally thinner, Lower Purple (a) being reduced near Aldbrough (Fig. 1, J) to thin lenticles lying in basins in the surface of the Upper Drab, and Lower Purple (b) thinning out north of Garton (Fig. 1, I). Lower Purple (c) contains numerous inclusions of Upper Drab clay at Aldbrough and Mappleton (Fig. 1, L). North of Hornsea the only occurrences of Purple are a thin reddish bed between gravels in the higher part of the cliff immediately north of Hornsea promenade (Fig. 1, N), and a similar bed between gravels at Atwick road end (Fig. 1, O).

There is little difference in erratics between the various beds of the Purple. The large igneous suite of the Drab has vanished except for a few small pebbles, and there are no Norwegian rocks. Triassic, Liassic and Magnesian Limestone material is abundant, and Carboniferous Limestone quite common. Chalk and grey flints are also common (though chalk is not so frequent as in parts of the Drab), but black flints are scarce. Greywackes and Cheviot porphyrites are plentiful. The igneous rocks that do occur are different from those in the Drab. Chalk is most abundant in Lower Purple (c), and Liassic shale in Lower Purple (a), the top of which further north is marked by shell fragments and gypsum. The brown Upper Purple has much fewer erratics than the Lower Purple beds.

SECTIONS NORTH OF BRIDLINGTON

During the late 1930's Bisat examined most of the coast section between Bridlington and the Tees estuary, and was able to trace northwards the main boulder clay horizons previously recognised in Holderness. He identified Basement Till on the foreshore at Sewerby, but failed to trace it in the cliff section there, which shows the sequence of Drab beds already mentioned overlying the chalk rubble (head), blown sand and interglacial beach deposits associated with the buried cliff. Various undated notes show that he recognised Basement Till also in the cliff section at North Landing, and on the foreshore at Reighton and opposite Flatt Cliff in Filey Bay. However, other notes on Filey Bay interpret the chalky bed with black flints and shell fragments exposed on Reighton foreshore as intercalated in the lower part of the Drab, and possibly equivalent to the Basement Drab.

The distinction between the Drab and Purple Clays in Filey Bay is as clear as in Holderness, the junction being readily observable between Reighton Gap and Hunmanby Gap on the southern side of the bay, and at Carr Naze on the northern side. In the centre of the bay, the Drab rises to the cliff top in many places, and the Purple is confined to isolated basins. 400 m south of Hunmanby Gap, a trough in the Drab extends below beach level, and is filled with varved silts (Bisat, 1940, Plate 17). Apart from this, the cliff foot throughout most of the bay is formed by a greyish brown bed; in the centre of the bay between Flatt Cliff and Filey, this is succeeded by a dark brown layer with large masses of Liassic shale near the top. A gravel and further Drab beds locally overlie the dark brown layer. Scandinavian erratics and the rich granite suite typical of the Lower Drab are generally absent from Filey Bay, and chalk is present only in the bed on Reighton foreshore, but black flints occur occasionally throughout the succession. However, the till overlying the Speeton Shell Bed on New Closes Cliff was identified as Lower Drab, because it contains larvikite. Elsewhere amygdaloidal porphyrites are common, which suggests correlation with lower parts of the Drab in Holderness, and Bisat concluded that in Filey Bay Middle and Upper Drab are probably absent. Middle Drab was identified, however, north of Scalby Ness, where a greyish brown bed containing rhomb porphyry overlies Lower Drab, which in turn rests on a thick sequence of gravels, sands and silts.

South of Hunmanby Gap and in other exposures further north (Carr Naze, Sandsend and Upgang) Bisat was able to recognise the main Upper and Lower divisions of the Purple, the junction often being marked by sands and gravels. The erratics are similar to those of the Purple in Holderness (Bunter sandstone, Liassic shale and limestone, Magnesian Limestone, Carboniferous Limestone, Jurassic sandstones, greywackes, Cheviot porphyrites, etc.), but chalk is absent. At Robin Hood's Bay (Fig. 2) and Upgang, an "intermediate bed" was recognised between the Purple and Drab clays; this was best exposed at Upgang, where it is about 8 m thick. It has some of the colour and erratic characteristics of each clay; Bisat never commented

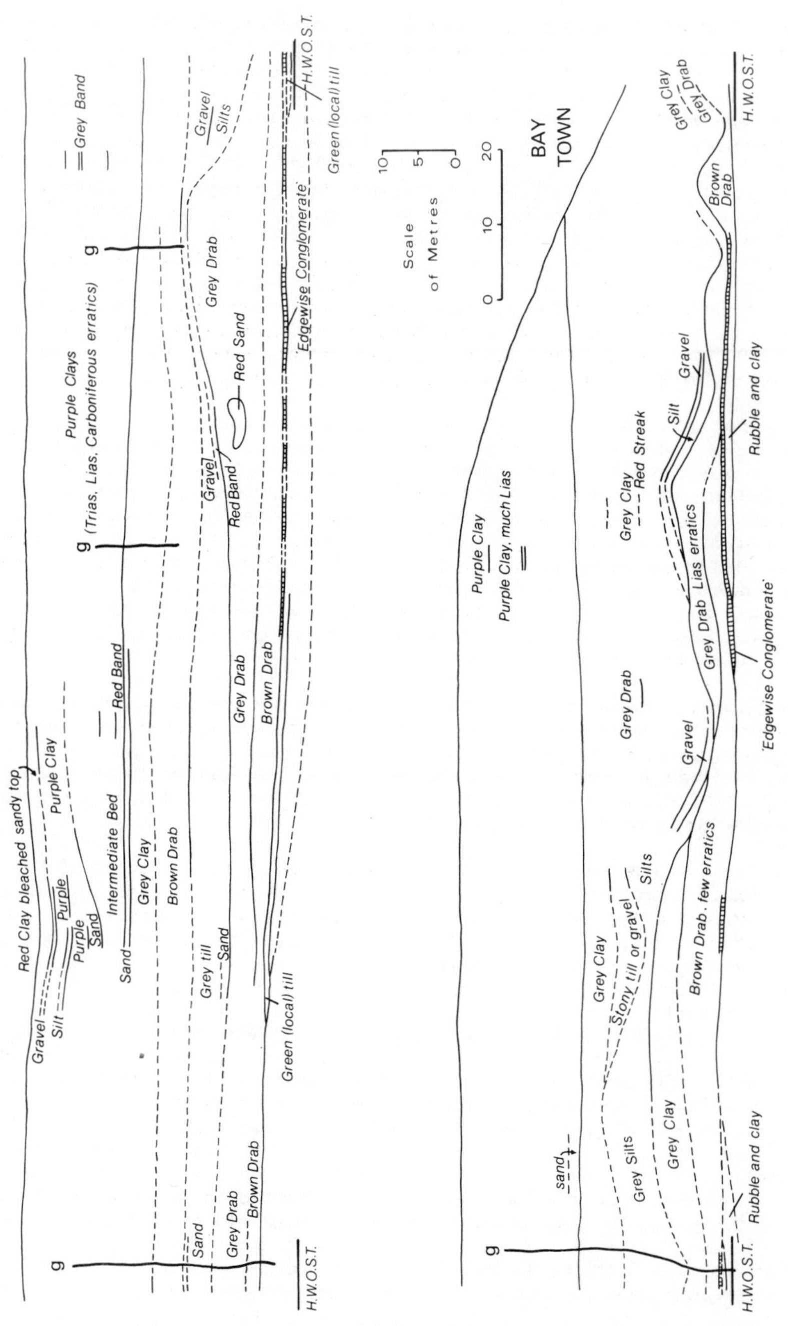

Fig. 2. Bisat's section of the drift deposits between Bay Town and the gully approximately 92m north of the headland north of Mill Beck, Robin Hood's Bay (g = position of gullies in cliff).

on its origin, but it could have resulted from mixing of the two main layers of the composite Devensian ice sheet, which we previously postulated invaded coastal areas of eastern Yorkshire south of the Tees estuary (Madgett and Catt, 1978).

Another new bed was recognised above the Purple Till at Sandsend. This is a buff-coloured clay packed with small pebbles of north British rocks (greywackes, porphyrites, quartzites and white quartz pebbles). The best exposure of this was at the top of the road section on the Whitby side of Sandsend, about 300 m east of the lane to Dunsley and on the west side of a small ravine (probably near NZ 870121), but it was also observed at the top of other ravines running inland between Sandsend and Whitby.

In Robin Hood's Bay the glacial sequence is best exposed in the cliffs between Bay Town and Mill Beck, though even here much is obscured by slipping. Bisat's sketch of this section is redrawn as Fig. 2. The bedrock surface (Lower Lias) is about high water level (spring tides), and rises to the small promontory at Mill Beck. A rubbly mixture of sandstone and clay (Liassic shale), up to 2 m thick, rests on the bedrock surface, and is locally overlain by a thin greenish till with sandstone and ironstone fragments in a matrix composed of disaggregated Liassic shale. Above this bed is a brown "edgewise conglomerate", a few cm thick, which is composed of broken pieces of laminated clay set at all angles. The succeeding bed is a similarly coloured till, up to 5 m thick and containing only a few, small, far-travelled erratics (porphyrites, rhomb porphyry, granites, quartzites, quartz) many of which are rounded beach pebbles. Bisat considered correlating this with the Basement Till, but was unsure of this and on the section referred it to the Drab. The grey till above this contains many erratics of Liassic and Triassic rocks, Magnesian Limestone, Carboniferous Limestone, greywacke, prophyrites and local sandstone; Bisat suggested correlating it with the Lower or Middle Drab of northern Holderness. Above this is a thick series of grey silts and sands, surmounted by gravels and another layer of brown till, which contains granite erratics, and was thought to be the source of the large pieces of Shap Granite on the foreshore of the bay. It also contains Liassic rocks, local sandstones, greywackes, Magnesian Limestone and porphyrites and was equated with the Lower Drab. A further layer of grey till, about 3 m thick, completes the Drab succession, and is capped by a thin sand from which many small springs arise.

The overlying intermediate bed is 5-8 m thick, and contains abundant Liassic shale and greywacke erratics, with some Triassic rocks and limestones. It is variable in colour, with reddish, brown and dark brown layers. The succeeding Purple Till consists of at least three beds with intervening silts and gravels, but was too poorly exposed to examine in detail.

PERIODICITY IN CLIFF EROSION IN HOLDERNESS

One of the draft papers among Bisat's notes has this title, and includes a summary of observations on coastal processes (movement of beach material and cliff erosion) between 1935 and 1952, mainly at Dimlington. During this period about 70% of the total length of Dimlington cliffs (exact limits not given) had at its foot a beach reaching to normal highwater level or above, thus precluding marine erosion. Under these circumstances, the cliff became an irregular slope of slumped clays, partially covered by vegetation, and useless for detailed examination of the till stratigraphy.

The beaches of Holderness do not form an even carpet fringing the cliffs, but consist of long banks separated by shorter troughs with little or no sand, in which the sole of the cliff and the foreshore clay adjacent to it are exposed for a distance of up to 1 km. It is only in these troughs that marine erosion of the cliffs becomes really effective, as the high sand banks are effective barriers to the waves, and for the time being prevent further marine erosion of the cliff.

Phillips (1964) described the same trough-like features, using the local dialect name "ord" for them. They seem to be unique to the Holderness coast, but only occur regularly south of Hornsea. She thought they are connected with the sheltering effect of Flamborough Head, only forming where the powerful waves caused by north or north-west winds first meet the coast with full force. At this point the rate of longshore movement of beach material would increase rapidly, and a gap would develop in the beach. Once they had been formed, they would not be filled in by southward drift, as the rate of longshore movement at any given time is approximately the same throughout the length of the south Holderness coast. She estimated that the ords move southwards at an average rate of about one mile per year.

However, Bisat's estimate of the average rate of movement at Dimlington over 15 years was only ¼ mile (400 m) per year, and he concluded that the cliffs at any point on the coast go through a cycle lasting approximately 12 years. Active erosion in a trough or ord may persist up to three years, and results in considerable steepening of the cliff by undercutting and the removal of the material that slips to the cliff base. The best cliff and foreshore exposures occur during this part of the cycle. The ord then moves south, usually as the result of a storm, and is replaced by a bank of sand, which is high enough to prevent the sea reaching the cliffs at most high tides. For several years the cliffs then become progressively less steep, masses of clay slipping down from the cliff top along arcuate surfaces, and remaining as back-tilted blocks at various levels because the cliff foot is protected from further erosion. Exposures then rapidly worsen, until the cliff foot again comes under attack through the arrival of the next ord.

Anyone interested in continuing the study of Holderness glacial stratigraphy from the point to which William Bisat and Lewis Penny have so painstakingly brought it, must therefore work within the confines of this geomorphological system. The casual visitor to the coast can often go away disappointed, disbelieving that even the broadest stratigraphic succession could ever be constructed from the tumbling, vegetated cliffs, boggy with mudflows and banked by beach deposits. But if the right time and place are chosen, the contrast is quite startling. To visit Dimlington at low tide after an equinoctial gale has just driven an ord southwards from Cliff Farm under the High Land is one of the most instructive experiences the Quaternary geologist can obtain in Britain. The moss silt basins first described by Bisat and Dell (1941) and later dated by Penny, Coope and Catt (1969) are clearly seen between the Basement Till and Bisat's Basement Drab, the blue Sub-Basement Clay with its abundant marine fauna is continuously exposed over half an acre of the foreshore, and all around are more rock specimens from northern England, Scotland and Scandinavia than in any museum. Frequent visits to relevant parts of the coast over a period as long as 15-20 years are thus necessary to see all the Holderness coast sections, which undoubtedly can still yield much useful information about the later glaciations of eastern England.

ACKNOWLEDGEMENT

We thank Dr W.H.C. Ramsbottom for allowing us to borrow Dr Bisat's notes.

REFERENCES

Bisat, W.S. (1939). The relationship of the 'Basement Clays' of Dimlington, Bridlington and Filey Bays. *Naturalist, Hull*, 133-135, 161-168.

Bisat, W.S. (1940). Older and newer drift in East Yorkshire. *Proc. Yorks. geol. Soc.*, 24, 137-151.

Bisat, W.S. and J.A. Dell (1941). On the occurrence of a bed containing moss in the boulder clays of Dimlington. *Proc. Yorks. geol. Soc.*, 24, 219-222.

Catt, J.A. and L.F. Penny (1966). The Pleistocene deposits of Holderness, East Yorkshire. *Proc. Yorks. geol. Soc.*, 35, 375-420.

Madgett, P.A. and J.A. Catt (1978). Petrography, stratigraphy and weathering of Late Pleistocene tills in East Yorkshire, Lincolnshire and North Norfolk. *Proc. Yorks. geol. Soc.*, 42, 55-108.

Mitchell, G.F., L.F. Penny, F.W. Shotton and R.G. West (1973). A Correlation of Quaternary Deposits in the British Isles. *Geol. Soc. Lond., Spec. Report.*, 4, 99 pp.

Penny, L.F., G.R. Coope and J.A. Catt (1969). Age and insect fauna of the Dimlington Silts, East Yorkshire. *Nature, Lond.*, 224, 65-67.

Phillips, A.W. (1964). Some observations of coast erosion: studies at South Holderness and Spurn Head. *Dock Harb. Auth.*, 45, 64-66.

Reid, C. (1885). *The Geology of Holderness.* Mem. geol. Surv. U.K., 177 pp.

Stubblefield, Sir J. (1974). William Sawney Bisat 1886-1973. *Biogr. Mem. Fellows R. Soc.*, 20, 27-40.

Dr J.A. Catt,
Rothamsted Experimental Station,
Harpenden,
Herts. AL5 2JQ.

Dr P.A. Madgett,
Department of Science,
Harrow College of Technology and Art,
Northwick Park,
Harrow HA1 3TP.

13. Major Contributions of North-east England to the Advancement of Quaternary Studies

F. W. Shotton

The name of Lewis Penny will always be associated with the north-east of England and with Quaternary studies. I felt, therefore, that a fitting tribute to him would be the discussion of a limited number of researches which have been accomplished in the region and which have, in my opinion, contributed in a major way to the development of our knowledge of the "Ice Age", far beyond their local implications. Since such a selection must omit many excellent pieces of work, especially perhaps recent ones which still have to pass the test of time, I tender my apologies to anyone who may feel neglected. A review of all that has been written about the Pleistocene of North-East England could not be encompassed within a single chapter.

It is necessary at this stage to define North-East England, as the term is used in this essay. Its eastern boundary is obviously the shore of the North Sea. Its western boundary marks the junction of low land (and therefore dominantly an area of glacial deposition) with the high land of the Cheviot and Pennines where erosional processes were more important than deposition. Conveniently it may be placed close to the 100 m contour line and this leads to a north-south trending boundary which, starting at Berwick-on-Tweed, runs through Wooller, Rothbury, west Newcastle, Darlington, Harrogate, Leeds, Rotherham, Mansfield and so to Nottingham. At this point we have to define an arbitrary southern limit, which is a west-east line to the Wash.

Not all the area so defined is low-lying, for it includes the impressive hills of the Yorkshire Moors (Cleveland Hills), whose highest summits rise above 400 m and which, with the Hambleton Hills flank the Vale of York to the west. Less imposing, but physiographically important to Quaternary geomorphology, are the Chalk hills of the Yorkshire and Lincolnshire Wolds.

The cases which are considered as deserving of special recognition for their importance, now follow in a rough chronological order.

KIRKDALE CAVE AND PROFESSOR W. BUCKLAND

Kirkdale Cave (SE 678856) is in Corallian limestone where Hodge Beck opens out on the north side of the Vale of Pickering. It was described and some of its contents figured by Buckland in his *Reliquiae Diluvianae* of 1823, though in the previous year he had announced the discovery to the Royal Society (Buckland 1822). Boylan (1972) has written an account of the scientific significance of the finds of hyaena bones, but he also names the persons who took part in the actual work of excavation and shows how Buckland was brought into the operation as a recognised expert.

Buckland listed 23 animals as present in the cave deposits, an incomplete and to some extent an erroneous list, as Boylan has shown, but nevertheless including species no longer living in this country. Not surprisingly, in view of the whole tenor of Buckland's book, these animals (particularly such mammals as hyaena, bear, elephant, rhinoceros and hippopotamus) were regarded as "antediluvian", i.e. destined to be subsequently exterminated in this country by the Noachian Deluge. That expression of Buckland's immature views is now of no significance. What he did convincingly show was that the bones of all the animals including some of the hyaenas, were crushed and gnawed by the hyaenas themselves. He also pointed out that the entrance to the cave was far too low for the larger animals to have entered alive and that they must therefore have been brought in by the carnivorous hyaenas as dismembered parts. Thus Buckland demonstrated the first "hyaena den" and laid a foundation for all subsequent work on the excavation of cave deposits which has proved to be so important in Pleistocene stratigraphy and the recognition of climatic cycles.

GEOLOGICAL SURVEY WORK IN THE SECOND HALF OF THE NINETEENTH CENTURY

What was written about the north-east in the thirty years or so which followed Buckland's account of Kirkdale Cave was largely of a general nature. It was in the second half of the century that real progress was made in establishing a stratigraphical succession in the Pleistocene. Much of this was due to the unspectacular but systematic mapping work of the Geological Survey, with Lamplugh standing out from amongst his contemporaries in the last quarter of the century. It was, however, Searles Wood and Rome in 1868 who were the first to subdivide the tills of Lincolnshire and South Yorkshire into units similar to those that are accepted today. They recognized three tills, Hessle, Purple and Basement. Since then these names have been in common use for more than a century, even though Wood and Rome's Basement Boulder Clay embraced the Drab Till and the underlying Basement Series of later workers. It is only very recently that Madgett (1975) has demonstrated that the Hessle Till is probably no more than the weathered top of the Purple Till. Meanwhile the Purple and Drab Tills have been named after the localities of Withernsea and Skipsea respectively.

Lamplugh's great contributions were the detailed observations he made on an ever-changing coastline and on excavations which he initiated and supervised at critical places, records which were to prove invaluable to later workers. He was responsible for early observations on the Bridlington Crag, though other people such as Forbes had recognized the deposit and commented on it before him. The great significance of the Bridlington Crag is that it occurs as erratics in the Basement Till and since it carries an arctic fauna, may well be all that is left as evidence of an Anglian glaciation in Holderness. What Bisat much later called the Sub-Basement Clay could be the equivalent of the Bridlington Crag in a more till-like form.

Lamplugh also conducted systematic excavation over several years on the famous 'buried cliff' at Sewerby, near Bridlington, and the value of his careful records was acknowledged by Catt and Penny when they reinvestigated this section (1966) and came to conclusions which are fundamental to British Quaternary stratigraphy.

It is strange that Lamplugh, throughout his long life (for his publications cover half a century, the latest being about 1925) remained a convinced monoglacialist. It is obvious too, from his writing, that Woolacott held a monoglacial viewpoint and it is to the credit of Trechmann, who was publishing papers at much the same time as Lamplugh and Woolacott, that he believed in a succession of glacials and interglacial

Another deposit which excited much attention in the latter half of the nineteenth century was the Kelsey Hill Gravel, underlain by Drab (Skipsea) Till and overlain by 'Hessle Clay' which is either weathered Purple (Withernsea) Till or a solifluction deposit. Since Prestwich in 1861 described its fauna, it has been repeatedly discussed during the succeeding century. The gravel contains an abundance of shells, both marine and fresh water with *Corbicula fluminalis* common. Since this bivalve is a diagnostic indicator of an interglacial it is not surprising that the Kelsey Hill Gravel has been from time to time accredited to the last major interglacial (the Ipswichian), even though the shells in the gravel are demonstrably derived. Since it has now been proved that the Drab Till is Late Devensian, the Kelsey Hill Gravels have to be derived from an earlier Ipswichian deposit. One supposes that the ice which carried the Drab Till eroded this deposit and that the gravels are the outwash or an esker emanating from this ice. Perhaps one day the parent body of these derived gravels may be located, if any of it now remains, and it should be an organically rich and very interesting deposit.

KENDALL AND THE GLACIAL IMPOUNDING OF WATER AGAINST THE CLEVELAND HILLS

When Kendall, in 1902, published his justly famous paper on the ice-damming of the Cleveland Hills, he was far from being a pioneer in the explanation of drainage diversions as the result of glacier fronts pressing up valleys and ponding lakes in their upper reaches. Nearly forty years earlier, in 1863, Jamieson had explained the Parallel Roads of Glen Roy, relating the three bench levels to the uncovering of three cols which acted as successive overflows to the ice-impounded lakes. Nevertheless Kendall's synthesis had the touch of genius which inspired later workers elsewhere, as for example, Trotter on the Pennine slope of the Eden Valley. The bare essentials of Kendall's reconstruction were that the Last Glacial ice, pushing from the north, was able to advance down the Vale of York and also along the east coast, but stagnated against the north side of the intervening Cleveland Hills. Here it impounded the north flowing streams into a number of lakes of which 'Lake Eskdale' was the largest and all this water then escaped from the top of the Ellerbeck, one of the Esk's tributaries, to cut the extraordinary valley known as Newtondale. Where this debouched into the Vale of Pickering, the flood water could not escape along an obvious eastern course to the sea because of the ice and moraine barrier of the North Sea ice which blocked it. The vast lake of about 380 km^2 which then formed found an escape point near Kirkham Abbey. It drained southward from here, cutting the Kirkham Abbey Gorge and so lowering the level of Lake Pickering that when the ice finally disappeared, the Lake had been virtually drained and the River Derwent retained its anomalous course. During the glaciation it must have fed another great glacier-dammed lake to the south - Lake Humber - but that was not part of Kendall's tale.

A very important aspect of Kendall's paper was that he defined the features that must be recognized to uphold any interpretation of ice-damming - lake floor deposits, inflow and overflow channels, and deltaic gravels in positions and at heights related to inflow channels and ice fronts. It is quite true that Kendall interpreted some channels as extra-glacial in origin, which we would now think of as tunnel valleys and sub-glacial in their mode of formation, (Gregory 1962). Such details in no way detract from the main Kendall thesis. Newtondale, from the point where it takes off from the head of Ellerbeck is a text book example of a great overflow channel. It has the typical flat floor and steep sides, an average gradient of only 1 in 1250, but at no point does it reverse its slope as is the wont of tunnel valleys. If one leaves Newtondale, to look over the vast expanse of alluvium in the Vale of Pickering, it needs little in the way of imagination to recreate the Lake and to picture our Mesolithic forbears living in their lakeside village at Starr Carrs when the ice but not yet the lake, had disappeared.

Reservations about Kendall's interpretation have certainly arisen and these have taken two different forms, Gregory (1962) postulates that the stagnating ice, liberating a massive amount of melt water, dispersed this mainly by means of eastward-draining, subglacial channels. Though this may be true, one cannot ignore a vast overflow down Newtondale - nor did Gregory attempt to do so. Catt (1977), whilst accepting both Newtondale and the Kirkham Abbey Gorge as overflow channels, believes that each is on too big a scale to have been cut solely by Upper Devensian melt water and would have them mainly determined by overflow from an earlier glaciation (?Wolstonian, ?Anglian). This is a matter of opinion only, since there is no stratigraphical evidence one way or another. How important it would be if the sediments of Lake Pickering were found to be of two different ages - post-Ipswichian and Pre-Ipswichian, thus giving the Derwent two opportunities to entrench itself in the Kirkham Abbey Gorge; but unhappily no such evidence has appeared.

TRECHMANN AND OTHER WORKERS IN THE FIRST HALF OF THE PRESENT CENTURY

During the present century work farther north in the old counties of Northumberlnad and Durham has produced results of more than local significance. As one moves into the north of England and into Scotland, the all-pervading presence of the Last (Late Devensian) Glaciation removes or conceals evidence of the earlier ice-invasions which assuredly affected Britain. Relics of the Middle and Early Pleistocene are very rare and correspondingly precious when they are discovered. We owe to C.T. Trechmann records of the few sites where such evidence has been found and a discussion of these can be found in Francis (1970).

It is a sad reflection upon the scant respect given to geological conservation (and to the preservation of the coastline) that one can only be told that under the mass of coal pit waste that spreads along the shore from Blackhall Colliery in County Durham, there are some very instructive, sediment-filled fissures in the Magnesian Limestone. Trechmann (1915, 1920) described them before they were covered. One fissure held a clay which carried macro plant remains, which were examined by Mrs E.M. Reid (1920) who suggested they indicated a Pretiglian age, i.e. the very beginning of what is now accepted as Pleistocene. They may, however, be earlier, and so not Pleistocene at all, according to Dr Pennington (1969). In a situation like this it is exasperating that, short of a major earth-shifting exercise, the deposit cannot be examined and sampled for further evidence.

Another open joint filling from the same locality, again described by Trechmann (1920), was rich with molluscs, plants including oak and pine wood, insects and vertebrate remains. The consensus of opinion favours a Cromerian age for this material, but whether it is Cromerian in the limited and correct British sense is uncertain. There is no doubt that it is the oldest proven Pleistocene in the north-east.

Trechmann (1920) also described from Warren House Gill (on the Durham coast about 10 km N.W. of Hartlepool) a section of "Scandinavian Drift" overlain by loess. It is unfortunate that once again colliery waste has covered that part of the section which exhibits loess, though a limited view of the Scandinavian Drift is still possible. There seems little doubt that loess was correctly diagnosed, probably resorted by water, but there is no reason to ascribe an interglacial origin to it as Trechmann did. No fauna was obtained from it and loess seems always to be associated with a cold, windy climate. As the Scandinavian Drift is now interpreted as a Wolstonian till impinging from the North Sea, the loess is regarded as a manifestation of the end of the Wolstonian, thus making it an 'older loess' as well as a unique example of this type of deposit in the north-east.

Woolacott is another name which appears in the first quarter of this century, particularly in regard to the splendid example of a raised beach at Easington, near to Warren House Gill (Woolacott 1920, 1922). Here a bench is cut into the Magnesian Limestone, about 18 m above mean sea level, and the beach gravels which rest upon it have marine shells which certainly do not indicate a cold sea. For several reasons, none of which is by itself conclusive, the beach is generally referred to the Ipswichian Interglacial and this may well be so; but the lack of unimpeachable evidence of age brings home to us how uncertain we still are in our understanding of the Ipswichian. If Easington does reflect an Ipswichian sea level of around +20 m, this has to be related to the undoubted Ipswichian beach at Sewerby, near Bridlington, 105 km to the S.E. and only 2 m or so above O.D.; and to the Ipswichian peats and silt at Tattershall, Lincolnshire, 112 km farther south but 25 km from the sea, where deposits with an abundant fresh water mollusc fauna lie just below mean sea level. These facts pose such questions as whether we are considering one period of time or several, whether there were large oscillations of sea level during the Ipswichian or if these conflicting beach heights are due to a post-Ipswichian warping of the North Sea basin. Our present state of knowledge does not allow us to give convincing answers to these three questions.

CATT AND PENNY (1966) AND OTHER RECENT WORK

In recent years several important papers have appeared from officers of IGS or from others prompted by them, on interglacial or Devensian deposits which bear upon the history and inter-relationship of the Vale of York end moraines of the last glaciation and Lake Humber. I do not intend to discuss this work, important though it is, for it is actively in progress and very much under discussion. I must, however, refer to the research of Lewis Penny himself, in collaboration with John Catt, on the Pleistocene of Holderness.

These authors had much in the way of earlier observations to take into consideration (particularly from Lamplugh and from Bisat) but from this and their own research they effected a remarkable clarification of several stratigraphical problems, notably:-

(1) The shingle and beach and blown sand which cover the Sewerby beach platform and bank up against the old cliff, and which in the past yielded Hippopotamus amongst other fossils, are of Ipswichian age.

(2) The Ipswichian beach planation cuts across from Chalk on to the Basement Till, thus proving the latter to be pre-Ipswichian. Catt and Penny ascribed it to the Wolstonian, though at that time they had to use the term "Saalian".

(3) Perhaps the major conclusion which resulted from their work was that the three tills, Hessle, Purple and Drab (to use the names which the authors did) were the product of one glaciation, the Devensian. It is true that Catt and Penny then regarded the Hessle Clay as a true till; but whether two or three tills is the correct explanation, there remains the concept of two or more ice masses from different source areas, co-existing as stratified layers of a single glacier, retaining their individuality and producing apparently distinct lodgement tills. It would also follow from this that the gravel layers which separate the tills farther north towards Scarborough, are of intra-glacial origin.

The Basement Till, which close to Sewerby is planed off below the old beach platform, farther south at Dimlington forms the foundation for the Devensian till sequence. Here the top of the Basement Till is weathered, presumably during the Ipswichian Interglacial. It is slightly undulating and in its hollows are the sediments of a few silty pools which contain moss. These were first recognised by Bisat and Dell (1941) as the Dimlington Silts. They obviously constituted a potential source for a radiocarbon date which would be older than the onset of the Devensian glaciation into this area.

Penny isolated sufficient of this moss to send to America for a radiocarbon dating. He was returned a figure of 18,500 ± 400 B.P. (I-3372). This was such an unexpected and incredibly low figure for something which antedated the advent of large glaciers into Holderness, that Penny was sceptical. With thoughts of contamination as an explanation of the figures, he requested an independent date from the laboratory of Birmingham University Geology Department. I well remember the laborious collection of about half a tonne of silt and the meticulous wet sieving of this to provide a few grams of moss. The labour was fully justified, for a date of 18,240 ± 250 B.P. (Birm-108) was obtained and it agrees perfectly with the American figure (Shotton, Blundell and Williams, 1969). Prior to these results, few people would have dared to predict that the Devensian glaciation of north-east England could have reached Holderness as recently as about 18,000 years ago, extended to the Wash very quickly, only to dwindle and disappear within about the next three millennia.

This date has taken on something of a magical significance, because 18,000 B.P. now often appears in international literature as the critical peak of the last glaciation. The Climap Project has taken this date when evaluating the physical conditions in the Atlantic at the height of glaciation in Europe and America. It is wise, I think, to express some reservation, not about the validity of the Dimlington figure, but as to whether the maximum extension of the Late Devensian ice was synchronous throughout the British Isles. In the Isle of Man there are sediments above Devensian till with dates a little older than 18,000 (though there may be an element of 'hard water error' in the measurements). At Four Ashes, near Wolverhampton, the locality of the Devensian stratotype, it is only possible to say that the two newest pre-glaciation dates, each of about 30,000 years, were obtained from organic lenses in gravels which admittedly were a few feet below the Irish Sea Till, but no younger organic lenses could be found. So it could well be that lobes of the Devensian Ice advanced to frontal positions which were not reached synchronously, but spread over perhaps a few thousand years, finishing last along the East Coast. It is interesting that Boulton and others (1977), arguing entirely from physical principles, came to the conclusion that the Holderness ice was to be explained by a late surge. This view accords well with the idea of non-synchroneity of behaviour of all parts of the glacier front. Certainly this is the way the Wisconsinan ice invading the U.S.A. from the Canadian Shield is interpreted.

DATING THE PRE-DEVENSIAN TILLS

Although I stated my intention to refrain from discussing work which is currently in progress, it is tempting to look at the varying views which are being promulgated at the moment on the age of the older tills. Much of the argument arises from East Anglia south of the Wash, where the Lowestoft Till (the "Great Chalky Boulder Clay") is unquestionably earlier than the Hoxnian Interglacial and belongs to the later part of the Anglian Glaciation. Chalky Boulder Clays which are found north of the Wash are considered to be of the same age by those who pin their faith to a reasonable similarity of grain-size analyses, heavy mineral content and an abundance of Chalk and flint erratics. Those who believe that tills which are Cretaceous-derived can be of two different ages, i.e. Wolstonian and Anglian, argue that if glaciers traverse the same or similar territory at different times, the mineralogy of their respective tills will be closely comparable - in other words, that a close mineral similarity between two tills does not necessarily mean that they are of the same age any more than that dissimilarity implies different ages irrespective of other considerations. In such controversial circumstances, reputable stratigraphy alone can provide unambiguous answers though it is often difficult to produce such when past events have removed so much of the evidence and the uncovering of what remains is largely a matter of chance.

The south Yorkshire and Lincolnshire Wolds may have played an important role in directing the flow of ice streams, with important implications to the nature of the tills produced. Those to the east of the Wolds would have an abundance of Chalk in the North Sea and on land to erode, whereas those to the west of the Chalk ridges would traverse extensive areas of Upper Jurassic clays and some Lower Cretaceous as well as Chalk. If, however, ice moved from the coastal region of north-east Yorkshire into the south-east of the county across the Chalk of Flamborough Head, there need not be any substantial difference between tills on either side of the Wolds.

There are a number of key sections, mostly in Lincolnshire. The first is at Kirmington, 15 km south of the Humber at Hull, where Watts (1959) on palynological grounds, correlated a thin peat in a borehole with the Hoxnian, (see also Boylan 1966). About 18 m below this peat, a grey clay with chalk and flint is interpreted as a till; this 'chalky boulder clay' can thus be correlated with the Lowestoft Till (Anglian) with no problem. Whether the till continues uninterruptedly into Norfolk is not important to the stratigraphical argument. However there is only one till above the Hoxnian peat, in the topmost 2 m of the borehole and this is said to resemble the Skipsea (Upper Devensian) Till. So if these correlations are correct, there would appear to be no till representative of the Wolstonian here, nor indeed any trace of the Ipswichian.

The next critical section is at Welton-le-Wold, near Louth (Alabaster and Straw, 1976), some 42 km south of the Humber at Hull and lying on the eastern slope of the Wolds. Here it is accepted that there are three tills, one Devensian, a Welton Till and a Calcethorpe Till. The Welton Till is covered by the Devensian Till on the lower ground, and is generally accepted to be pre-Devensian. The Calcethorpe Till, which is a chalky boulder clay, occurs on the higher ground to the west but slopes down eventually to cover the Welton Till.

Beneath the latter till occur gravels which have yielded a sparse vertebrate fauna, including the straight-tusked elephant, *Palaeoloxodon antiquus*, and three hand axes and a flake which are certainly Acheulian. This makes the gravels post-Hoxnian or at least late-Hoxnian, and the overlying Welton Till even later. So if the latter is pre-Devensian, it can only be Wolstonian. This view is reinforced by the similarity of the Welton Till to the Basement Till of Holderness, which is certainly pre-Ipswichian. So it seems that the Welton Till and the Basement Till can be correlated with each other and both with the Wolstonian.

It is very interesting, if this correlation of the Basement Till with the Wolstonian is accepted, that Wood and Rome as long ago as 1868 described the Basement Clay of Holderness as "a lead coloured clay abounding in Chalk debris accompanied by stones and boulders from all sorts of rocks; and it presents a close resemblance to that widespread Boulder-clay of the eastern and east-central counties...".

The position of the Calcethorpe Till is not so clear. It is a highly chalky till and may be correlated with what Straw called the Wragby Till; but the problem at Welton is whether the so-called Calcethorpe Till lying on the Welton Till is genuinely *in situ* or whether it is a solifluction sheet from higher up the slope, where *in situ* till certainly occurs. Straw acknowledges this ambiguity in his description of the site in Catt (1977), but this does not weaken the argument for a Wolstonian age for the Welton Till.

The other area of crucial importance is the low-lying ground west of the Lincolnshire Wolds, traversed by the rivers Witham and Bain down to the Wash. The basal Pleistocene deposit so far recognised throughout this area is a chalky-Jurassic till which Straw has called the Wragby Till and which, because of its lithology, has been correlated by some with the Anglian Lowestoft Till and hence with the till beneath the Hoxnian at Kirmington. However, throughout this region, no Hoxnian deposit has been found either above or below the Wragby Till. The successions at Tattershall and Kirkby-on-Bain typify much of this large area (Girling 1974, Catt 1977). Above the Wragby Till are Ipswichian peats and shell marls, followed by gravels (but no till) certainly of Middle and Late-Devensian age, and then Holocene sediments. Hence we have here a situation where a till derived from a glacier which must have moved southward down the valley west of the Wolds because of the high content of Upper Jurassic material in its till, can only be proved to be pre-Ipswichian. Not surprisingly, Straw and some others believe it to be a Wolstonian till.

This problem of the ages of pre-Devensian tills remains a major area of controversy which, I have no doubt, will be resolved in the future. It seems certain that south-east Yorkshire and Lincolnshire have relics of both Wolstonian and Anglian tills, but to solve the problem of the age of the Wragby Till (and this could have repercussions in East Anglia and the English Midlands), we must await the discovery of Hoxnian deposits either above or below the till.

REFERENCES

Alabaster, C. and A. Straw (1976). The Pleistocene context of faunal remains and artefacts discovered at Welton - le - Wold, Lincolnshire. *Proc. Yorks. geol. Soc.*, 41, 75-94.

Bisat, W.S. and J.A. Dell (1941). The occurrence of a bed containing moss in the boulder clays of Dimlington. *Proc. Yorks. geol. Soc.*, 24, 219-222.

Boulton, G.S., A.S. Jones, K.M. Clayton and M.J. Kenning (1977). A British ice-sheet model and patterns of glacial erosion and deposition in Britain. In F.W. Shotton (Ed.), *British Quaternary Studies: Recent Advances.* Oxford University Press. Chap. 17, pp. 231-246.

Boylan, P.J. (1966). The Pleistocene deposits of Kirmington, Lincolnshire. *Mercian Geol.*, 1, 339-350.

Boylan, P.J. (1972). The scientific significance of the Kirkdale Cave hyaenas. *Yorks. phil. Soc., Ann. Rep. for 1971*, 38-47.

Buckland, W. (1822). Account of an assemblage of fossil teeth and bones ..... discovered in a cave at Kirkdale. *Phil. Trans. R. Soc.*, 122, 171-236.

Buckland, W. (1823). *Reliquiae Diluvianae.* John Murray, London.

Catt, J.A. (1977). *Yorkshire and Lincolnshire.* Guide book for Excursion C7. X INQUA Congress, Birmingham.

Catt, J.A. and L.F. Penny (1966). The Pleistocene deposits of Holderness, East Yorkshire. *Proc. Yorks. geol. Soc.*, 35, 375-420.

Francis, E.A. (1970). Quaternary. In *Geology of County Durham.* Compiled by G.A.L. Johnson, edited by G. Hickling. *Trans. nat. Hist. Soc. Northumb.*, 41, No. 1. Chap. 12, pp. 134-152.

Girling, M.A. (1974). Evidence from Lincolnshire of the age and intensity of the mid-Devensian temperate episode. *Nature, Lond.*, 250, 270.

Gregory, K.J. (1962). The deglaciation of eastern Eskdale, Yorkshire. *Proc. Yorks. geol. Soc.*, 33, 363-380.

Jamieson, T.F. (1863). On the parallel roads of Glen Roy, and their place in the history of the glacial period. *Q. Jl geol. Soc. Lond.*, 19, 235-259.

Kendall, P.F. (1902). A system of glacier-lakes in the Cleveland Hills. *Q. Jl geol. Soc. Lond.*, 58, 471-571.

Madgett, P.A (1975). Re-interpretation of Devensian Till stratigraphy in eastern England. *Nature, Lond.*,253, 105-107.

Pennington, W. (1969). *The History of British Vegetation.* English Universities Press, London.

Prestwich, J. (1861). On the occurrence of *Cyrena fluminalis*, together with marine shells of Recent species, in beds of sand and gravel over beds of boulder clay near Hull. *Q. Jl geol. Soc. Lond.*, 17, 446-456.

Reid, E.M. (1920). On two pre-glacial floras from Castle Eden, County Durham. *Q. Jl geol. Soc. Lond.*, 76, 104-144.

Shotton, F.W., D.J. Blundell and R.E.G. Williams (1969). Birmingham University Radiocarbon Dates III. *Radiocarbon*, II No. 3, 263-270.

Trechmann, C.T. (1915). Scandinavian Drift on the Durham coast. *Q. Jl geol. Soc. Lond.*, 71, 53-80.

Trechmann, C.T. (1920). On a deposit of interglacial loess and some transported pre-glacial freshwater clays on the Durham coast. *Q. Jl geol. Soc. Lond.*, 76, 173-203.

Watts, W.A. (1959). Pollen spectra from the interglacial deposits at Kirmington, Lincolnshire. *Proc. Yorks. geol. Soc.*, 32, 145-152.

Wood, S.V. and J.L. Rome (1868). On the Glacial and Post glacial structure of Lincolnshire and south-east Yorkshire. *Q. Jl geol. Soc. Lond.*, 24, 146-184.

Woolacott, D. (1920). On an exposure of sands and gravels containing marine shells at Easington. *Geol. Mag.*, 57, 307-311.

Woolacott, D. (1922). On the 60-foot raised beach at Easington. *Geol. Mag.*, 59. 64-74.

Professor F.W. Shotton, F.R.S.,
111 Dorridge Road,
Dorridge,
Solihull B93 8BP.

14. The Quaternary Geology of the Sunderland District, North-east England*

Denys B. Smith

INTRODUCTION

This is an account of the drift deposits and related features of the area around Sunderland and is based on field work conducted at intervals from 1952 to 1975. The main physical features of the area are shown in Fig. 1 and comprise the sandstone escarpment of Gateshead Fell and the Tyne and Wear valleys in the west, the Wear Lowlands in the north-centre, the dissected East Durham Plateau (roughly coincident with the outcrop of the Magnesian Limestone) and a narrow coastal platform; the Cleadon and Fulwell hills north of the lower Wear are outliers of the plateau. The medium-scale relief of the areas below about 45 m O.D. is noticeably more subdued than that above this level.

In the preparation of this paper full account has been taken of the records of a large number of boreholes into and through the drift and of many sections that are now obscured; these included old quarries, gravel pits and clay pits that are now flooded or restored and many of the fleeting excavations associated with a host of construction and drainage schemes. There are, unfortunately, no records of many such earlier excavations, some of which were in key positions. The data are incorporated into 1:10,560 Geological Survey maps of the area and were summarised and briefly explained on 1:50,000 Geological Sheet 21 (Smith, 1978).

HISTORY OF RESEARCH

Despite its proximity to centres of learning, the excellence of its coastal exposures and the close relationship between the local geology and the social and industrial evolution of this populous district, there has been comparatively little detailed research into its drift deposits. The first major paper on the drift of this and the surrounding district was by Howse (1864), whose work has stood up remarkably well to the profusion of newer concepts and data, and regional syntheses have since been provided by Woolacott (1905, 1907, 1921), Raistrick (1931), Francis (1970) and Beaumont (1970). Important contributions were also made by Smythe (1908; 1912), Merrick (1909), Trechmann (1915, 1920, 1952) and Coupland and Woolacott (1926) but it is a feature of all this literature that it contains few details of individual

---

*Published by permission of the Director, Institute of Geological Sciences.

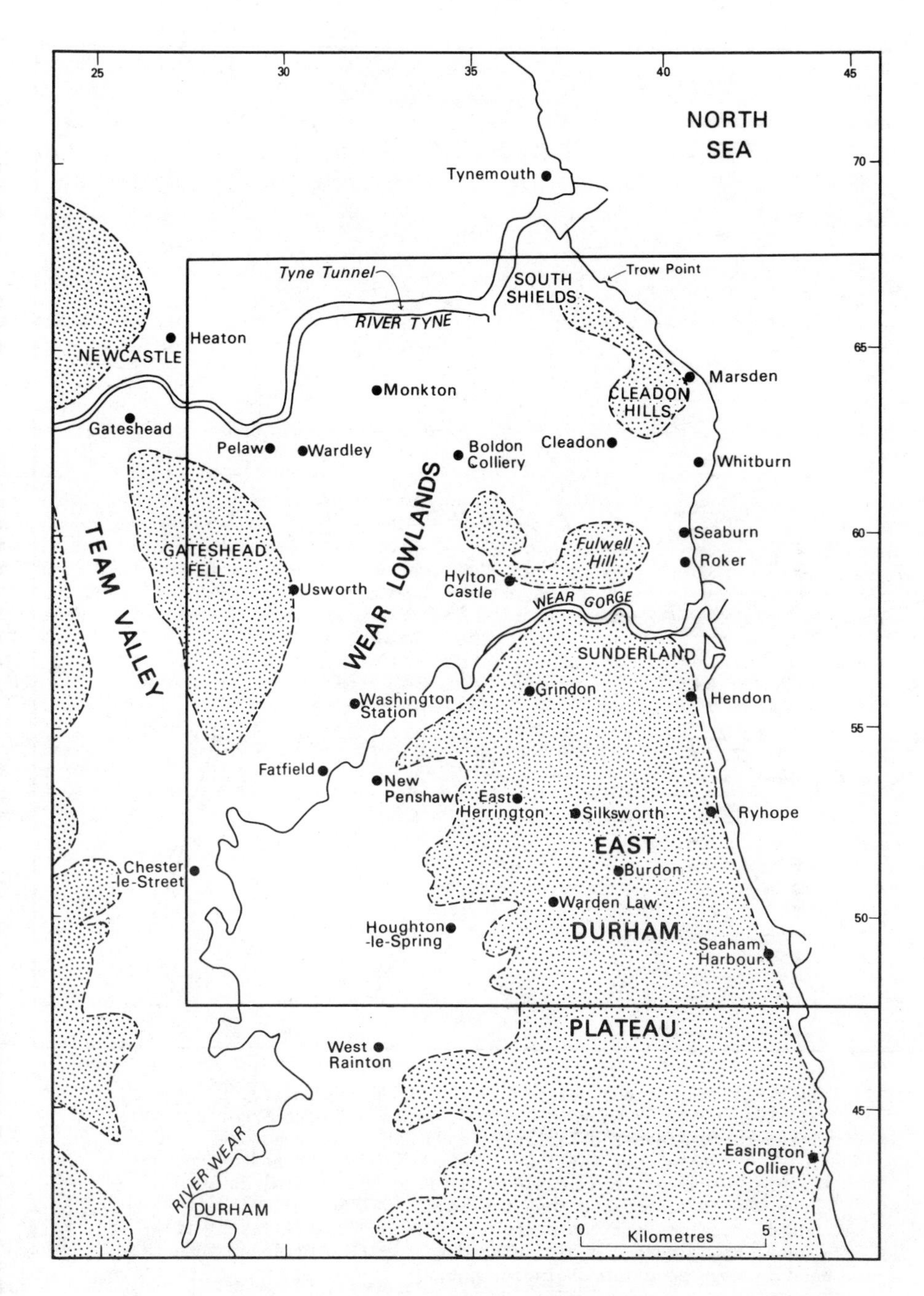

Fig. 1. Main physical features of the Sunderland district and its immediate surroundings showing places named in the text. Elevated areas are stippled.

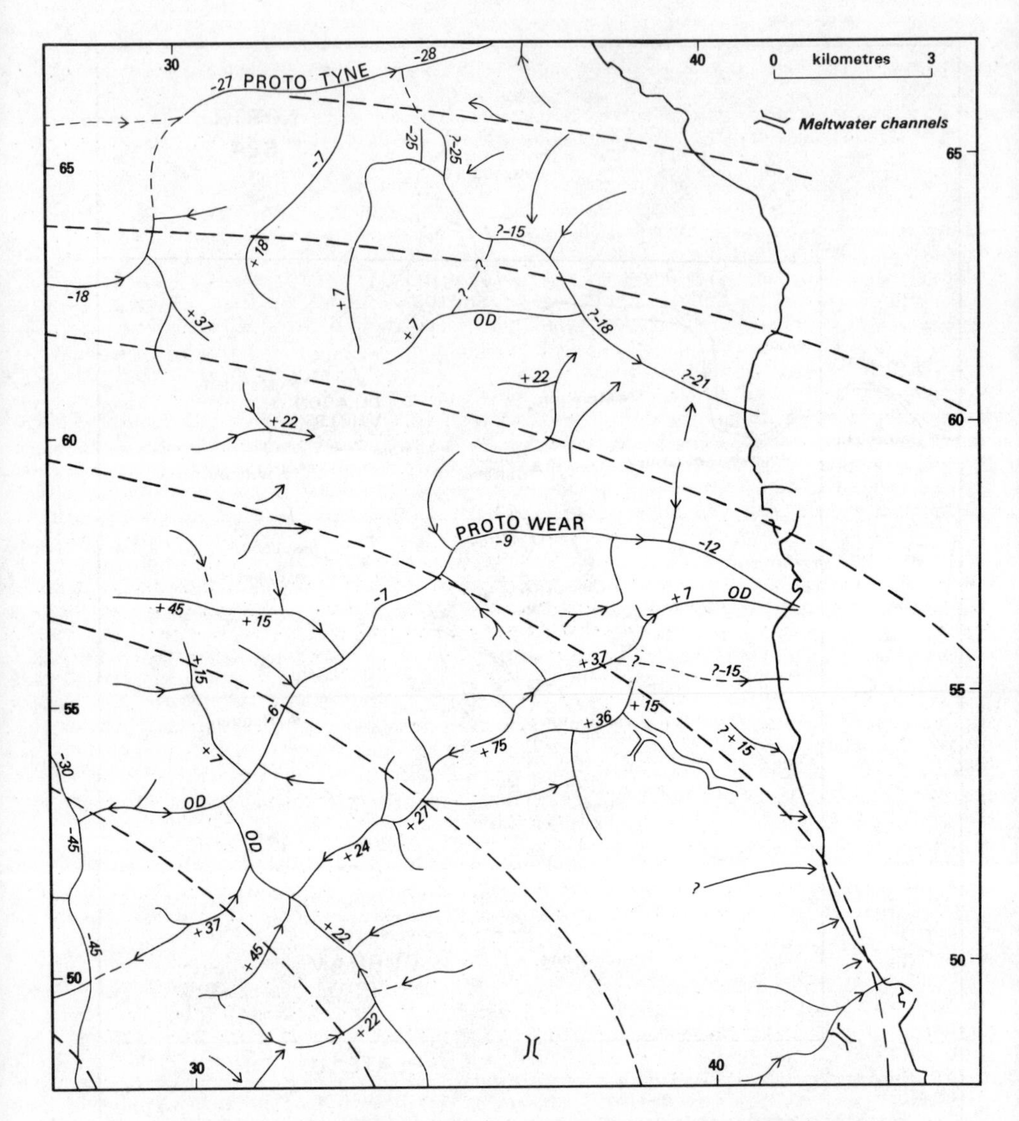

Fig. 2. Drift-filled (buried) valleys of the Sunderland district (levels in metres below or above O.D.), with second degree trend surface fitted to the inferred Tertiary landscape (from Beaumont, 1970, fig. 8.). Note that the lower course of both the proto-Tyne and the proto-Wear diverge considerably from the general pattern of consequent drainage and that the major WNW-ESE valley (north-centre) does not fit the pattern and is presumably later; it appears to be more closely related to the Tyne drainage than the Wear and may be of interglacial age.

sections; a rather more detailed account has been given for the district immediately to the south (Smith and Francis, 1967). A full history of Quaternary research in north-east England was given by Beaumont (1968) and need not be repeated.

THE MORPHOLOGY AND EVOLUTION OF THE SUB-DRIFT SURFACE

The main features of the sub-drift surface of the Sunderland district are similar to but more strongly marked than those of the present land surface. Drift is generally thin and locally absent on many higher parts of this surface and also in parts of the lowlands such as areas around Pelaw, Monkton, Boldon Colliery and Whitburn, but it is complex and up to 60 m thick in a network of valleys (Fig. 2) whose floors in many places lie well below Ordnance Datum. In general these valleys are wider, deeper and less steep-sided than their modern counterparts.

The work of several authors who have speculated on the evolution of the sub-drift surface of County Durham and adjoining areas was summarised by Beaumont (1970). These works, supplemented by Beaumont's own research, suggest that the present sub-drift landscape of the Sunderland district began to evolve during the late Mesozoic or early Tertiary when the area formed part of an extensive eastward-sloping peneplain; uplift and gentle doming of this peneplain, probably during the Miocene, was inferred by Bott (1968) to have resulted from the isostatic rise of the semi-rigid Alston Block and its underlying Weardale Granite. King (1967) and Beaumont (1970) applied trend surface analysis to this uplifted area and showed that the peneplain in the Sunderland district probably sloped gently to the east and north-east during the late Tertiary (Fig. 2).

Consequent drainage initiated on this surface was graded at some stage to a base level at about 30 m below present sea level (Beaumont, 1970, 35) and subsequent streams exploited belts of less-resistant bedrock to create strike features such as the Wear and Team valleys in the west; finally, obsequent streams up to 10 km long were initiated on the westward facing slopes of the subsequent valleys and of the eastwards-retreating Magnesian Limestone escarpment. Modification of this classic evolutionary pattern by a complex sequence of river captures has been inferred by several authors (see Beaumont 1970 for full references), and the present system of sub-drift valleys is the end-product of this evolution as locally modified by Pleistocene glacial effects. Most authors, however, have recognised that this interpretation is based on inadequate data and is almost certainly far simpler than reality; thus, for example, the inferred Tertiary peneplain of western Durham probably intersects and merges there with exhumed remnants of the more strongly tilted pre-Upper Permian peneplain and the general thesis takes little account of probable multiplicity of glaciations before the Late Devensian. Wide terraces cut in rock but overlain by drift exist at about 24 m and 42 m O.D. along the flanks of parts of the Wear Lowlands and at about 31 m O.D. along the coast but have not been satisfactorily explained by any authors and have been ignored or overlooked by most; they are necessarily pre Late-Devensian and could be related to inter-glacial or earlier high sea levels such as that evidenced by the ?Ipswichian raised beach at about 24 m O.D. at Beacon Hill, near Easington Colliery, just south of the district (Woolacott, 1920).

Because of the broad parallelism of the post-drift and sub-drift topography, late Devensian and Flandrian drainage was re-established on essentially the same lines as that on the underlying rock surface except in a few places such as East Herrington, South Shields, Whitburn and central Sunderland where sub-drift valleys were not reoccupied by major streams. Rapid downcutting gave rise to rock gorges in the lower courses of the Tyne and Wear valleys in places where the new drainage did not exactly follow the line of the buried valleys. However, the oft-repeated claim that the eastwards diversion of the Wear to Sunderland took place during the last glaciation is shown to be untrue by the large size, continuity, maturity and

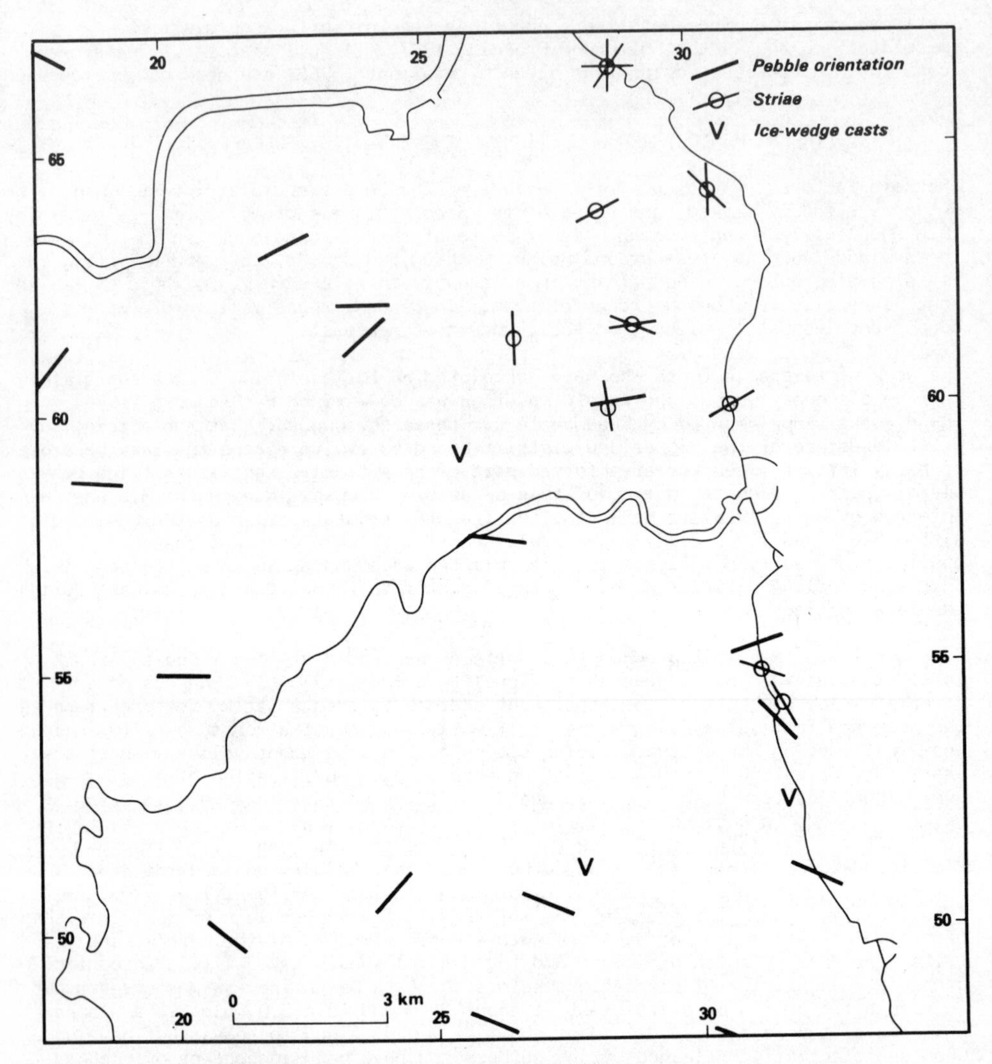

Fig. 3. Pebble orientation in the Durham Lower Boulder Clay, trends of striae and the position of known ice-wedge casts in the Sunderland district. Pebble orientation from Beaumont (1971, fig. 2) and the writer, striae from various sources (after Beaumont, unpublished).

steady eastwards fall of the buried valley through which the lower Wear runs for much of its course; this valley may have continued west of the diversion elbow at Chester-le-Street as the proto-Twizell Burn.

DETAILS OF THE SUB-DRIFT SURFACE

In detail the sub-drift surface ranges from uneven to smooth according to the lithology and coherence of the rock and the character of the drift, but in general the surface is gently rolling and with few eminences. A graphic account of such a surface was given by Howse (1864, 177-8) who, over a period of several years from 1854, saw several hectares of Magnesian Limestone laid bare during the course of quarrying at Trow Point (NZ 383664), South Shields. He described the surface as slightly irregular, with scattered *roches moutonées* and, in places, polished and strongly striated. Striae here trended E-W, NE-SW and N-S in different parts of the surface seen by Howse, who also recorded roughly E-W striae on the south side of the Cleadon Hills. The trends of these and other striae in the district have been summarized by Beaumont (unpublished) and are reproduced in Fig. 3 with his permission.

Where over-riding ice was not sufficiently erosive to remove all the pre-existing regolith, extensive traces of cryoturbation remain. Brecciated and cryoturbated rock is particularly well-exposed in the coastal cliffs east and south-east of Whitburn, where lenses and sheets of angular limestone debris with little or no matrix other than a partial cement of calcium carbonate are widespread. Ice-wedge casts occur to a depth of 6 m in the Westphalian mudstone at a quarry (NZ 31284750) near West Rainton, just south of the district, and clay-filled wedges up to 3 m deep were recorded in 1955 by R.H. Price (IGS files) in Permian sand (NZ 357591) near Hylton Castle, Sunderland. In many places the incorporation of local rock into the overlying drift is clearly evidenced by trains of angular debris streaming from torn-up substrate, and sheets of rock flour are locally abundant in crudely layered basal till; that the processes involved could be sharply selective is clearly shown at Cross Rigg Quarry (NZ 325539), New Penshaw, where a thin undisturbed coal seam forms rockhead for about 30 m but overlying mudstone is contorted and fragments stream out for at least 10 m to the SSW in overlying till.

THE DRIFT SEQUENCE

The many complexities of the sequence of drifts in the Sunderland district preclude simple summarization and the mutual age and relationships of some of the deposits remain uncertain. These uncertainties mainly concern higher members of the drifts, for the lowest widespread drift -- the Durham Lower Boulder Clay* -- blankets almost the whole district and is common to all the sequences. Although derived marine shells have been reported widely from the tills and gravels, none are stratigraphically informative and there is thus no direct evidence of the age of the Sunderland drifts. However, indirect evidence from other areas (see Francis, 1970, for summary) suggests that all the local Pleistocene sequence is Late Devensian and was probably formed between 20,000 and 13,000 years ago.

The type and thickness of drift is loosely related to the topography and relief of the rockhead surface, and in general the more elevated areas bear only a single stony clay (generally the Durham Lower Boulder Clay). In contrast, the Durham Lower Boulder Clay of the low-lying buried valleys is overlain there by up to 60 m of interbedded sands, silts, laminated and stony clays collectively termed The Tyne-Wear Complex (Smith, 1978), whilst in the narrow coastal belt the Lower Boulder Clay is overlain by a variety of deposits including an Upper Boulder Clay in the north and thick sand on the coast near Ryhope. A silty sparingly stony clay -- the Pelaw Clay -- widely caps the drift sequence in most low-lying areas. Lastly, some inland areas of intermediate altitude bear spreads of highly varied sands, gravels and stony clays whilst other areas, especially the Magnesian Limestone dip slope, bear spreads and mounds of fluvioglacial sand and gravel. The distribution of the main deposits is shown in Fig. 4 and their relationships are shown in Fig. 5.

---

* The term 'boulder clay' is used throughout for tough stony clay till with a preferred fabric; although regarded by some as a non-specialist term synonymous with "till", the author regards its continued use for this variety of till as useful and self-explanatory for most readers.

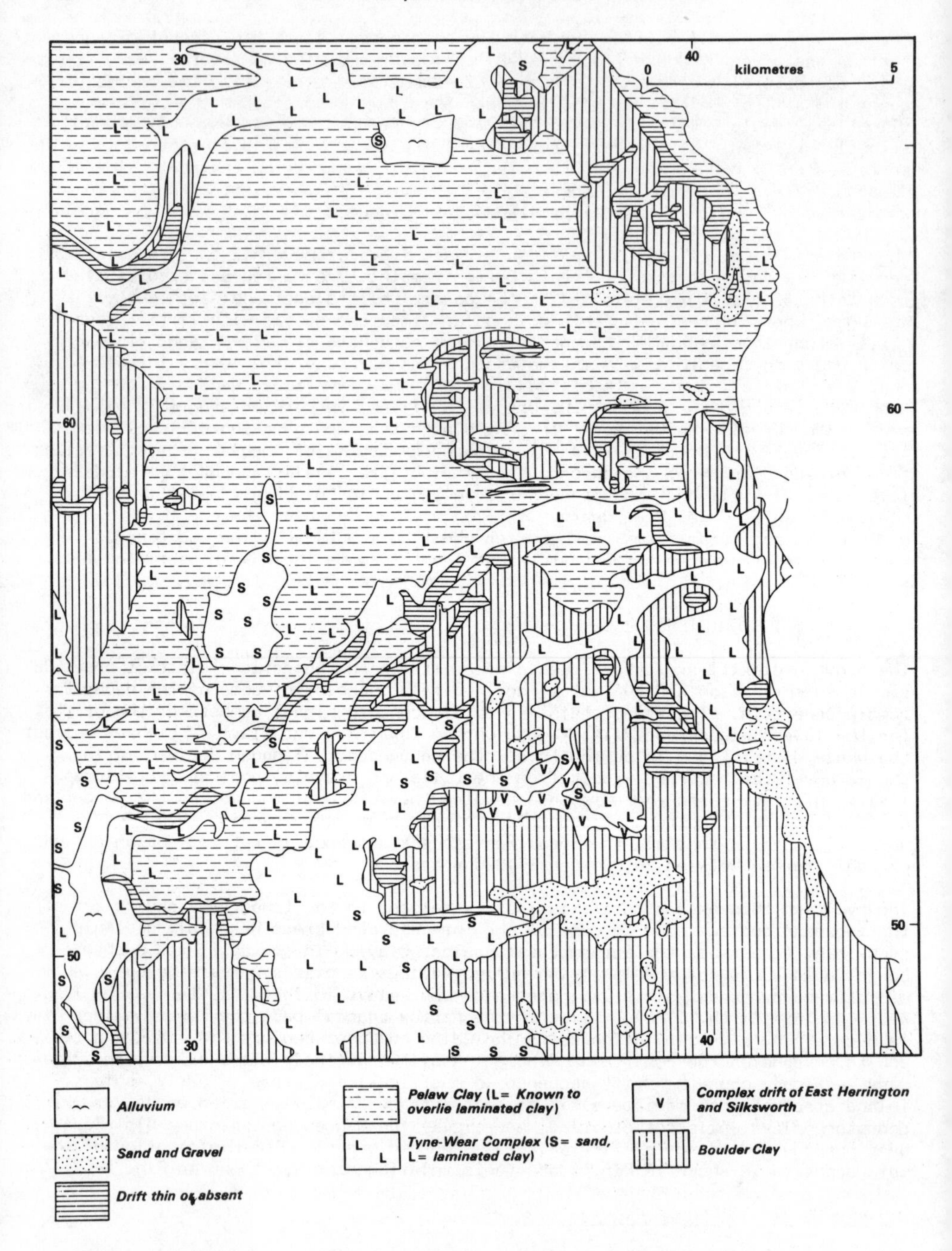

Fig. 4. Distribution of the main Quaternary deposits of the Sunderland district (simplified from Smith, 1978). The Prismatic Clay is not shown.

Basal Gravels.

Sand and gravel ranging from fine to very coarse is found in rockhead hollows below the Durham Lower Boulder Clay, and locally in the bottom of some buried valleys. Rarely exceeding 2 m in thickness, the gravels contain a high proportion of subrounded to subangular local rocks in addition to generally smaller better-rounded erratic pebbles of a suite similar to that in the overlying till (see below). Whilst some of the gravels may date from before the last glaciation, most were probably deposited by proglacial or subglacial streams.

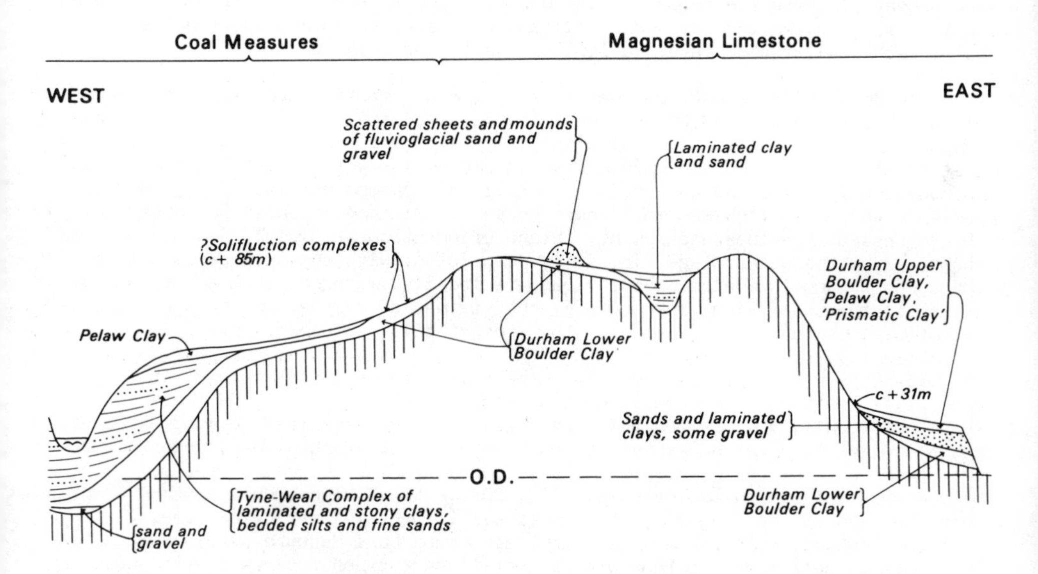

Fig. 5. Stratigraphical relationships of the main Quaternary deposits of the Sunderland district.

Durham Lower Boulder Clay

This till, recognised as the most continuous and uniform of the north-eastern drifts by Howse (1864) and all later authors, covers rockhead in most lowland parts of Northumberland, Tyne & Wear and County Durham and is also wide-spread on hills up to about 200 m O.D. It is continuous onto adjoining districts to the north (Land, 1974) and south (Smith and Francis, 1967), and, whilst generally 3 to 8 m thick over much of the Sunderland district, locally (as under the western part of Washington New Town) it exceeds 20 m.

The Durham Lower Boulder Clay is a dark brown, grey-brown or dark grey tough gritty overconsolidated stony clay. In much of the district it is indivisible apart from a thin crudely-laminated basal layer, but Merrick (1909) recognised two widespread main sub-units separated by a "leafy blue clay" on Tyneside and a twofold subdivision is general in the narrow coastal belt.

Boulders and stones in the Lower Boulder Clay are concentrated in its lower part but small stones abound throughout; they comprise a suite dominated by rocks of the local Coal Measures (everywhere) and Magnesian Limestone (in the east) but also include many rocks from south-west Scotland, the Vale of Eden, the Lake District and the northern Pennines; Carboniferous limestones (commonly well-rounded and polished), dolerite from the Whin Sill and Carboniferous ganisters form most of the large boulders at or near the base of the deposit. The importance of local influences in determining the erratic suite was demonstrated by Beaumont (1971a) who showed that Carboniferous sandstone, coal, shale and ironstone formed a high proportion of the stones in Lower Boulder Clay on Coal Measures of the Wear Lowlands but that the proportion declined progressively south-eastwards across the Magnesian Limestone where local dolomite becomes abundant. Later (1971b) Beaumont showed that the matrix (less than 2 mm fraction) of the Lower Boulder Clay (and also of all the higher clays in the sequence) was derived almost exclusively from broken-down Coal Measures rocks and comprises illite and mixed-layer minerals, together about half, with smaller amounts of kaolinite, quartz and chlorite.

Although Beaumont's results on the distribution of erratics are partly dependent on distance of transport and resistance to abrasion of the rock fragments, they clearly support the less analytical conclusions of earlier workers such as Woolacott (1905) that the ice from which the Durham Lower Boulder Clay was deposited flowed eastwards through the Tyne Gap and over adjoining hills and spread out in the Tyne and Wear lowlands where its flow was influenced by local eminences such as Gateshead Fell and the Magnesian Limestone escarpment. Stone orientation studies by Beaumont (1971a) augmented by the author (see Fig. 3) are consistent with the data derived from stone counts and indicator pebbles, but are locally different from the trends of striae on the underlying rock surface; the striae, of course, may have been caused by earlier ice than that from which the present till was deposited.

The bipartite nature of the deposit in the narrow coastal belt is clearly seen in many cliff sections from Sunderland southwards. The lower unit, commonly 1-2 m thick but locally 3 m, is generally dark grey or dark brownish grey and has a higher stone content than the dark brown thicker upper unit; it is probably equivalent to the lowest unit of the quadripartite Lower Boulder Clay in the coastal belt south of the district (Smith and Francis, 1967, 203). Local Magnesian Limestone comprises more than 70% of the stones in this lower clay, the remainder being a suite of western erratics, mainly small, and including many Coal Measures rocks; small fragments of soft red sandstone are present in many exposures. A smooth sharp surface separates the two units in the coastal cliffs of the Sunderland district, where the upper unit is generally similar to the single till of the Wear Lowlands, and carries a comparable suite of far-travelled erratics; weak bedding or poorly-defined lamination are not uncommon at the base of this upper unit, which also is less gritty than the lower and commonly possesses ill-defined coarse bedding in its upper part (Coupland and Woolacott, 1926, p. 4-5). In the cliffs about 1.5 km north of Seaham Harbour the 8 m upper unit is further subdivided by a slight discontinuity about 4.5 m above its base.

All authors since Woolacott (1905, 66) have accepted that the Durham Lower Boulder Clay (under its various historical names) is a ground moraine. In the coastal belt Coupland and Woolacott (1926, 3-4) further suggested that only the lower unit was a true ground moraine, the thicker upper unit being derived from englacial debris deposited when the ice melted; their view finds clear echoes in the observations of Boulton (1970, 1972) on the recent subglacial deposits of Spitzbergen where a "melt-out till" deposited from englacial debris in wasting stagnant ice widely overlies a thinner "lodgement till" plastered onto its substrate by the dirt-laden basal layer of active ice.

The Tyne-Wear Complex

Sediments of the Tyne-Wear Complex mainly comprise laminated and bedded clays and silts overlain by and locally interbedded with sand and with lenses and beds of stony clay. Deposits of this complex occupy most of the buried valleys of the Sunderland district and are extensive but patchy in the Wear Lowlands and on surrounding slopes up to about 70 m O.D.; above this, isolated but similar deposits at altitudes up to about 132 m may be related. In the Wear valley and its lower tributaries the deposits of the Tyne-Wear Complex merge with similar deposits described by Smith and Francis (1967) and Francis (1970) and in the north-west they extend far up the Tyne Valley; in the east they extend to the coast between Whitburn and Sunderland, and similar but strongly contorted drifts in the cliffs at Whitburn may form part of this complex. They generally rest sharply on the eroded surface of the Lower Boulder Clay but in places they rest directly on rock.

Laminated and bedded clays of the Tyne-Wear Complex are generally dark brown (weathering to grey-brown and purple-brown) and include films of grey and pale brown fine micaceous sand that locally thicken to form partial current ripples. Lamination is generally smooth, even and parallel, and distortion is uncommon except in surface layers affected by creep; the clays commonly dip gently towards the axes of the buried valleys, perhaps partly as a result of differential compaction. Small stones and a few boulders (?ice rafted) are sparingly scattered throughout but tend to be concentrated at certain levels, and small calcareous concretions are present locally. No undoubted organic remains have been reported in laminated clays of the Sunderland district but Woolacott (1905, 67) records tracks of fresh-water crustacea from a similar deposit in the Derwent Valley. The laminated clays have been widely worked for brick making, and the distribution of abandoned clay pits is an excellent guide to the position of many of the buried valleys. Their plasticity, however, and their content of water-bearing sand, cause considerable instability in civil engineering works. Where heavily loaded they are apt to flow, leading to the formation of pressure ridges like that around the northern fringe of the spoil from East Herrington Colliery (NZ 338535) and open trenches cut into it have been known to close overnight when left unshuttered.

Sands of the Tyne-Wear Complex are generally pale brown, fine- to medium-grained and locally rich in coal grains; pebbles are uncommon except in scattered thin beds and lenses of fine gravel where a suite of rock types similar to that of the Lower Boulder Clay is found. No organic remains have been reported. The sands occur as thin beds within the laminated clays and in several areas are most common towards the top of the sequence where they locally predominate. Sand is especially abundant in the south west of the district, from Chester-le-Street southwards, where it makes up much of the sequence, and it is also widespread between Usworth and Washington Station where it may be deltaic. Thick layered sand cropping out in the Wear Valley at Fatfield has been sculptured into isolated rounded hills, one of which (Worm Hill, NZ 311541) is the scene of the famous legend of the Lambton Worm.

Stony clays individually up to 7 m thick have been seen in some surface exposures and are recorded in many boreholes through deposits filling the lower Tyne buried valley. They are dark brown gritty sandy clays with abundant subangular to subrounded erratic pebbles, cobbles and some boulders and are lithologically similar to the Durham Lower Boulder Clay; they differ from the latter, however, in not being overconsolidated and in possessing little or no preferred fabric. Judging from borehole records these stony clays are discontinuous and probably lenticular, and in the few surface exposures seen they rest on eroded surfaces; of a number of possible mechanisms for their formation I favour an origin as mudflows, perhaps of soliflucted Durham Lower Boulder Clay from above the level of laminated clays on adjoining valley sides.

In the coastal belt at Whitburn, deposits in a similar stratigraphical position to the Tyne-Wear Complex comprise a thin but highly varied sequence of sands, gravels and laminated and stony clays. These lie mainly on Durham Lower Boulder Clay from which the thick upper unit has largely been eroded but which elsewhere is involved with them in widespread but not ubiquitous contortions and involutions; the lower unit of the boulder clay here is not involved in the deformation. Some impression of the range of relationships in the contorted sediments is given in Fig. 6. Undisturbed silty sparingly stony clay (?Pelaw Clay) lies on the truncated upper surface of the contorted deposits. Except for the deformation, these coastal deposits are typical of the Tyne-Wear Complex with which they may be equated with some reservation. Simple cryoturbation seems incompatible with the scale and pattern of the deformation which may have resulted from unequal loading, liquifaction and flowage under tundra conditions.

The suggestion by Lebour (1902) that laminated clay in the Derwent Valley was lacustrine was extended by Woolacott (1905, 67) to include deposits around Newcastle that are here included in the Tyne-Wear Complex. Woolacott inferred a late-glacial age and later (1913, 102) tentatively suggested deposition in a lake or lakes impounded by northern ice whose margin lay along the coast at Sunderland; in 1921 (p. 29) Woolacott reiterated this view but later (1921, 68-9 and 1922, 71) also suggested an origin as estuarine or marine deposits related to an inferred late-glacial sea level at about 45 m. This last interpretation has found little favour and a lacustrine origin was reiterated by Raistrick (1931, 287-289) who informally used the term lake Wear which was formalised to Lake Wear by Beaumont (1968, 9). The lack of collapse structures in the bedded clays and sands of the Tyne-Wear Complex in most of the district shows that the buried valleys were essentially free of ice when Lake Wear was formed, although the contortion at Whitburn may reflect the proximity of ice to the east.

Bedded Sands and Clays of the Coast at Ryhope

Much of the Durham Lower Boulder Clay has also been eroded off at the coast near Ryhope, where its channelled top is widely overlain by water-laid sand. The sand is thickest, up to about 20 m, for about 1.2 km south from the Crags (NZ 413545) at Hendon, Sunderland, where it is almost level bedded and contains sub-horizontal beds of silt and clay; an irregular lens of stiff brown gritty stony clay up to 6 m thick is traceable in the upper part of the sand for about 400 m mainly to the south of Salterfen Dene (NZ 415539) and contains a pebble suite similar to (but scantier than) that of the Lower Boulder Clay. Lenses of gravel are uncommon in the cliffs except near the base of the deposit where they occupy a N-S channel cut through into the Magnesian Limestone. Gravel, however, predominates south of Ryhope where it rests on the uneven surface of about 12 m of boulder clay, and is exposed both in the cliffs and adjoining workings (NZ 416522); here many of the larger clasts are of dolomite from the nearby Middle Magnesian Limestone reef and the gravel also contains rafts and rolled armoured balls of Lower Boulder Clay. Uppermost parts of the gravel are locally strongly cryoturbated, and ice wedge casts and pipes filled with clay from the otherwise flat-lying succeeding deposits are also present in the workings.

The distribution and high content of reef dolomite in the gravels at Ryhope suggest a strong genetic relationship with the major dry valley of Tunstall Hope (NZ 3953) into which the gravels persist for some distance before giving way to laminated clays at the north-west end (Fig. 7). I believe that they represent the remains of a delta fan and outwash plain radiating from the mouth of the Hope, with southwards currents predominating. I accept Woolacott's (1921, 30-31) view that Tunstall Hope was probably cut by overflowing lake waters, the implication being that it and the gravels date from the time when Tweed-Cheviot ice (see next section) was in the vicinity.

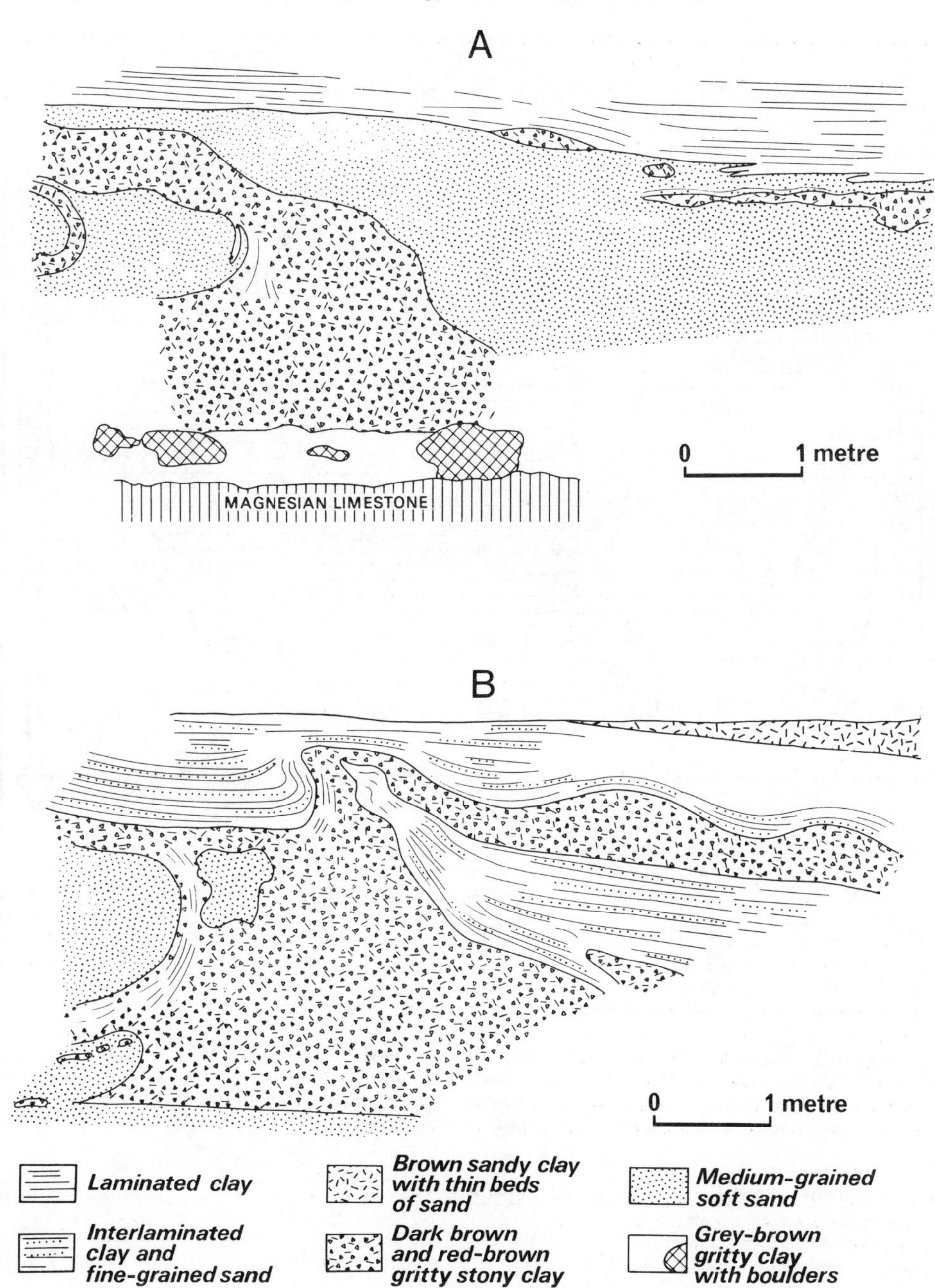

Fig. 6. Contorted drift in coastal cliffs near Whitburn, County Durham: A. At Grid Reference NZ 41056147; B. At Grid Reference NZ 41236173.

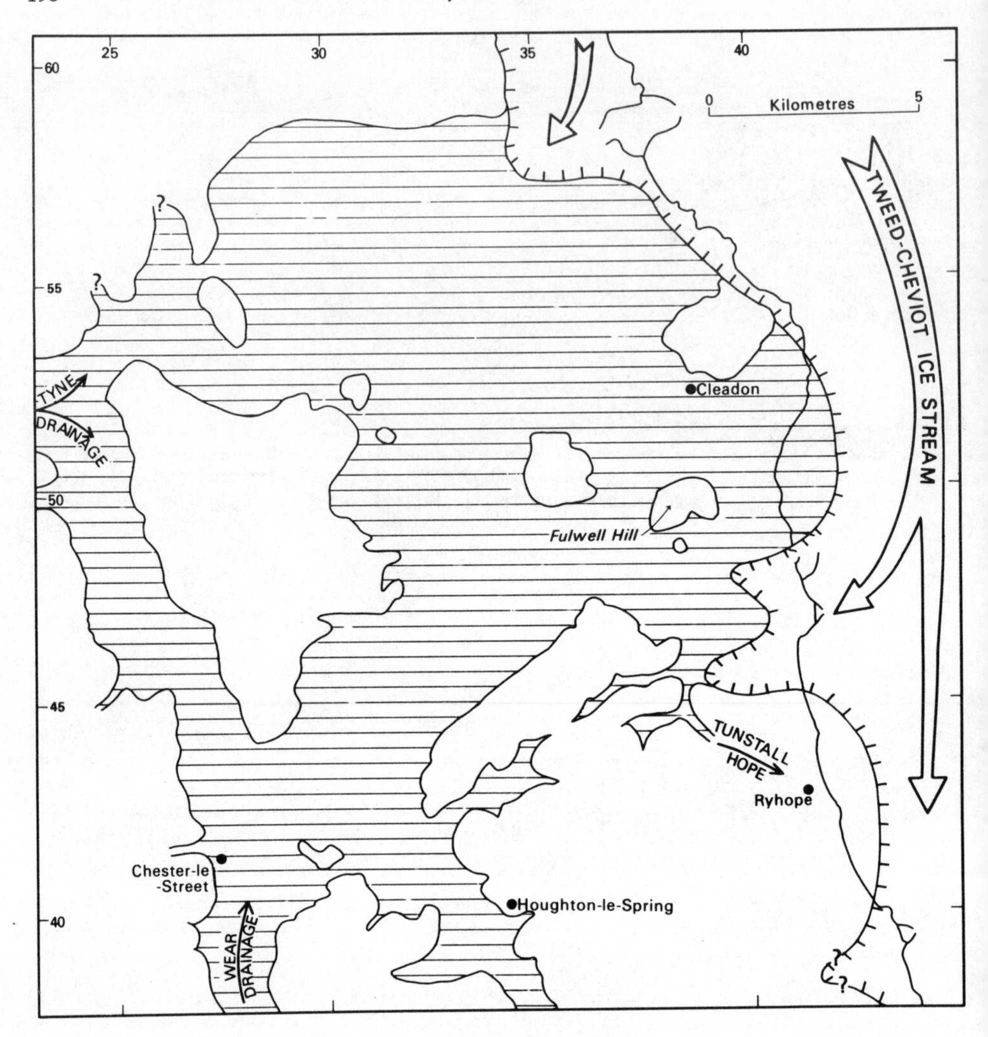

Fig. 7. Approximate configuration of Lake Wear during the inferred 43 m stand. The cutting of Tunstall Hope, unless it dates from an altogether earlier glacial episode, would have been initiated when the lake stood at perhaps 90 m.

Durham Upper Boulder Clay and Related Deposits

At Swaddles Hole, Tynemouth (NZ 369690), just north of the district boundary, about 3 m of red-brown on brown stiff stony clay rests on a thin bed of clayey sand that in turn overlies 6 to 9 m of typical Durham Lower Boulder Clay. It contains abundant pebbles, cobbles and some large boulders, comprising a suite dominated by Coal Measures rocks but with some Magnesian Limestone and Cheviot andesite and granite; the clay is lithologically similar to the Upper Boulder Clay of the Northumberland coast (Smythe, 1912) and the coast of south-east Durham with which it may be correlated with some confidence. From its lithology and stone content the clay probably represents a till deposited from ice from the Tweed valley and the Cheviot Hills.

Farther south, Cheviot and other distinctively northern rocks are abundant in scattered deposits of reddish-brown, stony clay (predominant), gravel and contorted sand that fill hollows eroded in the top of the bedded sand in cliffs east of Ryhope. These deposits were first recorded by Coupland and Woolacott (1926, 6) who also noted that many of the component pebbles were split. Their location and stone content suggest an affinity with the Upper Boulder Clay. Red-brown, stony clay (7.5 m) overlying gravel (2.3 m) and brown boulder clay (4m+) at a locality (NZ 419485) near Seaham Harbour may also be Upper Boulder Clay; if it were, appreciable penetration of Tweed-Cheviot ice into Dawdon Dene would be indicated.

Pelaw and Prismatic Clays

In almost all lowland parts of the district and locally on hills up to about 130 m, the youngest widespread drift deposits are brown or reddish-brown silty clays with scattered to abundant stones; they have been recognised and described by all previous workers, but uncertainty still surrounds their number, continuity and mode of origin. Two main separate deposits are recognised here - the Pelaw Clay of the Wear Lowlands and coast north of Sunderland and the Prismatic Clay mainly of the coast south of Sunderland - but some earlier authors have not made this distinction and have applied the term 'Prismatic Clay' (or its equivalents) to all the superficial clays.

The Pelaw Clay (type locality Pelaw Brick Pits, NZ 310625) is a blocky slightly reddish-brown to dark brown underconsolidated pebbly, silty clay that is widely 1 to 2 m thick in the Wear Lowlands but which amongst sections seen in recent years is up to 4.5 m thick; 6 m of uppermost stony clays was seen in 1923 (IGS records) at the Iris Brickworks Pit, Heaton, Newcastle (NZ 265648) and Woolacott (1905, 66) recorded a maximum thickness of about 9 m. Lenses and thin beds of fine sand, many contorted, are locally common in the Pelaw Clay and in places the deposit is weakly layered. Stones, generally less than 10 cm across, show no strong preferred orientation and comprise a suite similar to that of the Durham Lower Boulder Clay but are much less abundant. From Whitburn northwards to the Tyne the Pelaw Clay in the narrow coastal belt widely rests directly on rockhead, hollows in the surface of which contain gravel or Lower Boulder Clay (Howse, 1864, 177); pebbles in the clay here include chalk and flint. Small grey, calcareous, pedogenic concretions are common 1 to 1.5 m below the surface of the Pelaw Clay.

The base of the Pelaw Clay ranges from gradational where overlying some laminated clays to generally sharp and transgressive on most other deposits. On boulder clay the contact is commonly smooth and flat, but the two deposits are generally separated by a film or thin bed of sand, unevenly interlaminated sand and clay or a discontinuous layer of patinated small pebbles. Sub-vertical columnar joints, their faces discoloured with red, blue and purple hues, characterise upper parts of the deposit and, in combination with plasticity in its moist lower parts, lead to instability in excavations.

Howse (1864, 173) recognised that the deposit here called the Pelaw Clay (his "Scandinavian Drift") at Trow Point, South Shields differed from, and was younger than, "the true boulder clay" (i.e. the Lower Boulder Clay) and later (1880, 364) commented that it had been formed by different but unspecified processes. The continuity and uniformity of the Pelaw Clay over a wide range of other deposits, including Coal Measures and Magnesian Limestone, indicates that the deposit as a whole is not simply an *in situ* weathering product. Furthermore the widespread truncation of bedding and sedimentary structures in underlying deposits and the persistence of the clay far down the gentler flanks of many valleys may indicate that there is an erosional hiatus at the base. The possibility that the clay might be a till from a glacial readvance seems slight because of the lack both of a strongly preferred fabric and of derangement in the underlying clays and this lack of evidence of contemporary ice in the vicinity also argues against the suggestion by Francis (1970, 147) that the Pelaw Clay might be a flow-till. Woolacott's view (1905, 66-67) that the deposit was produced by late-glacial marine reworking of the glacial deposits likewise seems improbable in view of its poor grading and lack of marine fossils. Previous explanations of the origin of the Pelaw Clay all apparently having shortcomings, I advance tentatively the hypothesis that it is a congeliturbate formed by periglacial modification and redistribution of all pre-existing deposits (but especially of the plastic laminated clays, hence the coincidence of highest occurrences at 130-132 m) following the drainage of Lake Wear. The main processes I envizage are winter cryoturbation and slow solifluction combined with summer liquifaction, homogenization and mass flow in a saturated surface layer with its drainage impeded by underlying permafrost.

The Prismatic Clay - perhaps a geographical name such as the Ryhope Clay would be preferable - is a dull brown silty and sandy clay that caps the drift sequence south of Sunderland and in a few places inland. It is generally less than 1.5 m thick and commonly 0.4 m to 1 m, with pronounced columnar joints in its upper part grading to blocky below; in some places the Prismatic Clay is stoneless, but it generally contains scattered (locally abundant) small subrounded to rounded pebbles of a suite similar to that in the Upper Boulder Clay. It is weakly layered in some exposures and generally has a sharp relatively smooth base resting transgressively on all underlying deposits; where overlying till, a thin intervening lenticular layer of pebbly sand is common. As Coupland and Woolacott (1926, 7) noted, the Prismatic Clay maintains a roughly constant thickness for considerable distances, even down valley sides and onto bedrock. I accept their view that it is a relatively young deposit, formed not solely by weathering of underlying deposits but "the ordinary processes of denudation... in part rain wash, in part wind-formed and in part by weathering *in situ*"; it is probably younger than the Pelaw Clay and at Ryhope it may be genetically related to the well-marked coastal platform at about 31 m O.D.

Deposits of the Cleadon and Fulwell hills.

Drift is generally thin on the Concretionary Limestone of the Cleadon and Fulwell hills but most sections on the sides of the hills reveal several thin deposits of which the lowest in some places in undoubted Durham Lower Boulder Clay with an uneven (eroded) top. The remaining deposits comprise a highly variable complex of sands, laminated and unbedded stoneless clays, unbedded sandy stony clays and lenses of gravel, with a marked tendency for the uppermost deposit to be a red-brown, sparingly stony, sandy clay. Amongst these varied deposits are sediments lithologically comparable with the Tyne-Wear Complex and the Pelaw and Prismatic clays (in that order of superposition) but the sequence is too varied, and the evidence too fragmentary, to permit firm conclusions on correlation.

A deposit of sand (at a site now largely quarried away) at about 45 m O.D. on the northern slope of Fulwell Hill was carefully described and figured by Kirkby (1860) who recorded that the underlying Magnesian Limestone appeared to have been water-worn and bore many narrow cylindrical to conical sand-filled karst pipes up to 4 m deep; in places the sand was banked against a bed of gravel and elsewhere extended onto boulder clay which, he inferred, had earlier covered the whole site. Howse (1864, 174) interpreted the deposit as an ancient beach and this was also the preferred view of Woolacott (1897, 74; 1900, 256; 1905, 69-72) who recorded up to 3.5 m of gravel at about the same locality and from which he collected rolled flints, Cheviot andesites and a few fragments of *Cyprina* and *Littorina*; no bored pebbles were recorded. Almost all the sand and gravel had been removed by 1965 when I visited the site, but the small exposures then preserved revealed a varied sequence of sand, gravel and laminated and stony clays up to 2 m thick resting on a markedly uneven limestone surface that in places was badly fractured as if by frost action. The presence of laminated clay probably favours a lacustrine origin for these deposits, an alternative view allowed by Woolacott (1913a, 102) for the thicker deposit then visible.

An expanse of gravelly sand up to 12 m thick lying across the 30 m contour at Cleadon also was interpreted as a marine beach by Howse (1864, 174) and accepted as such by Woolacott (1897, 1900, 1905, 1906). The sand indeed contains fragmentary marine shells, but no more abundantly than most other Pleistocene sands and gravels in the region and I prefer Trechmann's view (1952, 174) that it probably is not marine. Similar gravelly sands, also with flint and chalk pebbles like the Cleadon sand are present at about the same level on the east side of the Cleadon Hills at Marsden (Howse, 1880, 164) and also form a spread beneath Whitburn village. The gravel at Marsden is no longer exposed but apparently lay between rockhead and Pelaw Clay; as with the Cleadon village deposit, Howse's view that the Marsden gravel was a raised beach is suspect in view of its stratigraphical position which suggests general equivalence to the lacustrine Tyne-Wear Complex.

Deposits at East Herrington and Silksworth

Temporary exposures (1966-7) in gently sloping and flat areas at East Herrington (NZ 3751) and Silksworth (NZ 3752) revealed the unsuspected presence there of a highly varied layered undisturbed sequence of sands, gravels and clays resting on an uneven eroded surface of Lower Boulder Clay and Magnesian Limestone; similar sequences presumably remain undiscovered in other parts of the district. Both areas lie at between 85 and 90 m O.D. in northwards-opening re-entrants leading to buried valleys filled with thick sands and laminated clays and appear to have been peripheral to late glacial or younger lakes.

The complex sequences at East Herrington and Silksworth are widely capped by brown sandy prismatic clay (?alluvium) and below this consist mainly of thinly interbedded and lenticular sands and sandy clays; laminated clay is uncommon at East Herrington but abundant at Silksworth. Most of the clays contain scattered to abundant small stones of the western suite, although some are silty, stoneless and contain laminae and thin beds of sand. Sedimentary structures other than generally parallel bedding are uncommon in much of the sand at East Herrington but northward sediment transport was indicated by imbrication in gravel at one exposure there. Of special interest at East Herrington was the presence in some sections of clayey gravels (grading to gravelly clays) containing an almost pure culture of angular Magnesian Limestone fragments, and streaks and tongues of locally-derived dolomite rock-flour; these gravels lack sedimentary structures or grading and may be mass-flow or solifluction deposits. Most of the other deposits appear to be water-laid, however, and probably are lake-marginal.

Other probable lake-marginal deposits have been noted (a) at about 48 m O.D. in temporary exposures (NZ 33464963) west of Houghton-le-Spring, where pieces of oak stems collected by the writer from a silty clay between layers of ?solifluction gravel had radiometric ages of 2,160 $\pm$ 110 and 3,400 $\pm$ 100 (analysis by two separate laboratories) and (b) at the entrance to Field House Sand Hole (NZ 353505), north-east of Houghton-le-Spring, where water-laid sands and laminated and unlaminated clays interfinger with wedges and streaks of gritty stony clay and tongues of redistributed Permian Yellow Sands and Magnesian Limestone flour; at about 132 m these are the highest lacustrine deposits noted in the district.

Fluvioglacial Sands and Gravels

These deposits overlie Lower Boulder Clay at Grindon (NZ 359546) and Warden Law (NZ 3701505) and scattered patches also occur elsewhere, mainly in the south of the district. The 23 m of sand at Grindon forms an elongate mound on a valley floor and contains pebbles mainly from western sources but with some Cheviot lavas (Woolacott, 1913b, 112) and Magnesian Limestone; abundant faults and large-scale contortions (Fig. 8) in the sand probably indicate contemporary ice-support and are consistent with interpretation of the deposit as an esker (Woolacott, 1905, 68) or a kame (Trechmann, 1952, 172). The massive deposit of gravelly sand capping Warden Law (the highest point in East Durham,197 m) appears to be a relic of a more extensive outwash spread; it contains a pebble suite similar to that at Grindon, including some Cheviot pebbles and some of Magnesian Limestone of types found only to the east and north. A gentle westerly sheet-dip and westwards-dipping cross-lamination in the Warden Law deposit are consistent with its content of eastern pebbles.

Alluvium, Peat and Wind-blown Sand

Downcutting of the Late-Devensian and early Flandrian River Tyne to at least 25 m below O.D. was followed by rapid silting of its lower course as sea level recovered to slightly above its present level. The resulting deposits have been proved by many boreholes to consist mainly of silt and fine sand with gravel present locally at the base but they are best known from boreholes, shafts and other excavations along the line of the Tyne Tunnel. Here Armstrong and Kell (1951) reported a thin pebbly layer with unworn *Mytilus* and with *Littorina* encrusted with unworn barnacles resting on laminated clay at about 13.5 m below O.D. and overlain by silt and fine sand containing abundant plant remains and, in places, a small range of vertebrate bones. Other bones, mainly of red deer, but with some *Bos primigenius* were found in association with plant and tree debris in alluvium excavated at Jarrow docks (Howse, 1861, 117-120). The presence of marine shells apparently *in situ* in the Tyne valley alluvium shows that sea level recovery after deglaciation initially outpaced both isostatic rebound and sedimentation and for a time the sea invaded the lower Tyne valley before receding as sedimentation again supervened; judging from sea level recovery curves (e.g. Gaunt and Tooley, 1974, fig. 6) this transgression was probably between about 9,000 and 7,500 years ago.

Comparable deposits have not been recorded from the Wear valley, where the lower course of the modern river is on resistant limestone and was not deepened to below modern sea level by Late-Devensian downcutting.

Peat is known in the Sunderland district only in a few shallow hollows and stream courses (mainly east and north-east of Washington), at South Shields and on the coast at Seaburn, Sunderland. The inland peats are thin and impure and are composed mainly of sedges; many are overlain by alluvium. The peat at South Shields was described by Howse (1861, 115) as about 0.3 m thick, 'very solid' and overlain by almost 4 m of alluvium; Howse considered it to be of drifted vegetation, and thought that the skeleton of an Irish Elk found in it was also drifted. The peat at Whitburn Bay, Seaburn is up to 2 m thick and lies on Durham Lower Boulder Clay mainly between tide marks. It is generally covered by beach sand. The Seaburn peat was already known to Howse in 1861 (p. 117), who later (1864, 170) recorded abundant fragmentary remains (some large) of alder, birch, hazel and oak in addition to bones of *Bos primigenius* and antlers of two species of deer. The report by Woolacott (1913b. 113) of tree roots in growth position was confirmed in temporary excavations (NZ 40516023) in 1965 and indicates that the peat is not a drifted accumulation. Woolacott's earlier observation (1897, 78) of a patchy layer of rounded stones which he believed to be littoral between the Seaburn peat and the underlying till is of considerable interest; if his interpretation is correct it indicates, as in the

Fig. 8. Normal faults and large-scale contortion in gravelly sand at Grindon Hill, Sunderland (Taken in 1953). IGS photographs NL141(top) and NL142(bottom).

Tyne valley, an early marine transgression followed by regression as the ensuing forest peat formed on land to below present sea level. This minor early transgression may correlate with that recognised in north-west England by Tooley (1974) and on Humberside (Gaunt and Tooley, 1974) and the peat is probably of much the same age (about 5,250 radiocarbon years according to Gaunt and Tooley) as the similar deposit at Hartlepool.

The youngest Flandrian deposits of the Sunderland district comprise recent stream alluvium and wind-blown sand. The former occurs in generally narrow belts along the minor streams where it rarely exceeds 1.5 m in thickness, but it is several metres thick in parts of the Wear valley between Fatfield and Sunderland (where the river was formerly tidal) and near Chester-le-Street. The wind-blown sand commonly contains frail gastropod shells and is restricted to a narrow coastal belt where it forms patches up to 2 m thick on the cliff tops at Ryhope, Hendon, Whitburn, Marsden and South Shields.

SYNTHESIS

Although all the glacial drift deposits of the Sunderland district are probably Late Devensian, it is probable that the area was covered by ice sheets during earlier cold periods such as the Wolstonian and Anglian; that no deposits from these presumed ice sheets have been recognised is an eloquent testimony to the efficiency of erosion during the long interglacials (of which likewise there are no recognised deposits) and of the erosive power of advancing Late Devensian ice. It is to be expected that successive glaciations of the area would follow generally similar courses but differ in detail with the interplay of lobes extending from different ice caps, and that many of the major glacigenic landscape features were shaped and modified on more than one occasion. It follows that it is unlikely that any simple reconstruction of the last glacial episode can satisfactorily explain all the glacial features of the district because some of them may date from earlier and different glacial episodes.

For the Late Devensian glaciation there is general agreement that the first ice to affect the area entered it via the Tyne valley, having accumulated to critical thickness on the high ground of the western Southern Uplands and the Lake District. Because of severe ice congestion in the Irish Sea area, this western ice streamed eastwards into the western part of the present North Sea, where it is believed (Kendall, 1902; Woolacott, 1907, 41), it was obstructed and deflected southwards by Scandinavian ice then covering much of the largely dry bed of the North Sea; this Scandinavian ice probably did not reach the present coast during the Late Devensian. The main surviving product of the western ice is the Durham Lower Boulder Clay and its equivalents in Northumberland, Cleveland and Yorkshire, and the thick outwash deposits of the middle Wear valley (Smith and Francis, 1967; Francis, 1970). Outwash may formerly have been more widespread but the extensive thinning and local removal of the Lower Boulder Clay and the absence of collapse structures in succeeding bedded deposits suggests that ice withdrew from most or all of the district and a phase of active erosion ensued before the Tyne-Wear Complex was formed. Ice-wedge casts in Lower Boulder Clay beneath 1 to 3 m of alluvium at Burdon (NZ 385509) may date from this phase.

The ice cap of the Cheviot Hills and eastern Southern Uplands appear to have reached critical mass later than that to the west, perhaps because of lower precipitation, but retained their vigour later. Ice from the Cheviots was initially unable to flow southwards because ground there was already occupied by the powerful western ice and it was deflected eastwards to merge with Scottish ice in the Tweed lowlands; there it spread eastwards into the adjoining area of the present North Sea before, like the Tyne Valley ice, being deflected southwards by the Scandinavian ice. Waning, downwasting and retreat of the western ice subsequently allowed Tweed- Cheviot ice to spread or be pushed westwards towards, and locally on to, the present coast as at Tynemouth, there depositing the Upper Boulder Clay and blocking eastwards drainage so as to create proglacial lakes in Northumberland (Smythe, 1908, 1912), the confluent valleys of the Tyne and Wear (Lake Wear, Fig. 7) and in lower Tees-side. Lake Wear became the repository of the Tyne-Wear Complex, at times reaching to about 132 m O.D. but generally lying below 90 m, whilst periglacial processes were active on higher ground and perhaps led to the observed interdigitation of lacustrine and inferred mass-flow deposits at lake margins. Surfaces recognised by Anderson (1940) and Westgate (1955) at about 96 m, 57 m, 43 m and 30 m above modern sea level must have been inundated by Lake Wear and may stem from periodic still-stands of the lake surface; that at 43 m is particularly well-marked and may be genetically linked to the sparingly shelly gravels on the north slope of Fulwell Hill and to caves at a similar level on Cleadon Hills (Howse, 1880). The alternative views of Anderson (1940) and Woolacott (1897 and later) that the surfaces and related deposits are marine is difficult to reconcile with the lack of evidence of high Late Devensian sea levels elsewhere in northern England and has found little favour.

The great channel of Tunstall Hope (Fig. 7) at Ryhope was probably cut by drainage overflowing south-eastwards from the early Lake Wear, the water then flowing mainly southwards (because of Tweed-Cheviot ice a little to the east) and depositing the Ryhope sands and gravels; the mixed but mainly stony clay deposits with Cheviot erratics that occupy hollows eroded in the top of the Ryhope sands and gravels perhaps indicate that the Tweed-Cheviot ice later spread westwards to reach or almost reach the present coast. Farther south, the stone content and sedimentary structures of the Warden Law sands and gravels suggest an origin as direct outwash from Tweed-Cheviot ice, but much of the evidence has been removed by late erosion and true relationships remain uncertain. The origin of the isolated Grindon sand mound is even more problematical; its reported content of Cheviot rocks (Woolacott, 1913b, 112) link it to the Tweed-Cheviot ice but its supposed ice-support features are at variance with the apparent lack of evidence that such ice ever reached Grindon; perhaps such evidence (in the form of Upper Boulder Clay or outwash deposits) was formerly more extensive but has since been removed along with much other evidence bearing on the glacial history of the district.

Drainage of Lake Wear was completed with the withdrawal of Tweed-Cheviot ice and was accompanied or followed by the cutting of the present valleys and gorges into and through the unconsolidated deposits of the Tyne-Wear Complex; contemporary base level was at least 28 m below modern sea level. Only at this stage could the Pelaw Clay have begun to form, although its formation presumably could have continued during remaining Late Devensian and early Flandrian time if the hypothesis that it is a congeliturbate is correct. Ice wedges in the gravels at Ryhope probably also were formed in these periglacial retreat phases, and pre-date the deposition of the Prismatic (?alluvial) Clay. Finally, with the recovery of sea level, fluviatile and marine alluvium filled the lower parts of the newly cut valleys and the 'submerged forest' at Whitburn flourished briefly; the thin wind-blown sands along parts of the coast are the most recent deposit of all, blown up from the modern beaches.

ACKNOWLEDGEMENT

This manuscript has greatly benefited from many helpful suggestions by D.I. Jackson.

RFFERENCES

Anderson, W. (1939). Possible late glacial sea levels at 190 and 140 feet O.D. in The British Isles. *Geol. Mag.*, 76, 317-321.

Armstrong, G. and J. Kell (1951). The Tyne Tunnel. *Geol. Mag.*, 88, 357-361.

Beaumont, P. (1968). A history of glacial research in Northern England from 1860 to the present day. *Univ. Durham Occas. Pap.*, No. 9, 21 pp.

Beaumont, P. (1970). Geomorphology. In J.C. Dewdney (Ed.), *Durham County and City with Teesside*. Br. Ass. Advmt Sci., Newcastle upon Tyne. Chap. II, pp. 25-45.

Beaumont, P. (1971a). Stone orientation and stone count data from the Lower Till sheet, eastern Durham. *Proc. Yorks. geol. Soc.*, 38, 343-360.

Beaumont, P. (1971b). Clay mineralogy of glacial clays in eastern Durham, England. In F. Yatsu and A. Falconer (Eds)., *Research methods in Pleistocene geomorphology*. Univ. Guelph. Geogr. Publn., No. 2, pp. 83-108.

Bott, M.H.P. (1968). Geophysical investigations of the Northern Pennine basement rocks. *Proc. Yorks. geol. Soc.*, 36, 139-168.

Boulton, G.S. (1970). The deposition of subglacial and melt out tills at the margins of certain Svalbard glaciers. *J. Glaciol.*, 9, 231-245.

Boulton, G.S. (1972). Modern Arctic glaciers as depositional models for former ice sheets. *J. geol. Soc. Lond.*, 128, 361-393.

Coupland, G. and D. Woolacott (1926). The superficial deposits near Sunderland and the Quaternary sequence in East Durham. *Geol. Mag.*, 63, 1-12.

Francis, E.A. (1970). Quaternary. In G. Hickling (Ed.), Geology of Durham County. *Trans. nat. Hist. Soc. Northumb.*, 41, pp. 134-152.

Gaunt, D.G. and M.J. Tooley (1974). Evidence for Flandrian sea-level changes in the Humber estuary and adjacent areas. *Bull. geol. Surv. Gt Br.*, No. 48, 25-41.

Howse, R. (1861). Notes on the fossil remains of some recent and extinct mammalia found in the counties of Northumberland and Durham. *Trans. Tyneside nat. Fld Club*, 5, 111-121.

Howse, R. (1864). On the glaciation of the counties of Durham and Northumberland. *Trans. N. Engl. Inst. Min. mech. Engrs.*, 13, 169-185.

Howse, R. (1880). Old sea caves and a sea beach at Whitburn. *Trans. nat. Hist. Soc. Northumb.*, 7, 361-365.

Kendall, P.F. (1902). On a system of glacier lakes in the Cleveland Hills. *Q. Jl geol. Soc. Lond.*, 58, 471-571.

King, C.A.M. (1967). An introduction to trend surface analysis. *Bulletin of quantitative data for geographers*, No. 12. Univ. Nottingham, Dept. Geogr., Nottingham.

Kirkby, J.W. (1860). The occurrence of "sand-pipes" in the Magnesian Limestone of Durham. *The Geologist*, 3, 293-298, 329-336.

Land, D.H. (1974). *The geology of the Tynemouth District*. Mem. geol. Surv. Gt Br. x + 176 pp.

Lebour, G.A. (1902). Note on a small boulder in the later glacial deposits in a 'wash out' near Low Spen in the Derwent Valley. *Proc. Univ. Durham phil. Soc.*, 2, 81-85.

Merrick, E. (1909). Glacial deposits around Newcastle upon Tyne. *Proc. Univ. Durham phil. Soc.*, 3, 141-152.

Raistrick, A. (1931). The glaciation of Northumberland and Durham. *Proc. geol. Ass.*, 42, 281-291.

Smith, D.B. (1978). *1:50,000 geological sheet 21 (Sunderland), including explanation*. Geological Survey of Great Britain.

Smith, D.B. and E.A. Francis (1967). *Geology of the country between Durham and West Hartlepool*. Mem. geol. Surv. Gt Br. xiii + 354 pp.

Smythe, J.A. (1908). The glacial phenomena of the country between the Tyne and the Wansbeck. *Trans. nat. Hist. Soc. Northumb.*, 3, 79-109.

Smythe, J.A. (1912). Glacial geology of Northumberland. *Trans. nat. Hist. Soc. Northumb.*, 4, 86-116.

Tooley, M.J. (1974). Flandrian sea-level changes during the last 9,000 years in north-west England. *Geogr. J.*, 140, 18-42.

Trechmann, C.T. (1915). The Scandinavian Drift of the Durham coast and the general glaciology of South-East Durham. *Q. Jl geol. Soc. Lond.* 71, 53-82.

Trechmann, C.T. (1920). On a deposit of interglacial loess and some transported preglacial freshwater clays on the Durham coast. *Q. Jl geol. Soc. Lond.*, 75, 173-203.

Trechmann, C.T. (1952). On the Pleistocene of Eastern Durham. *Proc. Yorks. geol. Soc.*, 28, 164-178.

Westgate, W. (1955). *The geomorphology of the Wear Valley*. Ph.D. thesis, University of Durham.

Woolacott, D. (1897). *Geology of north-east Durham*. Hills and Co., Sunderland.

Woolacott, D. (1900a). On the boulder clay, raised beaches and associated phenomena in East Durham. *Proc. Univ. Durham phil. Soc.*, 1, 247-258.

Woolacott, D. (1900b). On a portion of a raised beach on the Fulwell Hills, near Sunderland. *Trans. nat. Hist. Soc. Northumb.*, 8, 165-171.

Woolacott, D. (1905). The superficial deposits and pre-glacial valleys of the Northumberland and Durham coalfields. *Q. Jl geol. Soc. Lond.*, 61, 64-95.

Woolacott, D. (1906). On an exposure of the 100 feet raised beach at Cleadon. *Proc. Univ. Durham phil. Soc.*, 2, 243-246.

Woolacott, D. (1907). The origin and influence of the chief physical features of Northumberland and Durham. *Geogrl J.*, 30, 36-54.

Woolacott, D. (1913a). The geology of north-east Durham and south-east Northumberland. *Proc. geol. Ass.*, 24, 87-107.

Woolacott, D. (1913b). Report on an excursion to Sunderland and Tynemouth. *Proc. geol. Ass.*, 24, 108-114.

Woolacott, D. (1920). On an exposure of sands and gravels containing marine shells at Easington, Co. Durham. *Geol. Mag.*, 57, 307-311.

Woolacott, D. (1921). The interglacial problem and the glacial and post glacial sequence in Northumberland and Durham. *Geol. Mag.*, 58, 26-32, 60-69.

Woolacott, D. (1922). On the 60-ft raised beach at Easington, Co Durham. *Geol. Mag.*, 59, 64-74.

Dr Denys B. Smith,
Institute of Geological Sciences,
5 Princes Gate,
South Kensington,
London SW7 1QN.

15. Status and Relationships of the Loch Lomond Readvance and its Stratigraphical Correlatives

W. G. Jardine

INTRODUCTION

The major climatic events that occurred in Britain between c. 13,500 and 10,000 B.P., the time frequently termed the 'Lateglacial' interval in British literature (e.g. Gray and Lowe, 1977, xiii), are considered by a number of Quaternary workers to comprise an earlier 'Windermere Interstadial' and a later 'Loch Lomond Stadial' (*cf*. Coope, 1977, 325-326 & 328-331; Sissons, 1979, 199). The proposed selection of a standard reference section for the Windermere Interstadial (Coope and Pennington, 1977), together with the recent renewal of interest in, and intensification of study of the limits of the glaciers of, the Loch Lomond Readvance suggests that selection of stratotypes for the Stadial and the Readvance may be imminent. Before stratotypes are selected, however, the terminology relating to, and the classificatory relationships between and among, the Loch Lomond Readvance, the Loch Lomond Stadial and their chronostratigraphical correlatives require to be considered. In the discussion that follows, the terminology of both the American Commission on Stratigraphic Nomenclature (1961) and the International Subcommission on Stratigraphic Classification (Hedberg, 1976) is used because, although the terminology of the latter body is to be preferred, the terms 'stadial' and 'interstadial' are not included in the International Stratigraphic Guide. It is arguable that adherence to strict definitions of the terms stade (stadial) and interstade (interstadial) may lead to further difficulties rather than elucidation. On the contrary, however, the writer believes that much of the confusion that exists in relation to the terminology and subdivision of the Quaternary sub-era and its deposits has been caused by misunderstanding of, and slackness in the use of the terms of, litho-, bio- and chronostratigraphy. The better-known (informal) term 'Loch Lomond Readvance' is used here rather than 'Loch Lomond Advance' although perhaps the latter term is more 'correct' because Britain probably was entirely ice-free immediately prior to the formation of the 'Loch Lomond' glaciers (*cf*. Sissons 1977, 58).

STADIAL IN RELATION TO READVANCE

The term 'glacial readvance' does not appear in any recognised scheme of stratigraphical classification, but it and the term 'secondary advance of ice' describe the same geological event (Jardine, 1972, 49). A secondary advance of ice is an essential feature of the Stade (Stadial), a geologic-climate unit of the American Code of Stratigraphic Nomenclature (American Commission on Stratigraphic Nomenclature 1961, 660). It follows that, adhering to the same Code, the Loch Lomond Readvance was an essential feature of the Loch Lomond Stadial, ' ... a climatic episode ... which perforce included a secondary advance of glaciers, but which was not necessarily exclusively represented by the deposits of such a readvance' (Jardine, 1972, 51).

The Readvance and the Stadial were geological events, and geological events are recorded only if traces of their occurrences are preserved in rocks. The lithological record of the Loch Lomond Readvance is preserved in a number of glacial and fluvioglacial deposits present at several localities in northern Britain. In due course, the sedimentary record of the Readvance present at one of the localities may be chosen as the stratotype of the Loch Lomond Readvance and, following the rules of lithostratigraphical nomenclature (Hedberg, 1976, 40-43), the term used to denote the sediments concerned should be one such as 'the *type* Loch Lomond Formation'. 'Loch Lomond' is used here for the purposes of discussion but, when the definitive choice is made, another distinctive geographical feature rather than Loch Lomond may be chosen to name this unit (*cf*. Holland and others, 1978, 10). It should be noted that the time-span of the *type* Loch Lomond Formation may be substantially shorter than the total period during which the Loch Lomond Readvance occurred (Fig. 1). This is because the Readvance probably was not synchronous throughout the area where the sedimentary evidence of its existence has been traced in Britain.

STADIAL IN RELATION TO CHRONOSTRATIGRAPHY

The American Code of Stratigraphic Nomenclature, unlike many schemes of stratigraphical classification, provides explicitly for the subdivision of Quaternary events on the basis of geologic-climate units. Quaternary deposits, which record geological events, may be subdivided, however, on bases additional to those of climate. One of these bases is chronostratigraphy, within the hierarchy of which the lowest-ranking division is the chronozone (Hedberg, 1976, 69). As discussed elsewhere, within a limited geographical region, such as the British Isles, the equation of two divisions, one based on climate, the other chronostratigraphical in character, is valid and useful (Jardine, 1972, 49). Thus, the Loch Lomond Stadial and a corresponding (unnamed, as yet) chronozone ' ... may be equated, provided it is recognised that the time interval of the Stade is related to an actual geological event (a secondary advance of ice) within the area concerned, whereas the corresponding time interval of the ... Chronozone, being chosen arbitrarily by stratigraphers (like all chronostratigraphical divisions), is selected so as to equal the time interval of the Stade' (Jardine, 1972, 49).

A chronozone defined in this way would be of the second kind described by Hedberg (1976,69-70) in the International Stratigraphic Guide, because the known time-span of such a chronozone would vary as information concerning the duration of the Loch Lomond Stadial increased. Also, following the rules given by Hedberg (1976, 70), and keeping in mind the necessity for the Loch Lomond Readvance to be defined on the basis of lithostratigraphy in the near future, the chronozone concerned would require to be designated formally 'the chronozone of the Loch Lomond Formation'. Perhaps for brevity the term 'Loch Lomond Chronozone' might be used informally.

The time-span of the Stadial is determined largely, but not necessarily entirely, by the duration of the Loch Lomond Readvance. Because the Readvance probably was not synchronous throughout the area where the sedimentary evidence of its existence has been traced in Britain, the time-span of the Stadial, and therefore of the equivalent chronozone, cannot be determined at any individual location where the Readvance is well defined. One method of defining the time-span of the Stadial and the equivalent chronozone would be to combine the evidence from the location where the earliest traces of the Readvance occur with that from the location where the latest traces of the Readvance are to be found (Fig. 1).

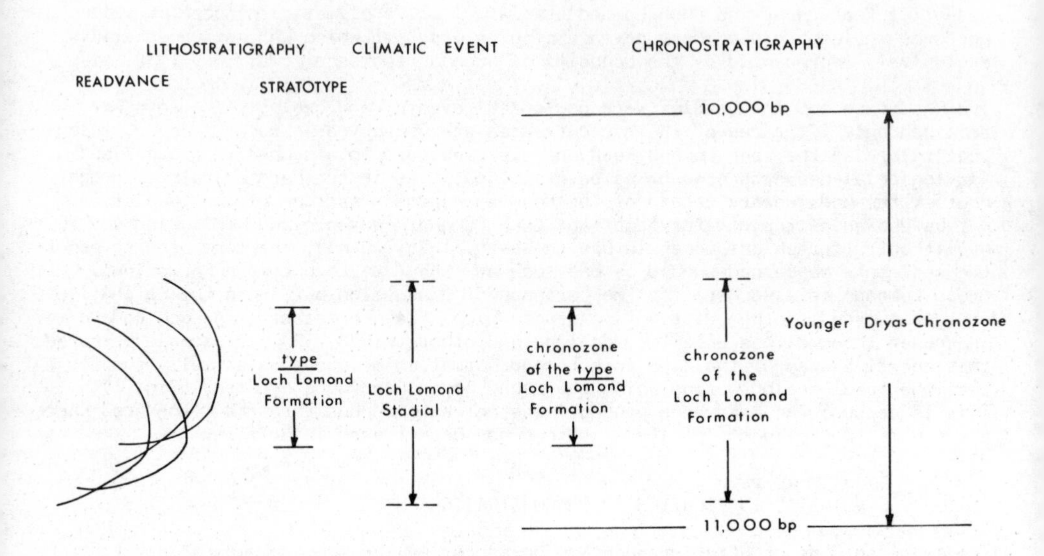

Fig. 1. Diagrammatic representation of the Loch Lomond Readvance, Loch Lomond Stadial, their chronostratigraphical correlatives and the 'Younger Dryas Chronozone' of Mangerud and others (1974). On the left four possible (time) ranges of the Readvance (in different parts of Britain) are shown. One method of defining the (currently-known) time-span of the Stadial, by combining evidence from the location where the earliest traces of the Readvance occur with that from the location where the latest traces of the Readvance occur, is represented diagrammatically. The time-spans of the 'chronozone of the *type* Loch Lomond Formation' and the 'chronozone of the Loch Lomond Formation', the former being determined by the span of the stratotype of the Loch Lomond Formation, the latter being chosen to correspond with the span of the Loch Lomond Stadial, are represented by horizontal lines. The horizontal broken lines defining the Loch Lomond Stadial and the chronozone of the Loch Lomond Formation indicate that the time-span of these divisions is liable to change as research progresses. The horizontal full lines defining the *type* Loch Lomond Formation and the chronozone of the *type* Loch Lomond Formation indicate that the time-span of these divisions will remain unaltered unless and until a new stratotype for the Loch Lomond Formation should be chosen.

COMPARISON WITH CHRONOSTRATIGRAPHICAL DIVISIONS PROPOSED BY OTHER AUTHORS

'The chronozone of the Loch Lomond Formation' considered above is a different kind of stratigraphical division from that proposed for Norden by Mangerud and others (1974) for approximately the same period. The 'Younger Dryas Chronozone' of Mangerud and others is broadly (but not precisely) the same as the first kind of chronozone described by Hedberg (1976, 69), taking its name from a previously-recognised biozone, 'the Younger Dryas period or pollen zone III of Jessen and DR3 of Nilsson' (Mangerud and others, 1974, 118). The 'Younger Dryas Chronozone' of Mangerud and others differs from the first kind of chronozone described by Hedberg in that its span is defined on the basis of a time interval measured in radiocarbon years (11,000 to 10,000 B.P.) rather than on the span of the stratotype of a lithostratigraphical unit (Hedberg, 1976, 89).

A problem related to the difference in nature between chronozones on the one hand and stadials and interstadials on the other hand was illustrated by Pennington (1975, 1977a) and Gray and Lowe (1977). The problem arises because there are indications that termination of (Loch Lomond) stadial conditions occurred in the Lake District of England and the NW Highlands of Scotland c. 10,500 B.P. (Pennington, 1975, 170) or c. 10,400 B.P. in the NW Highlands of Scotland (Pennington, 1977a, 141) and it is 'awkward' that the later part of the 'Younger Dryas Chronozone' of Mangerud and others (1974) should be markedly different climatically from the earlier part (cf. Gray and Lowe, 1977, 179-180). As seen in Figure 1, and as implied in the discussion above, no such problem arises where a chronozone is selected so that its time-span equals that of a stadial or interstadial. In the case considered here, 'the chronozone of the Loch Lomond Formation' is the time-equivalent of the Loch Lomond Stadial, and an additional chronozone may be inserted between 'the chronozone of the Loch Lomond Formation' and the beginning of the Holocene epoch, without the rules of chronostratigraphy being violated.

SELECTION OF STRATOTYPES

One method of defining the time-span of the Loch Lomond Stadial and the equivalent chronozone was suggested above (p. 169). Another, and equally valid method, would be to define the beginning of the Stadial and the equivalent chronozone 'at the locality where the earliest indications of climatic changes are to be found. This need not be the locality at which the earliest manifestations of a glacial readvance are found.... Indeed, the time limits of the stade may be defined on any reliable evidence of climatic changes, glacial deposits being only one type of such evidence. Hence the beginning of the stade may be defined, and is best defined, at a locality ... which is outside the area which has been glacially affected by the readvance. Similarly, the end of the stade (or rather the beginning of the succeeding interstade, because in stratigraphic practice it is the base of any unit which should be defined ...) is best defined' at a locality outside the area that has been glacially affected (Jardine, 1972, 51).

Applying these principles to the given case, it may be seen that (paradoxically) the Loch Lomond Stadial and the equivalent 'chronozone of the Loch Lomond Formation' are best defined outside the limits of the Loch Lomond Readvance. The most suitable site (or sites if Stadial and equivalent chronozone require separate type locations) for the stratotype(s) for the Stadial and the equivalent chronozone will be in organic deposits that bear indications of climatic changes related to the onset and termination of the Loch Lomond Readvance. Such a site or sites may be chosen at any suitable location(s) in northern Britain where a continuous sequence of organic deposits covers the duration of the onset, climax and termination of climatic conditions that promoted the Loch Lomond Readvance. Possible sites would be those studied by Pennington in the Lake District (1975, 1977b) or in the NW Highlands of Scotland (1977a), but it is suggested that it would be preferable if a site much closer to the limit of the Loch Lomond glacier were to be chosen, partly because stratigraphical divisions related to the 'Loch Lomond' event are being defined and partly because the Loch Lomond glacier was one of the largest valley glaciers involved in the readvance of ice (Sissons 1979, 199). A site that appears to satisfy these last requirements is that at Muir Park Reservoir near Drymen (Vasari and Vasari, 1968, 35-38; Vasari 1977, 155-156). It is suggested that further investigation of this site be made with a view to the site being considered as the stratotype of the Loch Lomond Stadial and/or the equivalent 'chronozone of the Loch Lomond Formation'.

Selection of a stratotype for the Loch Lomond Readvance presents different problems from those involved in choosing a stratotype for the Loch Lomond Stadial and the equivalent chronozone. As mentioned elsewhere (Jardine, 1972, 51-52), no rules are laid down specifically for the naming and field definition of glacial readvances, but the rules of lithostratigraphy may be applied to the inorganic sediments that are deposited by advancing ice. On this basis, 'the deposits of ... [the Loch Lomond] readvance should be given a name of a geographical locality where the deposits are typically developed, well exposed and readily identifiable, and the locality concerned should be designated the type locality of ... [the] readvance' (Jardine, 1972, 52; *cf*. Hedberg, 1976, 37 & 40).

No formal proposal has been advanced, as yet, for the naming of the sediments laid down by the ice of the Loch Lomond Readvance or for selection of a stratotype for these deposits. The term 'Loch Lomond Formation', therefore, is used here in the meantime in reference to the deposits laid down by the glaciers of the Loch Lomond Readvance. It is possible that the apellation 'Loch Lomond' (well-known internationally) will remain acceptable as the necessary geographical designation of the Formation, provided a stratotype is selected on, or close to, the (southern) shore of Loch Lomond but, as mentioned above, when a stratotype is chosen eventually, it may be necessary to apply another geographical name to the Formation. When selectio of a stratotype has been made, it follows that the span of 'the chronozone of the *type* Loch Lomond Formation' will be definable by the span of the relevant deposits at the selected site. This chronozone will be of the first kind defined by Hedberg (1976, 89-90).

ACKNOWLEDGEMENTS

Professor T. Neville George and Dr J.D. Lawson read the text and made helpful suggestions for its improvement.

REFERENCES

American Commission on Stratigraphic Nomenclature. (1961). Code of Stratigraphic Nomenclature. *Bull. Am. Ass. Petrol. Geol.*, 45, 645-665.

Coope, G.R. (1977). Fossil coleopteran assemblages as sensitive indicators of climatic changes during the Devensian (Last) cold stage. *Phil. Trans. R. Soc.*, B. 280, 313-340.

Coope, G.R. and W. Pennington (1977). The Windermere Interstadial of the Late Devensian. *Phil. Trans. R. Soc.*, B. 280, 337-339.

Gray, J.M. and J.J. Lowe (Eds). (1977). *Studies in the Scottish Lateglacial Environment.* Pergamon Press, Oxford. 197 pp.

Hedberg, H.D. (1976). *International Stratigraphic Guide; A Guide to Stratigraphic Classification, Terminology, and Procedure.* John Wiley, New York. 200 pp.

Holland, C.H. and others (1978). A guide to stratigraphical procedure. *Geol. Soc. Lond. Spec. Rep.* No. 10. 18 pp.

Jardine, W.G. (1972). Glacial readvances in the context of Quaternary classification. *Proc. Int. geol. Congr.*, 24 (12), 48-54.

Mangerud, J., S.T. Andersen, B.E. Berglund and J.J. Donner (1974). Quaternary stratigraphy of Norden, a proposal for terminology and classification. *Boreas*, 3, 109-128.

Pennington, W. (1975). A chronostratigraphic comparison of Late-Weichselian and Late-Devensian subdivisions, illustrated by two radiocarbon-dated profiles from western Britain. *Boreas*, 4, 157-171.

Pennington, W. (1977a). Lake Sediments and the Lateglacial Environment in Norther Scotland. In J.M. Gray and J.J. Lowe (Eds), *Studies in the Scottish Lateglacial Environment*, Pergamon Press, Oxford. pp. 119-141.

Pennington, W. (1977b). The Late Devensian flora and vegetation of Britain. *Phil. Trans. R. Soc.*, B. 280, 247-271.
Sissons, J.B. (1977). The Loch Lomond Readvance in the Northern Mainland of Scotland. In J.M. Gray and J.J. Lowe (Eds), *Studies in the Scottish Lateglacial Environment*, Pergamon Press, Oxford. pp. 45-59.
Sissons, J.B. (1979). The Loch Lomond Stadial in the British Isles. *Nature, Lond.*, 280, 199-203.
Vasari, Y. (1977). Radiocarbon Dating of the Lateglacial and Early Flandrian Vegetational Succession in the Scottish Highlands and the Isle of Skye. In J.M. Gray and J.J. Lowe (Eds), *Studies in the Scottish Lateglacial Environment*, Pergamon Press, Oxford. pp. 143-162.
Vasari, Y. and A. Vasari (1968). Late- and Post-glacial macrophytic vegetation in the lochs of northern Scotland. *Acta bot. fenn.*, 80, 1-120.

Dr W.G. Jardine,
Department of Geology,
The University,
Glasgow G12 8QQ.

16. Ice-Damned Lakes in Glen Roy and Vicinity: A Summary

J. B. Sissons

INTRODUCTION

Although there was much controversy last century on the ice-dammed lakes of Glen Roy and vicinity, the views of Jamieson (1863, 1892) have long been generally accepted and have been repeatedly summarised. Geomorphic research in the area during the present century has been negligible. The writer recently attempted a partial rectification of this neglect. Owing to editorial resistance to long papers the results have been published in various journals (Sissons, 1978, 1979a, 1979b, 1979c, in press) and a coherent summary is needed. This contribution seeks to provide such a summary. The following pages do not contain detailed evidence or comprehensive arguments for the proposed interpretations, while the maps and diagrams are generalised. The reader requiring details should consult the papers cited above.

THE LOCH LOMOND ADVANCE

The ice that dammed up lakes in glens Gloy, Roy and Spean accumulated during the Loch Lomond Stadial, approximately equivalent to the Younger Dryas of Scandinavia. Major glaciers flowed eastwards into the Great Glen where they coalesced, completely occupying the glen southwards from the southern end of Loch Ness (Fig. 1). The ice crossed the Great Glen in the Spean Bridge area (Wilson, 1900; Peacock, 1970) and flowed up glens Gloy and Roy. In the Spean valley the ice of western derivation merged with glaciers nourished south of the Ben Nevis range. The most important of these was the Treig glacier, which spread out as a large piedmont tongue (Fig. 1). The ice that occupied Glen Spean received only a modest contribution from the northern slopes of the Ben Nevis range and this only in the west: in the east some corries failed to nourish glaciers. Only a few glaciers, mostly very small, developed in the high ground east of Glen Roy.

Such differences in glacier size primarily reflect differences in former snowfall which, as Jamieson (1892) emphasised, were a major factor in the very existence of the ice-dammed lakes. Some 50 km SW of the area shown in Fig. 1 equilibrium firn lines for the stadial glaciers were as low as 300 m, yet in the NE part of Fig. 1 firn lines averaged over 800 m (Sissons, 1980). Such values imply limited snowfall in the NE and consequent intrusion of glaciers from the south and west where snowfall was heavy.

In many parts of the area illustrated in Fig. 1 the limit of the Loch Lomond Advance is accurately defined by landforms, although in other parts morphological evidence is lacking and the limit has been interpolated. The extent of the Treig piedmont tongue is recorded by an end moraine that is almost continuous for 15 km. Several end moraines occur on the northern flanks of the Ben Nevis range, the broadest having a width as great as 250 m. A small end moraine that marks the ice limit in Glen Roy is succeeded immediately southwards by a massive drift barrier, which

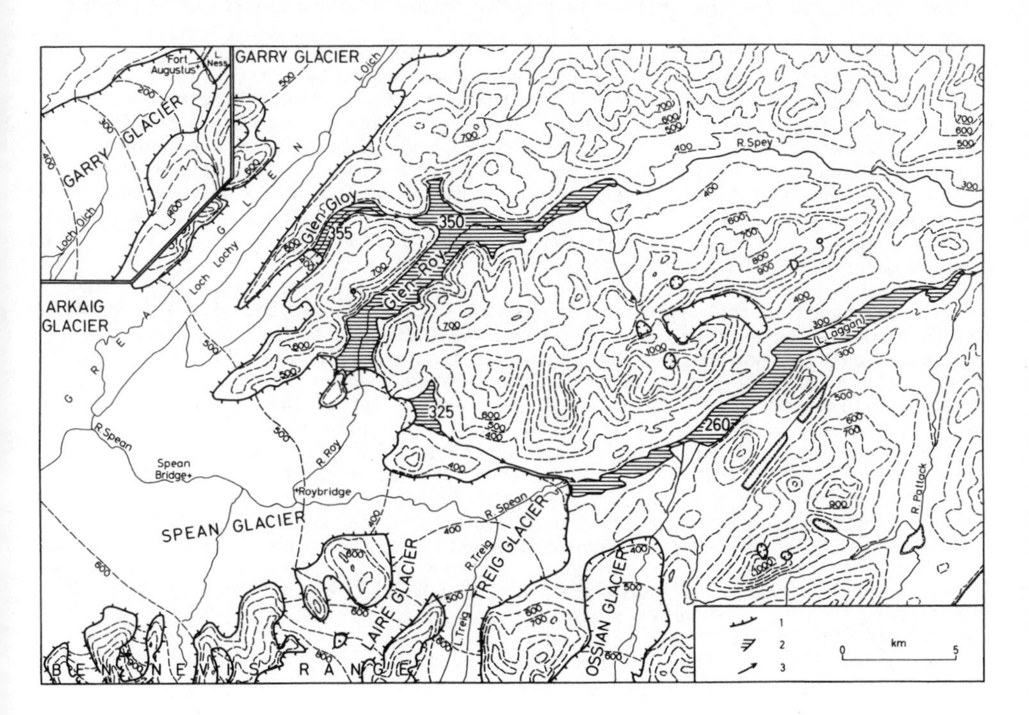

Fig. 1. Maximal extent of the Loch Lomond Advance in Glen Roy and the surrounding area (1). Contemporaneous ice-dammed lakes (2) and their overflow routes (3) are shown. Contours are at 300, 600 and 900 m.

probably attains 80 m in thickness and is composed of glacial till, fluvioglacial sand and gravel, and fine lake sediments. In Glen Gloy, however, the morphological evidence is poor and the former ice limit is tentatively placed in the vicinity of a few mounds of drift.

THE PARALLEL ROADS

As Agassiz (1840) first recognised, the parallel roads are the shorelines of former ice-dammed lakes. Exceptions are a few short stretches of road that are kame terraces, deposited by streams flowing between glacier ice and a valley side. There is one obvious road in Glen Gloy (355 m) and three in Glen Roy (260, 325 and 350 m), the lowest Roy road being continued along the slopes of Glen Spean. Each of these roads corresponds approximately in altitude with the floor of a col. Of various much fainter roads mentioned in the literature two in Glen Roy (located between the 325 and 350 m roads) certainly exist but do not accord with cols.

In some places the roads are backed by an obvious fossil cliff and at a few places a road can be seen to truncate bedrock. Numerous measurements show that over-steepening of the slope behind each road, as well as of that in front, is wide-spread, implying that the roads are largely the result of erosion at the back and deposition at the front. Erosion is attributed to severe frost action in the critical environment of the slightly varying lake level and to wave action, the former accounting for the dominantly angular nature of the deposits of the roads.

Deltas sometimes occur where streams entered the ice-dammed lakes, a particularly large one, which was supplied with glacial outwash, being located by Loch Laggan. Where obvious deltas are not present roads are often wider near streams and the deposits are less angular than normal. In part of Glen Roy measurements of road widths indicate that wave action carried more stream debris along the roads up the glen than down it, pointing to the importance of SW winds blowing up the glen. There is a strong statistical correlation in the area as a whole between road volume (a measure of the amount of material eroded or deposited to form a road) and fetch from SW, again leading to this conclusion, although the funnelling of winds by deep, narrow glens has to be borne in mind.

It is very probable that different lengths of time were available for the formation of the roads: for example, the 325 m road was formed while the ice advanced and retreated only 1.5 km whereas the comparable figure for the 260 m road is 8 km. Hence different average dimensions for the various roads might be anticipated. Yet average volumes for the 350, 325 and 260 m roads differ little, being respectively 9.1, 9.0 and 8.6 m^3 (per metre length of road). The time available for the formation of any one road varied along part of its length: for example, outside the Loch Lomond Advance limit in Glen Roy the 350 m road had a longer time in which to be formed than within the limit, the time available within the limit diminishing down-valley to zero at the position occupied by the ice-dam when the 350 m lake level ceased to exist. Better developed roads might therefore be expected outside the former ice limits than within them. Yet the average widths of the four principal roads in glens Roy, Spean and Gloy (recalculated to eliminate the influence of hillslope angle, with which road width is highly correlated) range between 10.0 and 11.3 m outside the former ice limits and between 9.4 and 11.7 m within them. These rather surprising results appear explicable only if each road was at first formed very quickly and, thereafter, developed very slowly.

SEQUENCE OF PRINCIPAL LAKE LEVELS IN GLEN ROY

As glacier ice advanced eastwards across the Great Glen into Glen Spean a lake at 260 m became established in glens Roy and Spean, its overflow being at the eastern end of present Loch Laggan. Later, as the ice advanced up Glen Roy, the 325 m lake level came into existence, its outlet leading down to a much reduced 260 m lake in the Spean-Laggan valley. With further glacier advance a small 325 m lake remained in a valley tributary to Glen Roy (Fig. 1) while the Glen Roy lake level rose to 350 m with overflow at the head of the glen. During ice retreat this sequence was reversed.

INITIAL DRAINAGE OF THE 260 m SPEAN/ROY LAKE

The maximal length achieved by the 260 m Spean/Roy lake during glacier retreat, as indicated by the western termination of the parallel road on each side of the Spean valley, was 35 km, lake area being 73 km^2. The maximal depth of the lake, adjacent to the 7 km long ice-dam, was 200 m and lake volume was 5 km^3. This volume is twice that of Loch Lomond, the second largest Scottish lake, and two-thirds that of Loch Ness, the largest. This great volume of water cannot have drained away

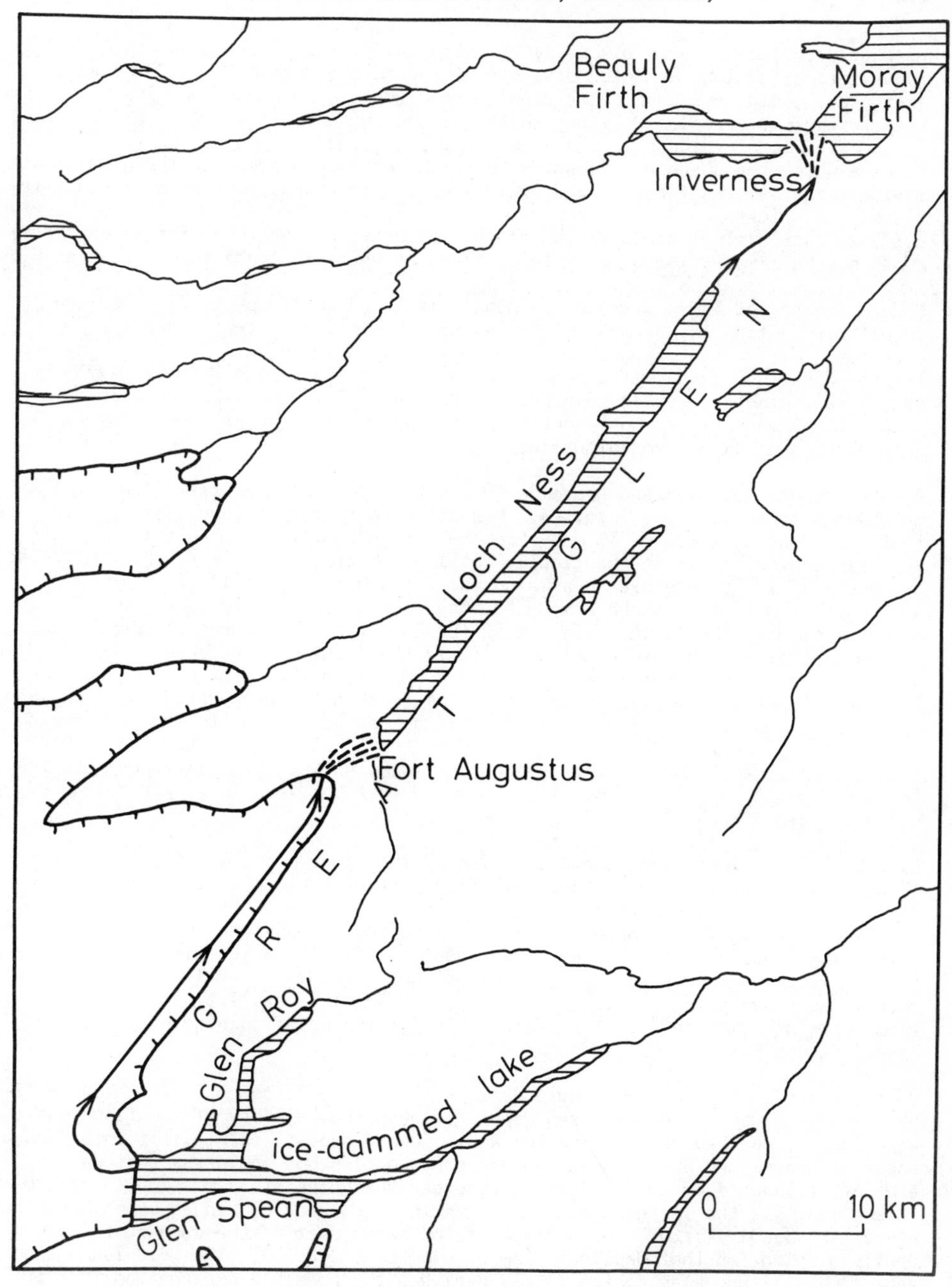

Fig. 2 Extent of the Spean/Roy ice-dammed lake immediately before initial drainage, contemporaneous glacier limits, route of the major jökulhlaup, and jökulhlaup deposits (broken lines).

gradually or escaped supraglacially or englacially, thus leaving no trace of its route for, by analogy with the drainage of modern ice-dammed lakes, drainage, once initiated, would have been catastrophic and a route over or through ice would have been very rapidly replaced by one on the ground surface. The only available route for initial lake drainage is the Spean gorge, several km long and up to 30 m deep, through which the Spean now flows NW to the Great Glen across its former drainage divide (Fig. 3), its former broad valley leading SW being occupied by a tiny misfit stream. That the Spean gorge was not simply excavated subaerially as the river cut down during final deglaciation of its vicinity, but had been formed earlier, is shown by the composition of a suite of terraces where it leads into the Great Glen.

Modern ice-dammed lakes often drain subglacially at the deepest point and are completely emptied in, at most, a few days. The Icelandic term "jökulhlaup" is used for the resultant floods. Such catastrophic subglacial drainage is envisaged for the Spean/Roy lake. If an equation derived by Clague and Mathews (1973) is valid, peak discharge would have been c. 22,500 m^3s^{-1}. The 7-km-long ice-dam across the Spean valley, no longer supported by the lake water, would have collapsed into a chaos of blocks and pinnacles.

A major deposit of sand and gravel occupies the floor of the Great Glen for 3 km southwards from Loch Ness (Fig. 2). Dissected to form a terrace, its surface rises gently from an altitude of 31 m by Loch Ness to slightly above 40 m towards its southern end, where it is chaotically kettled. Certain aspects of the feature are anomalous if it is interpreted as a normal outwash terrace: (i) it begins well within the limit of the Loch Lomond Advance whereas outwash terraces normally commence at or near the limit; (ii) it is far larger than features of comparable age in the central and eastern Highlands; (iii) the large area of chaotic kettling is unique for such features; (iv) the terrace ends 15 m above the present surface of Loch Ness. These anomalies disappear if the feature is interpreted as a jökulhlaup deposit. In particular, the terminal altitude can be related to the very rapid raising of the surface level of Loch Ness by the jökulhlaup: since the highest raised shoreline at the southern end of the loch, backed by a cliff cut into the terrace deposit, is at 22.5 m, it is inferred that the loch level was raised by 8.5 m (31 minus 22.5 m).

Loch Ness is drained along the Great Glen by the River Ness, which enters the sea (Beauly Firth) at Inverness (Fig. 2). This part of the Great Glen contains abundant drift deposits, especially sand and gravel laid down during ice-sheet decay. The River Ness follows a broad valley cut into these deposits, the eastern side of this valley usually being a steep, prominent bluff. It is believed that this bluff and the width of the valley are a result of the jökulhlaup. A less capacious valley is envisaged as having existed previously, its initial inability to cope with the flood waters having been partly responsible for the brief raising of the surface level of Loch Ness.

On the above interpretation debris derived from the enlargement of the Ness valley by the jökulhlaup would have been deposited when the velocity of the flood waters was suddenly checked on entering the sea. Such a deposit exists. It comprises a sheet of gravel and sand (including abundant cobbles and boulders) that has an area of at least 7 km^2 and a maximal recorded thickness of 39 m. The layer underlies almost all the low ground of northern Inverness and continues beneath the waters of the Beauly Firth. The surface of the major part of the deposit forms a gently sloping fan that declines from c. 6 m to c. 2 m O.D. The gravel and sand lacks any silt and clay content: this excludes the possible explanation that it accumulated in successive increments through intermittent floods of the River Ness, for fine estuarine sediments (that are accumulating along the Inverness coastline today, as in the past) would inevitably have been incorporated. Thus the absence of silt and clay implies accumulation of the gravel and sand in a very short time.

Lines of boreholes across the narrows at the entrance to the Beauly Firth show that the gravel and sand layer continues across the firth to the northern shore, where its surface is almost at sea-level. Yet the gravel surface descends to -12 m beneath the channel that exists near the axis of the narrows. Since a considerable channel must always have existed hereabouts, if only to evacuate the run-off from the 3000 km^2 of the Highlands that drain to the Beauly Firth, the gravel and sand layer could only have been built up to sea-level on the northern side of the channel if it were produced by a catastrophic event. It is therefore interpreted as having been deposited by the jökulhlaup. It may also be noted that the altitude of the gravel fan accords approximately with the sea-level in the Beauly Firth at the time the fan was deposited (this having been determined from independent evidence).

INITIAL DRAINAGE OF THE GLOY LAKE

At the entrance to Glen Gloy the River Gloy leaves its former valley, turning abruptly through 100° to enter a rock gorge that leads northwards towards Loch Lochy. There are no meltwater channels or belts of water-swept rock on the adjacent hillslopes; hence it is inferred that the gorge is the subglacial route by which the Gloy ice-dammed lake was abruptly drained.

THE LAKE THAT IMMEDIATELY SUCCEEDED THE 260 m LAKE

When modern ice-dammed lakes have emptied the tunnel normally closes and re-filling begins. Since the escape route for the Spean/Roy lake passed beneath some 300 m of glacier ice such tunnel closure is inferred. If precipitation in the Spean/Roy catchment was comparable to that of today 3 years would have been required to restore the lake to its original level. Such total restoration was not possible, however, since the original ice-dam had collapsed and deglaciation was in progress. Furthermore, the drainage of modern ice-dammed lakes often becomes an annual event, normally occurring in summer. Hence it is suspected that subsequent jökulhlaups were much smaller than the one that drained the 260 m lake. It appears necessary to postulate these additional jökulhlaups since, at the time end moraine A (Fig. 3) was formed, ground immediately to the east, at altitudes as low as 60 m, would have been largely or entirely ice-free, whereas at the time the 260 m lake was drained the glacier surface in the same area must have been above 260 m: the removal of such a thickness of ice would have taken a considerable time.

Along the floor and lower slopes of the Spean valley is an extensive suite of fluvial terraces that accumulated very rapidly (see below). Hence, if the lake that succeeded the 260 m lake stood at constant levels for even limited periods, associated terraces should exist. The absence of such terraces implies a continually varying lake level (except probably during winter freeze-up when inflow would have been negligible or would have ceased), lowering by jökulhlaup alternating with raising by inflow.

LAKE DRAINAGE BY THE LUNDY CHANNEL

The main (western) part of the Lundy channel (Fig. 3) is a conspicuous feature a kilometre long, part of its southern side being a cliff of metamorphosed limestone up to 30 m high. The channel floor slopes down westwards. The eastern part of the channel has been partly modified by later stream action. Kames occur immediately north of the western part of the channel and the kame sands and gravels mantle its northern slope. A large section at the base of this slope showed the sands and gravels to be clearly bedded and *in situ*. Thus the channel was cut subglacially (or, at least, cleaned out by subglacial meltwaters) shortly before the kames were

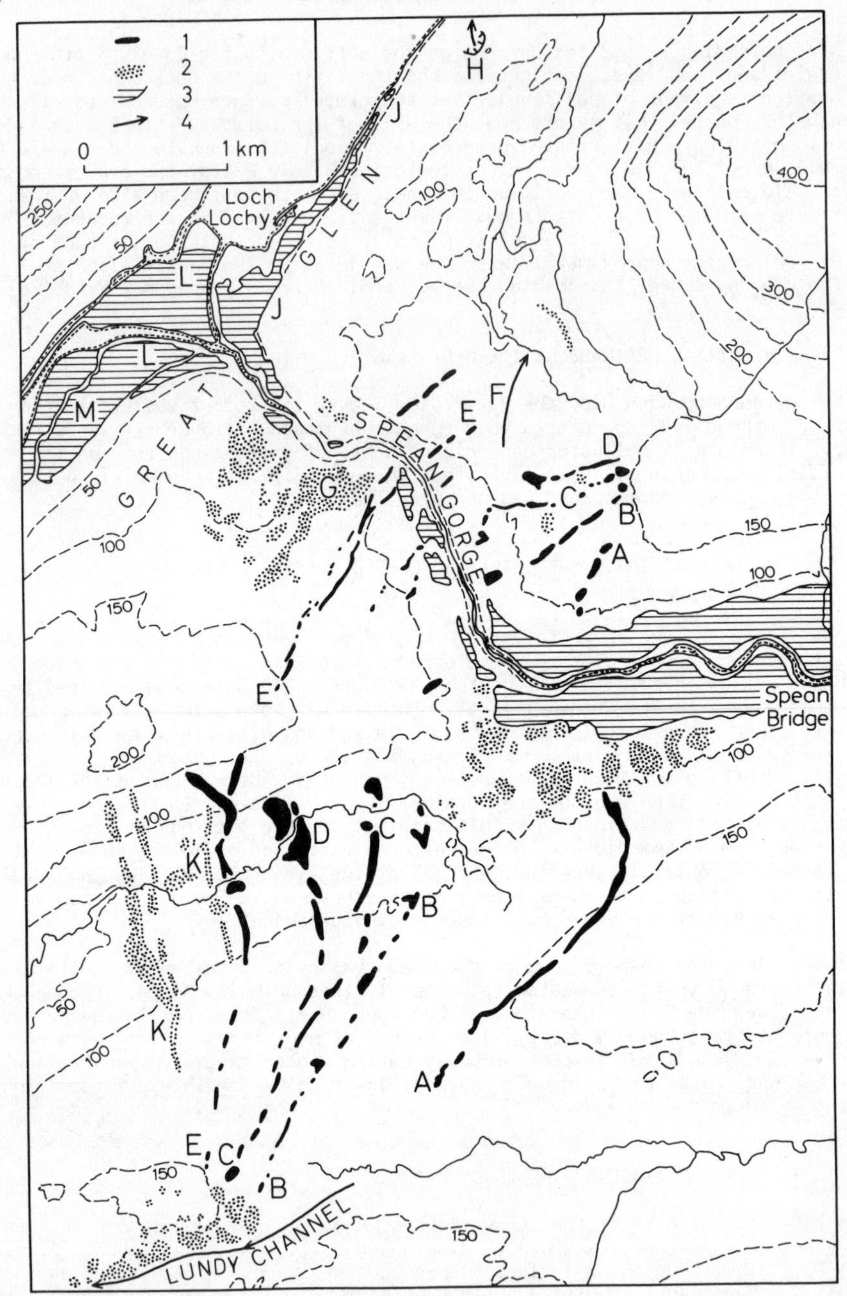

Fig. 3. Some landforms of the area around Spean Bridge. 1, End moraines. 2, Kames. 3, Terrace belt, including present floodplain. 4, Meltwater channels. Contour interval 50 m. For explanation of letters see text.

deposited, glacier ice collapsing into the channel after meltwater flow ceased. Lake drainage through the channel is therefore implied.

Five end moraines 1-6 m high and up to 5.5 km long terminate near or point towards the Lundy channel, suggesting a relationship between them and the channel. End moraine formation could not have occurred when a high ice front, subject to collapse on lake drainage, existed. The sudden appearance of end moraines can be attributed to the change in glaciological conditions caused by the replacement of the Spean gorge by the Lundy channel as the jökulhlaup route: the Spean lake could no longer drain completely, the entrance to the Lundy channel being 70-80 m higher than that to the Spean gorge. The sudden cessation of end moraine formation resulted from a further change in glaciological conditions when the Lundy gorge ceased to function as the lake outlet and considerable areas of dead ice developed (see below).

THE FINAL LAKE LEVELS IN THE SPEAN AND ROY VALLEYS

The rivers Treig, Spean and lower Roy are bordered by numerous terraces almost all of which are very clear features even though many are peat-covered. All the terraces have been accurately levelled at closely spaced intervals, the peat having been bored through where present.

The terrace sequence in the Roy valley (Fig. 4) commences some 60 m above present river level at the major drift barrier that was deposited just within the limit of the Loch Lomond Advance. This means that, after the major Spean/Roy lake ceased to exist, a small lake was impounded in the mid-part of Glen Roy by the drift barrier.

The terraces in Glen Roy, which have surfaces composed of sand and gravel, decline in altitude quite rapidly down valley to Roybridge, where they merge with sand and gravel terraces sloping with considerable gradient down the Spean valley. Westwards from the vicinity of Roybridge, however, the higher terraces become very gently inclined or horizontal and are composed almost entirely of sand resting on fine lake sediments. This sand was deposited in the final Spean ice-dammed lake, which occupied several levels before being wholly infilled with sand.

At H in Fig. 3 is an isolated abandoned waterfall site cut in bedrock to a depth of 30 m, from which a channel 80-100 m wide leads NW and north. Since the channel is a clear feature despite being cut in drift, formation very late in the deglaciation of the area is implied for otherwise it would have been modified or destroyed by glacier movement. The waterfall site is interpreted as having been formed by the final Spean jökulhlaup, at which time the Lundy channel ceased to operate. The jökulhlaup ended when the Spean lake had fallen to the level of the col at F in Fig. 3, this causing the lake level to stabilise.

Lake overflow through col F is demonstrated by a channel cut into it, the highest point of the channel floor being 111.5 m. Allowing for water depth in the channel, this altitude accords with the terminal altitude of 112-115 m for the highest Spean terrace (Fig. 4), which occurs on both the north and south sides of the valley near Roybridge. Just west of moraine E (at G, Fig. 3) there was formerly a very large, flat-topped kame bordered on all but its eastern side by a steep ice-contact slope. The surface of the kame was at 113 m. The bedding revealed by sand and gravel working that destroyed the feature showed it to be a delta deposited by meltwaters from the wasting glacier ice. Thus an ice-dammed lake with its surface at c. 113 m existed in the Spean valley. The kame-delta and a large area of kames and kettles that adjoins it demonstrate ice stagnation immediately following the formation by active ice of moraine E. Glacier stagnation is similarly demonstrated by sand and gravel crevasse fillings immediately behind moraine E in the former valley of the

Spean (K, Fig. 3) and by the kame area by the Lundy channel.

The terminal altitudes of gently sloping and horizontal terraces of sand in the Spean valley show that the lake level at c. 113 m was succeeded by others at 99, 96.5 and 90.5 m (Fig. 4). The lake meanwhile shrank in size westwards owing to infilling with the sand until it was completely extinguished (indicated by terraces A and B (Fig. 4), which terminate by the Spean gorge). The infilling was very rapid for the margin of the dead ice that held up the lake (to the extent that there was a definite margin) meanwhile retreated no more than a kilometre. Outflow from the lake when it stood at the 99 m and lower levels was alongside the Spean gorge (not through the gorge) and then marginally or submarginally to the ice along the side of the Great Glen by Loch Lochy. The passage of the water across the latter area

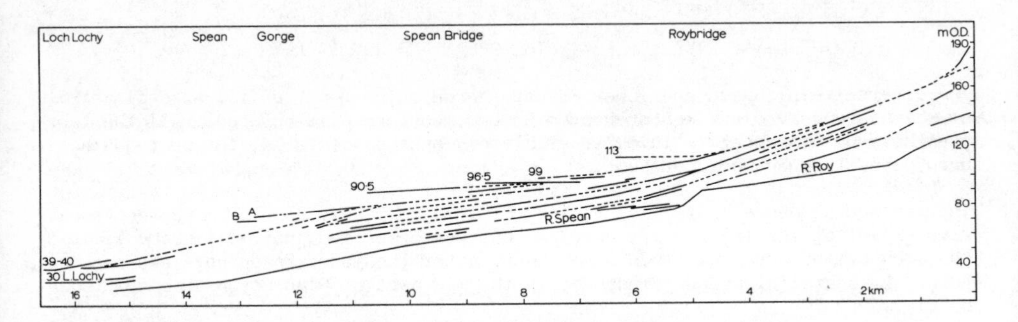

Fig. 4. Height-distance diagram of terraces in lower Glen Roy, part of the Spean valley, and thence alongside Loch Lochy.

is demonstrated by bedrock swept free of drift.

FORMER LEVELS OF LOCH LOCHY AND REMOVAL OF THE ICE DAM

With further decay of the ice the waters of the Spean were able to utilise the Spean gorge, removing any dead ice that remained in it. This in turn resulted in the dissection of the sand infill of the former lake and the production of a series of terraces with much steeper gradients than those related to that lake (Fig. 4). Initially glacier ice still occupied the SW part of the Great Glen, its northern terminus being in Loch Lochy. This is shown by a fluvial fan built out against the ice from the gorge exit, this fan (and a higher feature) being continued alongside Loch Lochy by kame terraces (J, Fig. 3). The terrace terminations show that the surface level of ice-dammed Loch Lochy was 39-40 m (compared with the present artificial c. 30 m or natural c. 26 m), overflow being along the floor of the Great Glen towards Loch Ness.

With further decay glacier ice in the Great Glen became restricted to the part SW of the Spean gorge. A raised shoreline bordering Loch Lochy shows that the loch stood at just below 34 m. A large fan of sand (L, Fig. 3) related to this level of the loch was deposited at the gorge exit. South-westwards the fan merges imperceptibly into a large terrace (M. Fig. 3) that slopes down towards SW, thus indicating flow in this direction. This terrace therefore records the establishment of the present south-westward flow to Loch Linnhe of the waters of the Spean and Loch Lochy, damming by glacier ice having ceased.

REFERENCES

Agassiz, L. (1840). On glaciers, and the evidence of their having once existed in Scotland, Ireland and England. *Proc. geol. Soc. Lond.*, 3 (72), 327-332.

Clague, J.J. and W.H. Mathews (1973). The magnitude of jökulhlaups. *J. Glaciol.*, 12, 501-504.

Jamieson, T.F. (1863). On the parallel roads of Glen Roy, and their place in the history of the glacial period. *Q. Jl geol. Soc. Lond.*, 19, 235-259.

Jamieson, T.F. (1892). Supplementary remarks on Glen Roy. *Q. Jl geol. Soc. Lond.*, 48, 5-28.

Peacock, J.D. (1970). Glacial geology of the Lochy-Spean area. *Bull. geol. Surv. Gt Br.*, 31, 185-198.

Sissons, J.B. (1978). The parallel roads of Glen Roy and adjacent glens, Scotland. *Boreas*, 7, 229-244.

Sissons, J.B. (1979a). Catastrophic lake drainage in Glen Spean and the Great Glen, Scotland. *J. geol. Soc. Lond.*, 136, 215-224.

Sissons, J.B. (1979b). The limit of the Loch Lomond Advance in Glen Roy and vicinity. *Scott. J. Geol.*, 15, 31-42.

Sissons, J.B. (1979c). The later lakes and associated fluvial terraces of Glen Roy, Glen Spean and vicinity. *Trans. Inst. Br. Geogr.*, 4, 12-29.

Sissons, J.B. (1980). Palaeoclimatic inferences from Loch Lomond Advance glaciers. In J.J. Lowe, J.M. Gray and J.E. Robinson (Eds.), *Studies in the Lateglacial of North-West Europe*, Pergamon Press, Oxford. pp. 31-43.

Sissons, J.B. (in press). Lateglacial marine erosion and a jökulhlaup deposit in the Beauly Firth. *Scott. J. Geol.*

Wilson, J.S.G. (1900). In *Summ. Prog. geol. Surv. Gt Br.* for 1899, pp. 158-162.

Dr J.B. Sissons,
Department of Geography,
The University,
High School Yards,
Edinburgh. EH1 1NR

17. Evidence for Late Devensian Landslipping and Late Flandrian Forest Regeneration at Gormire Lake, North Yorkshire

Anne Blackham, Clive Davies and John Flenley

INTRODUCTION

Gormire Lake (SE 503832) lies at the north eastern edge of the Vale of York (or, more particularly, that part of it which is known as the Vale of Mowbray), just below the scarp of the Hambleton Hills at Whitestone Cliff (Fig. 1). This position is strategic in Quaternary terms, for it is precisely on the Late Devensian glacial limit (Penny, 1974). Ice is believed to have banked up against the cliffed escarpment, and the lake lies in what was probably a marginal drainage channel (Kendall and Wroot, 1924).

The lake and its environs are of special interest for at least two reasons. Firstly, how and when did the lake originate? The explanation given by Kendall and Wroot is that the channel became dammed by a landslip from the escarpment. Probably the channel drained from north to south, as do the smaller channels parallel to it, Butterdale (with its man-made fish ponds) and the lesser valley to the west. The dam is probably therefore the col to the south-east of the lake. The basal deposits of the lake should therefore provide a minimum age for the landslip. A maximum age will be provided by the date of deglaciation of the area. The date of the landslip will be of interest because major landslips of this type could be the catastrophic response to a sudden environmental change or to a prolonged climatic phase favouring unstable conditions. Relevant changes within the time scale available would include the disappearance of ice, changes in temperature and variations in precipitation.

A second reason for interest is the fact that some of the woodland surrounding the lake may have some claim to be regarded as ancient. In fact Garbutt Wood on the north-eastern side (Fig. 1) is a Yorkshire Naturalists' Trust nature reserve containing six species which are regarded by Peterken (1974) as 'primary woodland' indicator at least in Lincolnshire. The species in question are *Oxalis acetosella, Chrysosplenium oppositifolium, Lysimachia nemorum, Adoxa moschatellina, Conopodium majus* and *Primula vulgaris*. One of these, *O. acetosella,* was regarded as *confined* to primary woodlands in the Lincolnshire survey. Peterken defines primary woodlands as those which 'occupy sites that have been continuously wooded throughout the historical period and comprise much-modified fragments of the former natural forest cover'. Of course, the primary woodland species (i.e. those which tend to be confined to such sites) will not be the same everywhere. For instance, in the Gormire area *O. acetosella* is not confined to primary woodlands; it occurs also on open moorland under heather. Despite this, the size of the list is impressive. In Lincolnshire Peterken found that the commonest number of primary woodland species present in such a wood was only seven. There is thus a distinct possibility, on floristic grounds,

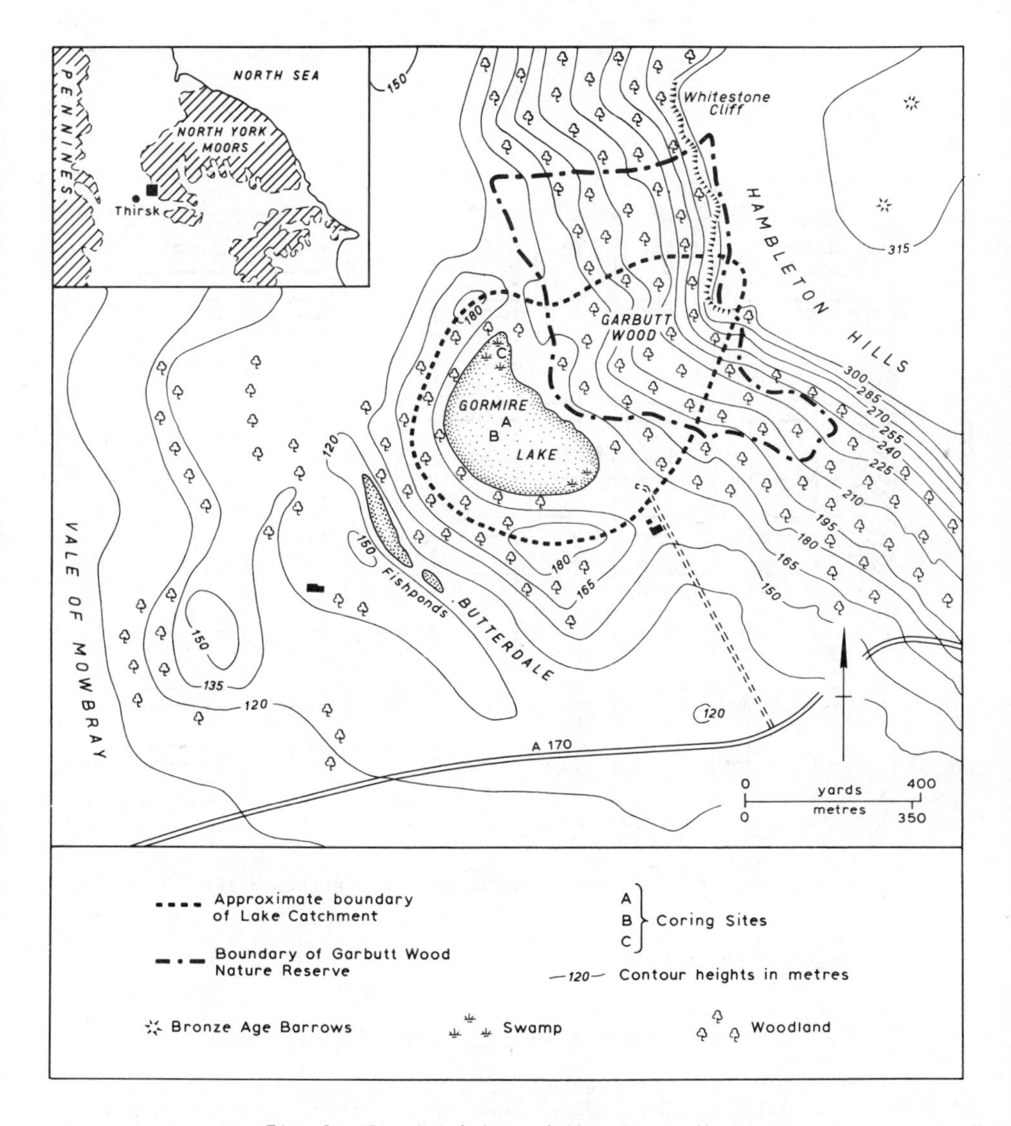

Fig. 1. Gormire Lake and its surroundings.

that Garbutt Wood is ancient. It therefore seemed worthwhile to test this possibility by investigating the vegetational history of the Gormire catchment through pollen analysis of the lake sediment.

METHODS

A brief bathymetric survey showed that the central part of the lake had a fairly flat bottom at 5 - 6 m depth.

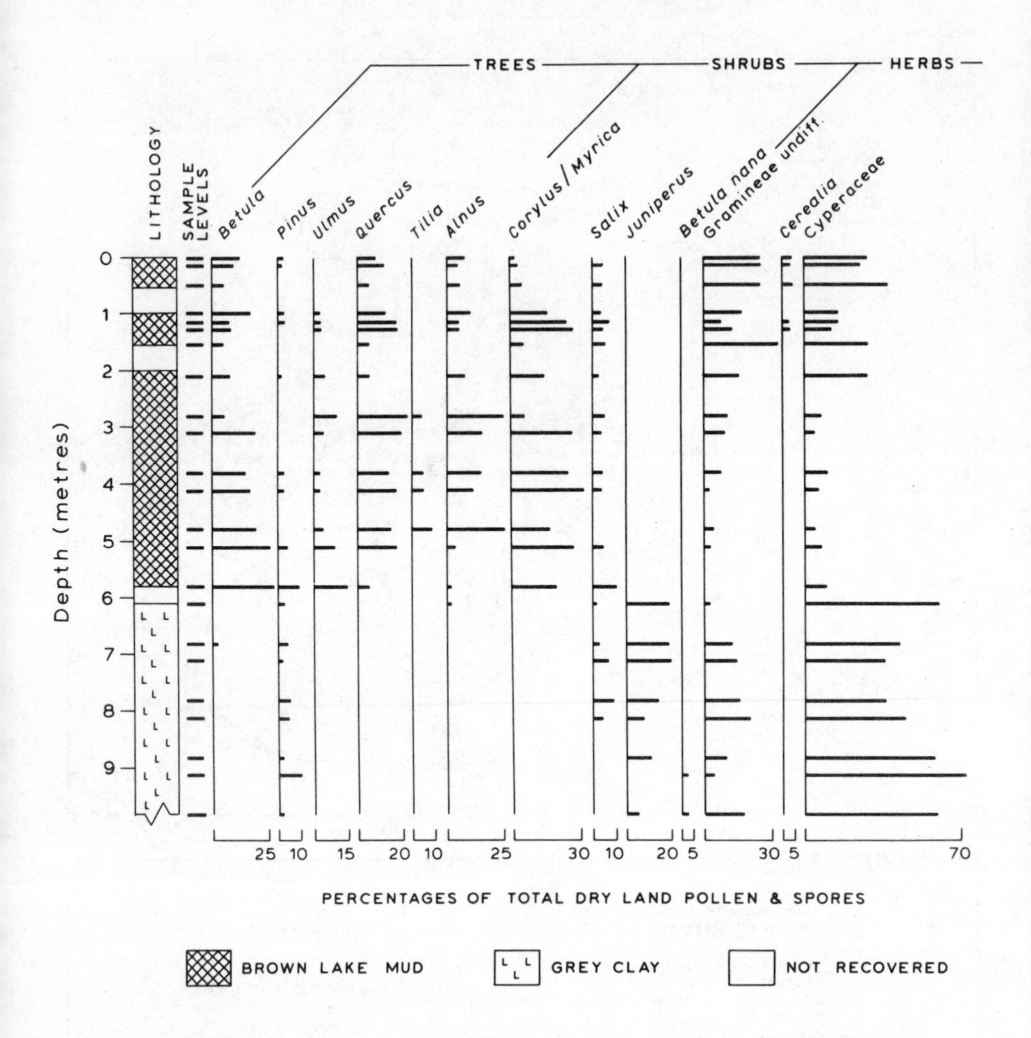

Fig. 2. Stratigraphy and outline pollen analysis of Core A -

The sediments in the lake were investigated at three points (Fig. 1). Core A was obtained with a piston sampler (Walker, 1964) used from a raft, Core B with a Mackereth sampler (Mackereth, 1958) and Core C (in the marginal swamp) with a D-section sampler (Jowsey, 1966). Core A was analysed for pollen and (from 200 cm depth) for plant macrofossils and Core B for pollen and palaeomagnetic susceptibility. Core C was recorded only stratigraphically.

RESULTS

(a) Core A The core was collected below 6.0 m of water and consisted of 10.0 m of sediments obtained in 1 m units. Core recovery was good except at three points where it is thought that relatively hard material plugged the entrance of the core

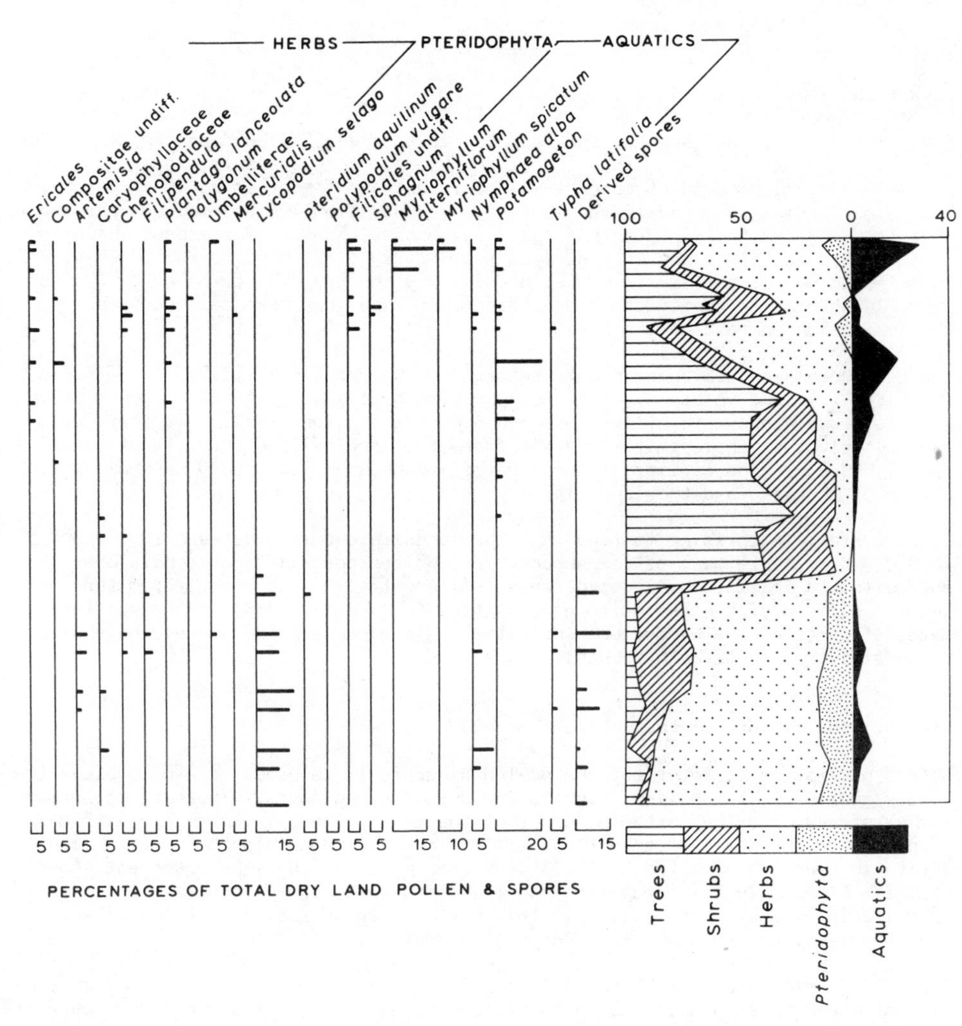

- Analysis by A. Blackham

tube. The stratigraphy and pollen analyses are shown in Fig. 2; the plant macrofossil analyses are listed in Table 1.

(b) Core B The core was continuous and 2.73 m in length. It was collected beneath c. 6 m of water and was believed to be more-or-less complete except that, due to the fluid nature of the top sediment, some of the surface deposit may have been lost. The stratigraphy, pollen analyses and palaeomagnetic analyses are shown in Fig. 3.

(c) Core C The core consisted of half-metre increments collected successively downwards from the surface of the swamp (which was approximately the lake surface level). The stratigraphy is shown in Fig. 4.

DISCUSSION

Age of the lake and the landslipping.

A minimum age for the lake and the landslip is provided by the basal analyses from Core A. Although this was probably not the base of the sediments, it was the deepest material which could be obtained without the use of heavier boring apparatus. Palynology indicates a Late Devensian age. The basal pollen assemblages are probably to be correlated with the RB3 pollen assemblage zone in Holderness which is dated by Beckett (1975 and in press) as from 11,220 ± 220 B.P. to 10,120 ± 180 B.P. The minimum age of the lake and the landslip may therefore be taken as c. 10,000 B.P. An attempt to confirm this by radiocarbon assay has been delayed because of the low carbon content of the clay.

The Late Devensian ice sheet had retreated from Holderness by 13,045 ± 270 B.P. (Beckett, 1975 and in press), from Seamer Carrs by 13,042 ± 140 B.P. (Jones, 1976) and from Lake Windermere by 14,623 ± 360 B.P. (Pennington, 1977). It therefore seems likely that the Gormire area was freed from the grip of the ice between 15,000 and 13,000 B.P. The earlier of these dates may thus be taken as a maximum for the origin of the lake and the landslip.

It is scarcely surprising to find that the landslip occurred between 15,000 and 10,000 B.P. The removal of ice support from Whitestone Cliff could have been sufficient to initiate a landslip there. Alternatively, the change from the continental climate of the Devensian glacial time (e.g. Bell, 1969) to the moister climates of parts of the Windermere Interstadial (e.g. Coope, 1977) could have been responsible. Perhaps both mechanisms operated.

Vegetational history of the lake catchment.

Before attempting to interpret the pollen diagrams it is necessary to consider the likely sources of the pollen which is found in the sediments. There is no permanent inflow stream into the northern corner of the lake. There is therefore little input of pollen to the lake by streamflow. It is, however, possible that pollen might arrive in overland flow (Peck, 1973) from some parts of the lake catchment shown in Fig. 1. Presumably the steep slope on the north-east side of the lake would be the most likely to contribute pollen in this way. There is also no outflow of water from the lake, and thus no export of pollen from it.

The atmospheric component of the pollen rain may be slightly unusual. The escarpment of the Hambleton Hills is characterized by powerful upward convection currents which have made it a popular area for gliding, and gliders frequently pass directly over the lake. Presumably convection cells exist over the lake and/or the adjacent escarpment, especially on warm days when pollen might be released.

The precise effects of all these factors in determining the origins of the pollen rain are difficult to assess. Overland flow could bring pollen only from within the catchment, but is not likely to be a major contributor. Atmospheric pollen entering a lake of this size could well consist to a substantial extent of the trunk-space component (Tauber, 1965, 1967), which may travel for as far as 1 km in the case of light grains. The convection cells may well encourage such travel but they presumably influence chiefly the canopy component. The cells are however a distinctly limited development, restricted to the area of the escarpment, and it seems possible that their effect on the pollen rain is simply to mix and distribute the extra-local component, without bringing in additional regional pollen.

Depth	Macrofossil
200cm	Many leaf fragments, plus monocot remains. *Betula* bud scale.
250cm	Many monocot remains, *Quercus* bud scale, *Hypnum cupressiforme*.
300cm	Many plant fragments, plus monocot remains, leaves with reticulate venation, *Quercus* bud scale, *Salix* bud scale, *Betula* fruit scale, several moss fragments including *Pohlia delicatula* and *Hypnum cupressiforme*, capsule of *Rumex* sp.
350cm	Many plant fragments, plus leaves with reticulate venation, *Phragmites* remains, *Salix* leaf fragments, *Betula* fruit scale, *Hypnum cupressiforme*.
400cm	Many plant fragments, plus monocot leaves, *Salix* bud scale, *Betula* leaf fragment and bark.
450cm	Many plant fragments, leaf remains with reticulate venation, *Betula* leaf fragments and fruit, *Quercus* bud scale, *Phragmites* remains.
500cm	Many plant fragments, monocot remains, *Betula* fruit scale, *Chara* oospores.
550cm	Many fragments of leaves with reticulate venation, *Salix* leaf fragment, *Betula* fruit case and bud scale, *Chara* oospores.
600cm	No macrofossils were recorded from 600cm downwards.

Table 1. Plant macrofossils from Core A.

In general, then, it is reasonable to conclude that the pollen recruited to the lake sediment is derived mainly from the lake catchment and from not far outside it, and that the vegetational history as revealed by pollen analysis is therefore relevant to these areas.

The depth of the lake (c. 6 m) is probably sufficient for some temperature stratification to develop in summer, with consequent lake overturn in autumn. This is likely to lead to redistribution of sediment and the pollen it contains. This effect is important but is unlikely to invalidate the generalized vegetational history (Bonny, 1976, 1978).

In a general sense, the vegetational history, as suggested by Figs. 2 and 3 and Table 1, parallels that at other sites in the region, such as Star Carr (Walker and

Godwin, 1954), Tadcaster (Bartley, 1967) and Seamer Carrs (Jones, 1976). The earliest vegetation indicated, at about 11,000 B.P., is essentially herbaceous. Gramineae, Cyperaceae and *Lycopodium selago* were apparently important. *Betula nana* and *Juniperus* were the only shrubs present. *Pinus* pollen, occurring at this time, was probably wind-blown from far to the south, or possibly derived. *Juniperus* became gradually more important before the change to forest cover which occurred at the time represented by 6 m depth in Fig. 2, probably about 10,000 B.P. The pollen analyses are too widely spaced to permit a detailed reconstruction of the changes in forest composition, but there is no evidence that these were unusual.

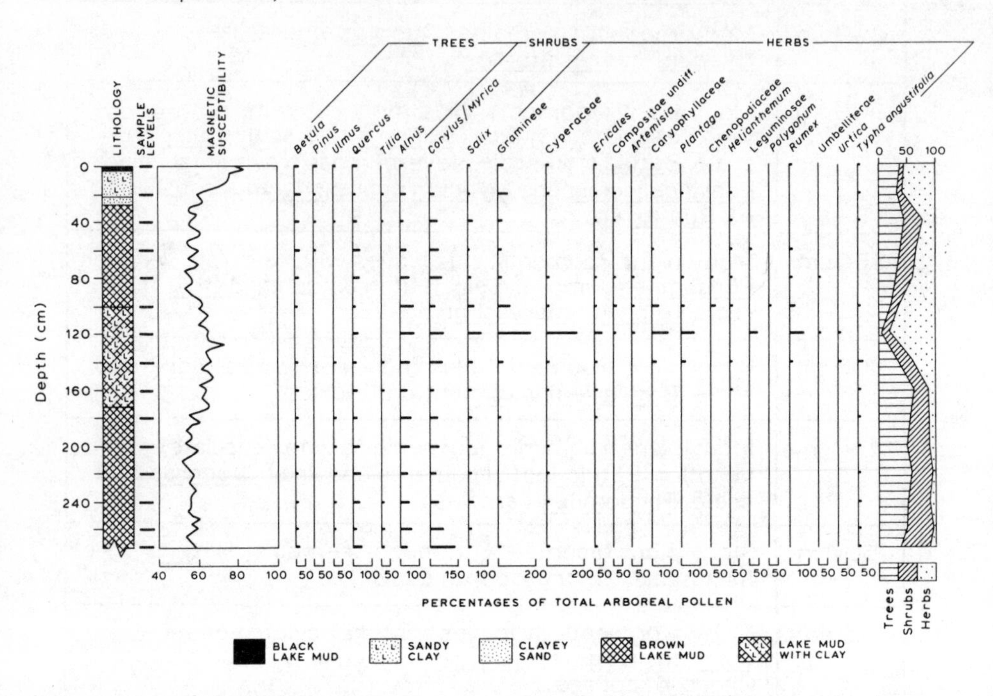

Fig. 3. Stratigraphy, palaeomagnetic susceptibility and outline pollen analysis of Core B. Analysis by C. Davies.

It is clear from both pollen diagrams that a major forest clearance occurred in the lake catchment. In Fig. 3 there are increases from about 140 cm depth in the values for many herbaceous taxa, especially Gramineae, Cyperaceae, Caryophyllaceae, *Rumex* and *Plantago*. The last in particular is a well-known indicator of forest clearance. The summary diagram confirms the dramatic relative increase in herbs at the expense of trees and shrubs at this time. In Fig. 2 there are increases, from about 2.10 m depth, in the values for Gramineae, Cyperaceae, *Plantago lanceolata*, Chenopodiaceae and Ericales. The summary diagram also shows the dramatic reduction in tree pollen which was seen in Fig. 3.

The lowest levels of tree pollen reached in this phase fell below 10% of the total dry land pollen in both Figs. 2 and 3. Such a low value is consistent with a more or less totally deforested landscape, and must imply that the lake catchment was virtually without forest at the time.

The idea of a major deforestation is strikingly supported by the stratigraphy. In

Fig. 3 there is a change from brown lake mud (gyttja) to a clay-rich gyttja at c. 170 cm, and a reversal of this change at c. 100 cm. The clay-rich band correlates with slightly higher values of susceptibility, which are presumably due to the iron contained in the inwashed material. It is noteworthy that the peak value for susceptibility occurs just below the minimum value for forest pollen. In Fig. 2 the relevant stratigraphic evidence is absent, probably because the presence of inwashed inorganic material prevented effective operation of the piston sampler. The marginal core (Fig. 4) also contains an allochthonous layer, the grey clay at 85 cm to 60 cm depth. Inwashed clay has been shown, on the North York Moors, to be the result of soil erosion following forest clearance (Simmons and others, 1975; Curtis, 1975).

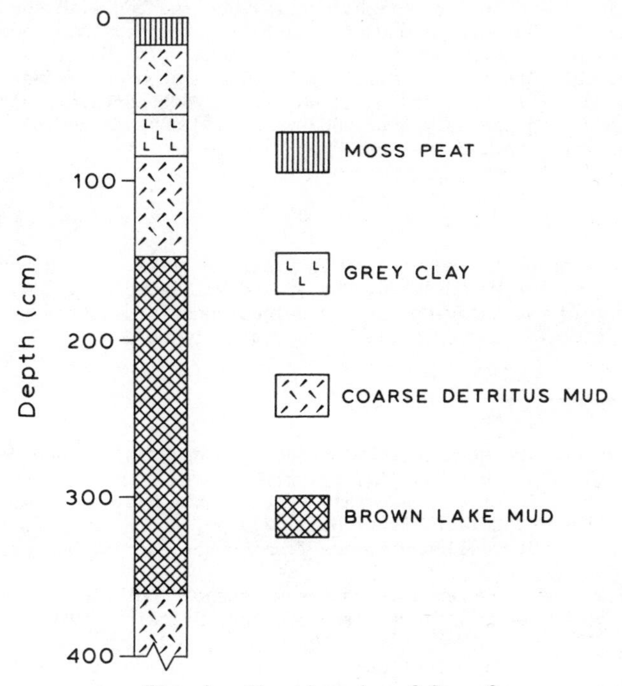

Fig. 4. Stratigraphy of Core C.

The location of the marginal core suggests that clearance was taking place on the slopes to the north and east of the lake, and possibly within what is now Garbutt Wood.

It has not been possible to determine the date of this clearance. Radiocarbon assay was judged unsuitable, because the allochthonous material, assuming it contained humus, would itself contaminate the sample with older carbon. The pollen diagrams are not sufficiently detailed to permit dating, although the clearance is clearly well after the rise of *Alnus* and may therefore be judged to be late Flandrian. An attempt to date the clearance phase palaeomagnetically was not successful, because the magnetic intensity was too low.

Both pollen diagrams suggest that there was a dramatic recovery of the forest after deforestation. This regeneration occurred to an extent rarely found in England. Forest cover was substantially restored.

Figure 3 provides evidence of more recent disturbance. From 27 cm depth the deposit becomes much more sandy, and susceptibility values rise to their highest of the whole core. This probably indicates renewed human activity, although there is only a slight indication of reduction in tree pollen. Figure 2 shows a greater reduction in tree pollen in samples near the top of the core. Detailed correlation of the two cores is not possible, but both are topped by organic sediment, perhaps suggesting some recent relaxation of disturbance.

The Flandrian vegetational history therefore suggests that there has been at least one major phase of forest clearance, followed by regeneration, followed by renewed disturbance, perhaps followed by a relatively recent decline of that disturbance. This modern decline in disturbance might possibly be equated with the regeneration of woodland (excluding Garbutt) around the rest of Gormire. This woodland is definitely modern, with sycamore canopy and a poor flora. The major forest clearance probably involved complete deforestation of the lake catchment. The 'primary woodland' indicators are therefore growing in what is almost certainly old secondary woodland, although it is probably over 2000 years old and has therefore existed throughout the historic period, as predicted by Peterken (1974).

ACKNOWLEDGEMENTS

We thank Mr. N. Hetherton for permission to visit the lake. Assistance in the field was provided by S. Clark, M. Marsh, R. Murray and N. Sutherland. We are grateful to Professor F. Oldfield for carrying out the palaeomagnetic measurements, and to Dr. M. Wyon for providing a species list from Garbutt Wood.

REFERENCES

Bartley, D.D. (1962). The stratigraphy and pollen analysis of lake deposits near Tadcaster, Yorkshire. *New Phytol.*, 61, 277-287.

Beckett, S.C. (1975). *The Late Quaternary Vegetational History of Holderness, Yorkshire.* Ph.D. thesis, University of Hull, 275 pp.

Beckett, S.C. (in press). Pollen diagrams from Holderness, N. Humberside. *J. Biogeogr.*

Bell, F.G. (1969). The occurrence of southern, steppe and halophyte elements in Weichselian (Last Glacial) floras from Southern Britain. *New Phytol.*, 68, 913-922.

Bonny, A.P. (1976). Recruitment of pollen to the seston and sediment of some Lake District lakes. *J. Ecol.*, 64, 859-887.

Bonny, A.P. (1978). The effect of pollen recruitment processes on pollen distribution over the sediment surface of a small lake in Cumbria. *J. Ecol.*, 66, 385-416.

Coope, G.R. (1977). Fossil coleopteran assemblages as sensitive indicators of climatic changes during the Devensian (Last) cold stage. *Phil. Trans. R. Soc.*, B 280, 313-340.

Curtis, L. (1975). Landscape periodicity and soil development. In M. Chisholm and P. Haggett (Eds.), *Processes in Physical and Human Geography. Bristol Essays.* pp. 249-265.

Jones, R.L. (1976). Late Quaternary vegetational history of the North York Moors. IV. Seamer Carrs. *J. Biogeogr.*, 3, 397-406.

Jowsey, P.C. (1966). An improved peat sampler. *New Phytol.*, 65, 245-248.

Kendall, P.F. and H.E. Wroot, (1924). *Geology of Yorkshire.* Reprint 1972, Scolar Press, Menston, Yorks. 973 pp.

Mackereth, F.J.H. (1958). A portable core sampler for lake deposits. *Limnol. Oceanogr.*, 3, 181-191.

Peck, R.M. (1973). Pollen budget studies in a small Yorkshire catchment. In H.J.B. Birks and R.G. West (Eds.), *Quaternary Plant Ecology*, Blackwell Scientific Publications, Oxford. pp. 43-60.

Pennington, W. (1977). The Late Devensian flora and vegetation of Britain. *Phil. Trans. R. Soc.*, B 280, 247-271.
Penny, L.F. (1974). Quaternary. In D.H. Rayner and J.E. Hemingway (Eds.), *The Geology and Mineral Resources of Yorkshire*. Chapter 9, pp. 245-264.
Peterken, G.F. (1974). A method for assessing woodland flora for conservation using indicator species. *Biol. Conserv.*, 6, 239-245.
Simmons, I.G., M.A. Atherden, P.R. Cundill and R.L. Jones (1975). Inorganic layers in soligenous mires of the North Yorkshire Moors. *J. Biogeogr.*, 2, 49-56.
Tauber, H. (1965). Differential pollen dispersion and the interpretation of pollen diagrams. With a contribution to the interpretation of the elm fall. *Danm. geol. Unders.* II Series, No. 89, 1-69.
Tauber, H. (1967). Investigations of the mode of pollen transfer in forested areas. *Rev. Palaeobotan. Palynol.*, 3, 277-286.
Walker, D. (1964). A modified Vallentyne Mud Sampler. *Ecology*, 45, 642-644.
Walker, D. and H. Godwin, (1954). Lake-stratigraphy, pollen-analysis and vegetational history. *Star Carr*. by J.G.D. Clark, Cambridge University Press, Cambridge. Chapter 2, pp. 25-69.

Miss Anne Blackham,
14 Rookery Close,
King's Lynn,
Norfolk. PE34 4EH.

C. Davies, Esq.,
149 Sheffield Road,
Warmsworth,
Doncaster,
South Yorkshire. DN4 9QX.

Dr. J.R. Flenley,
Department of Geography,
The University,
Hull. HU6 7RX.

18. Beach Development and Coastal Erosion in Holderness, North Humberside

Ada W. Pringle (née Phillips)

INTRODUCTION

The Holderness coast, (Fig. 1) bounded to the north by the chalk promontory of Flamborough Head and about 60 kms to the south by the Humber Estuary at Spurn Head is well known for the very rapid rate of erosion of its Pleistocene till cliffs, which Valentin (1954) calculated as 1.20 m on average per year along its entire length. This rapid erosion is constantly revealing fresh exposures of the till which is now recognised as having three main divisions. The Basement Till (Catt and Penny, 1966) is found intermittently in the base of the cliff between Kilnsea and Holmpton and at Bridlington; the overlying Skipsea Till (Madgett and Catt, 1978) forms the base of the cliff along the entire coast except where the Basement outcrops in the south-east and between Holmpton and Tunstall where the uppermost Withernsea Till is exposed. All the tills are predominantly composed of boulder clay, with colour and erratic content helping to distinguish one from another. Lenses of sands, silts and gravels are found especially at the junctions of the different tills (Penny, Coope and Catt, 1969).

Along this moderate to high energy coast, in terms of marine processes, the eroded till is rapidly broken down into its constituent parts. The sands, gravels and erratic boulders provide the main source of beach material whilst the silt and clay fractions are quickly taken up into suspension by the waves and are carried away from the beach. The aim of this essay is to examine the form of the beach development resulting from the erosion of these till cliffs and to demonstrate how the beach form itself controls the detailed rate of erosion.

BEACH FORM AND SEDIMENT CHARACTERISTICS

Below the cliffs which rise to a maximum height of about 35 m O.D. at Dimlington, but are more commonly about half this height, the beaches run without interruption along the whole length of the Holderness coast. The form of the beach however changes from north to south. In Bridlington Bay, it is over 300 m wide at mean low water and has a gentle overall gradient of about 1.5°. It is characterised by a well developed ridge and runnel system, with three of four ridge and runnel units usually present. The beach material consists almost solely of sand and shell fragments, with a small amount of shingle close to the cliff foot.

Southwards from Barmston to the tip of Spurn Head the beach takes on its characteristic Holderness form which in cross-section can be divided into two distinct parts.

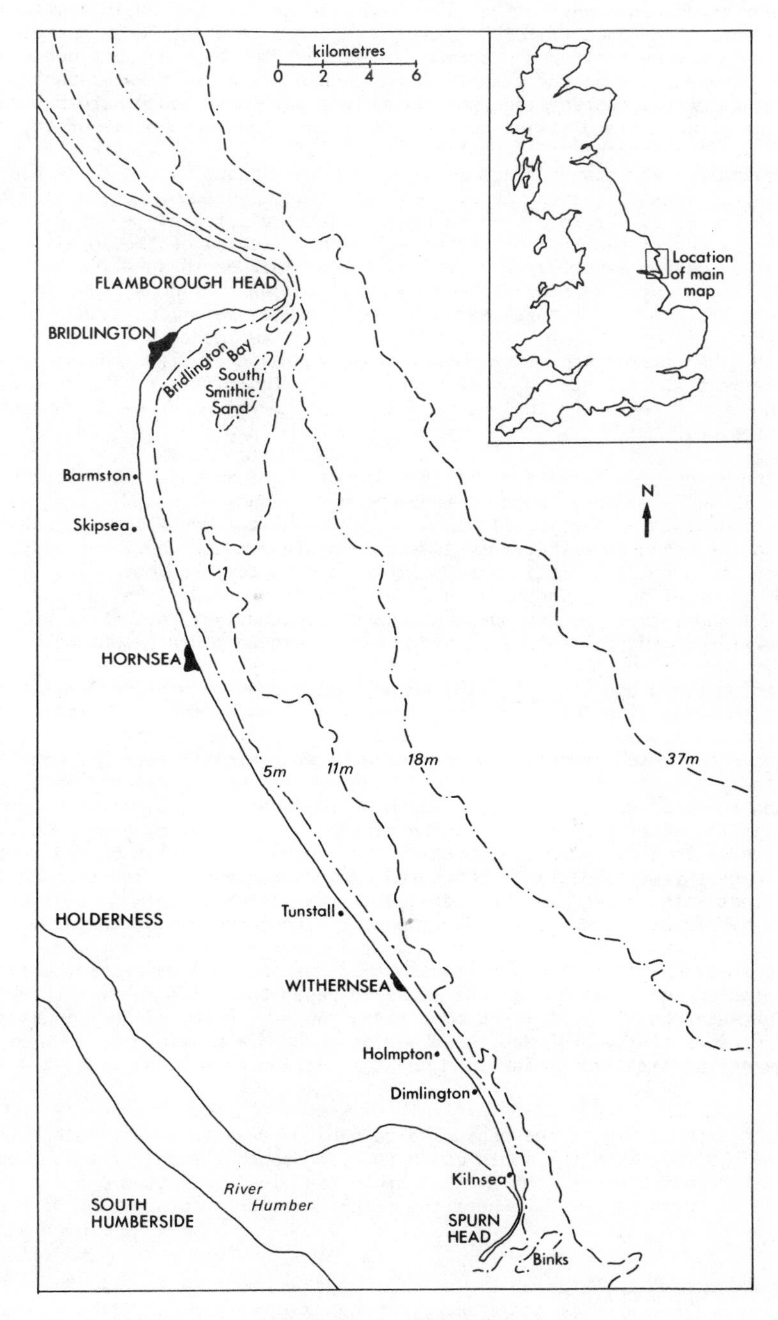

Fig. 1. Location Map

The upper beach, adjacent to the cliff foot, is usually convex in profile and is composed of coarse sand (1-1.15 mm) and shingle up to about 20 cm diameter. Whilst the shingle may be scattered over much of the surface of the upper beach it is frequently concentrated on the steeper lower slopes. The beach water-table comes to the surface at the foot of the upper beach and the lower beach extends from here to low water mark. This division between upper and lower beaches is often, although not always, a clear line along the beach and in profile lies at a marked change in beach surface angle. In local dialect it is termed the 'grope'. The lower beach is characterised by a gentle, even slope. Much of it is covered by a thin film of surface water at low tide, but in places patches of sand may rise above the water-table, in dialect they are 'cranches' or 'crauches'. The material of the lower beach usually grades from medium sand (0.5-0.25 mm) near the upper beach junction to fine sand (0.25-0.125 mm) at low water mark but a small amount of fine shingle may be present after storms. At the southern end of the Holderness coast, especially south of Withernsea, the beach contains a larger volume of sediment, which is probably the result of the general accumulation of material from the cliffs to the north plus the presence in the Dimlington area of particularly high cliffs which are capable of yielding more material to the beach. As a result, the upper beach becomes a much larger feature.

Along the Holderness coast from between Barmston and Skipsea southwards beach features known as 'ords' are found. Figure 2 shows a generalised diagram of an ord, which is basically a section of beach where the normal upper/lower beach form breaks down and the beach is either very low or completely absent near the cliff foot, exposing the underlying till shore platform basement. From both field investigations and the study of aerial photographs (Scott 1976) it appears that on average ords are between 1 and 2 km in length and that there are usually 10 ords between Barmston and the tip of Spurn Head. Over time they can be seen to migrate southwards.

Near the northern end of an ord the upper beach becomes lower and narrower southwards and the junction with the lower beach swings in towards the cliff foot. The lower beach conversely becomes wider and is raised slightly above the water-table. At the ord centre the beach reaches its lowest level at the cliff foot and beach material is commonly absent here for a distance ranging from only a few metres to over 100 m, exposing the till shore platform. South of the centre the upper beach begins to reform at the cliff foot but the shore platform is usually exposed along its seaward margin for a long distance southwards. The lower beach within an ord often develops into a high asymmetrical ridge above the beach water-table. Its steeper landward slope traps water along the seaward edge of the shore platform in a water filled runnel which drains seaward at low tide at the southern end of the ord.

In between ords, where the upper/lower beach form is well developed, only extreme high water spring tides enable the waves to reach the cliff foot. At the site of an ord high water on all spring and neap tides reaches the cliff foot increasing the amount of wave energy expended there and on the shore platform by 600% or more. The average spring tidal range is 5.0 m and the average neap tidal range 2.4 m at Bridlington.

As can be seen in Fig. 1 the submarine morphology seaward of the Holderness coast is relatively simple with the depth contours lying roughly parallel to the coast. Only two offshore banks exist, the South Smithic Sand in Bridlington Bay and the Binks to the east of Spurn Head. Elsewhere the coast receives no protection from offshore features.

BEACH CHANGES

Replicate surveying of beach cross-profiles at Spurn Head in the early 1960's (Phillips 1962) and with particular reference to ords along the Holderness coast in the mid 1970's (Scott 1976) has revealed that both short and long term changes are taking place on the beaches. The short term changes are those which are seen most frequently as a result of constantly varying wind, wave and tidal conditions and they are on a relatively small scale. The long term changes tend to be on a larger scale and result from a specific wind, wave and tidal relationship which produces the most marked effect on this coast.

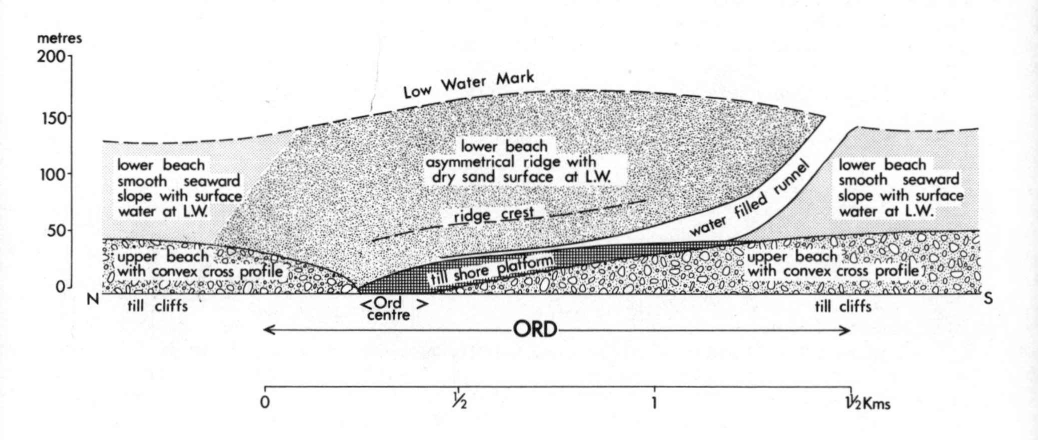

Fig. 2. A generalised plan of an Ord

(a) Short Term Changes

(i) Beach build-up at Spurn Head. During a year of weekly surveys at Spurn Head on beaches between ords and where ords were present the building up of the beach was the most frequently observed change. Figure 3 (a) shows the typical built-up beach profile. This was produced by the removal of material from the lower part of the profile and its addition to the upper part most commonly in the form of a swash bar, with its steep slope shorewards and its gentle slope seawards. Usually only one swash bar was present at about high water mark, but on occasions when tides were decreasing in range from springs to neaps several parallel swash bars were observed ranging in age from the oldest highest up the beach to the newest lowest down the profile. The presence of a swash bar indicates that the waves are steepening the profile in order to attain a state of equilibrium. When a swash bar was absent and yet building up processes predominated the upper beach was markedly convex in profile with a clearly defined crest, sometimes etched with beach cusps. Another morphological feature associated with the building up of the beach was the sand rise (cranch or crauch) on the lower beach. These took the form of either large continuous features or small irregular, oblique ones rising above the beach water-table.

As the inter-relationship between wind, wave and tidal conditions is a very complex and involved one it is no easy task to determine the precise conditions which lead to any particular type of change on a beach. Table 1 attempts to relate loss and gain of material on the upper and lower beaches to weekly periods of onshore and offshore winds of all velocities. It is apparent that there is a tendency for offshore winds and their associated waves to raise the upper beach level and to lower the height of the lower beach. A more detailed examination of the weekly beach changes in relation to wind direction and speed revealed that building up occurred under all wind conditions except when strong winds above 15 knots blew from between north-west, north-east and south-east, although the most marked build-up occurred after an offshore wind had been blowing for several days. Then, in addition to the redistribution of material landwards up the beach profile an overall net gain was usual. The offshore winds lessened the wave steepness (ratio of wave height/wave length) to an average of 0.0063 and they produced an offshore movement of water close to the sea surface which was compensated by a shoreward movement near the sea-bed. The relatively low steepness of the waves together with the shoreward drift of water over the sea-bed helped to build up the beach. The energy content of the waves observed under these wind conditions and calculated according to the formula:-

$$E = 2000\ H^2T^2$$

where E is wave energy in joules per metre of wave crest per wave length

H is wave height in metres

T is wave period in seconds

	Onshore Winds (Winds from between North and South East)				Offshore Winds (Winds from between South and North West)			
	Upper Beach		Lower Beach		Upper Beach		Lower Beach	
	Gain	Loss	Gain	Loss	Gain	Loss	Gain	Loss
S.E. Spurn	5	6	5	1	17	4	6	9
E. Spurn	6	11	9	8	16	6	12	8
N.E. Spurn	9	9	10	5	23	3	8	15
TOTALS	20	26	24	14	56	13	26	32

Table 1. Loss and gain of material on Upper and Lower Beaches in relation to onshore and offshore winds of all velocities. Figures refer to number of periods between consecutive weekly observations from April 1960 to April 1961.

was less than 88,970 joules/m of wave crest/wave length. The wave period was less than 8.5 seconds. The maximum recorded angle at which these waves approached the coast and which influences the rate of longshore drift, was 20° southwards, with an average of 4.3°. When building up processes predominated, the only disturbance of the predicted tide heights was a depression of the water level when strong winds blew from between south and west, causing wave action to occur at an abnormally low level and to be absent from the higher part of the beach.

(ii) Beach destruction at Spurn Head. On five occasions during the 12 months of observations at Spurn Head marked beach destruction occurred. Material was combed down the beach from high water mark, any swash bars previously present were obliterated, as also were sand rises from the lower beach. Figure 3(b) shows the typical combed-down profile. This gave an almost even slope to the profile from high water mark to low water mark and also produced a blurring of the boundary between upper and lower beach divisions. The material combed down from the higher part of the beach was sometimes deposited on the lower part, but on other occasions was carried out below low water mark. One distinctive beach form was constructed under these

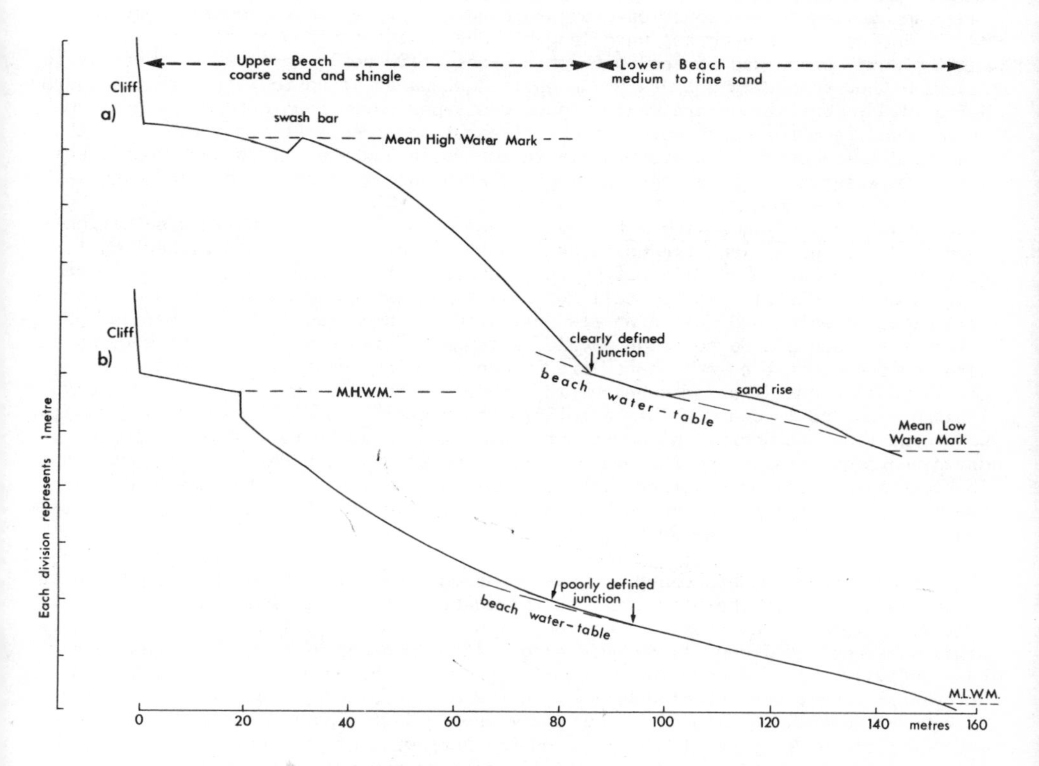

Fig. 3. Short term changes in profile of the High Beach between Ords.

a. Built-up beach form especially associated with offshore wind conditions.

b. Combed-down beach form associated with onshore wind conditions especially winds from the northerly quarter over 15 knots velocity.

conditions, namely a tongue of coarse sand and shingle which was drawn out south-eastwards from the upper beach to lie at a low angle to the main beach alignment. This form suggests that a rapid longshore movement of material was taking place. After these periods of marked beach destruction rapid re-building commonly took place. Less marked beach destruction was associated with the cutting by the waves of a low cliff in the upper beach, up to about a metre in height, facing seawards. Below this there was moderate combing of material down the profile.

Whilst Table 1 indicates that onshore winds tend to be associated with a loss of material from the upper beach together with a gain on the lower beach and this helps to identify conditions leading to small scale destruction with beach cliffing, a much more detailed examination of the observations was needed to determine the specific wind, wave and tidal conditions which led to the five periods of major destruction. They were in fact associated meterologically with the presence of a depression centred over southern Scandinavia which resulted in strong north to north-west winds blowing over the northern North Sea. Winds from this direction blow over the longest fetch towards Holderness of over 2,500 km from the north-east Atlantic Ocean and can lead to the development of the largest waves which can affect this coast. The observations showed that when winds of over 15 knots from this direction had been blowing for at least several hours waves with an energy content greater than 88,970 joules/m of wave crest/wave length broke on the east coast of Spurn Head. The maximum wave period was 12 sec, with an average of 10 sec. Despite considerable refraction they reached the shore at an angle southwards of up to 40°, with the average being 18.3^{o}. Whilst onshore winds produced steeper waves than offshore winds, the storm waves from the north were not the steepest recorded. Strong north to north-west winds also produced a storm surge in the North Sea and if its peak coincided with high water the high energy storm waves were able to reach abnormally high levels.

(iii) Short term changes within the Holderness ords. In investigations specifically concerned with ords along the Holderness coast, Scott (1976) distinguished short term changes relating to distinct parts of these features and also relating to variations in the distribution of till mud balls. Whilst these observations on ords at Atwick, Holmpton and Easington are not strictly comparable with the earlier observations at Spurn and do not span as wide a range of wind conditions, the same short term tendencies on the lower beach are apparent. With westerly, southerly and light and variable winds, the lower beach ridge characteristic of an ord, was most commonly lowered or moved landward. This parallels the overall beach build-up during similar wind conditions at Spurn. With northerly winds up to 20 knots, the lower beach ridge moved seawards paralleling the beach destruction at Spurn, under similar conditions. The most marked change occurred with northerly winds over 20 knots when the lower beach ridge was raised and moved landwards. This was probably part of a longer term change and will be considered further below.

The northern and southern upper beaches bounding the centre of an ord were found to be not as dynamic in the short term as the lower beach. The swash bar, the characteristic feature of build up at Spurn on the upper beach, was found after comparable wind, wave and tidal conditions here also. After periods of moderately destructive wave activity a low seaward-facing cliff was cut in the upper beach, as at Spurn. An overall change was detected however in the ords during a long non-stormy period of several months. A slight raising and widening of the northern upper beach took place whilst there was a slight lowering and narrowing of the southern upper beach. This appears to be related to a relatively slow southward longshore movement of sediment affecting both the ords and the higher beaches between them.

"Armoured mud balls" form a distinctive and variable feature of ords. They originate as angular blocks of till which have been eroded from the cliff face and commonly contain small erratics. As they are worked over and rolled about by the waves, small beach pebbles become embedded in their spherical surfaces. After they have been affected by several tides they are rapidly reduced in size and are usually destroyed in less than a week. The mud balls tend to collect on the exposed till platform within the ord or in hollows in the sand, towards the cliff foot. Their presence reflects the absence of a built-up upper beach to protect the cliff foot from wave attack and also the occurrence during the immediately preceding days of storm conditions with steep powerful waves, mainly from the north, producing very marked cliff erosion.

Overall it is possible therefore to correlate the short term, small scale changes within the ords with particular wind and therefore wave conditions. During a period of off-shore winds from between west and south the features within an ord become subdued, a narrow upper beach may form along the cliff foot at the ord centre, reducing the area of exposed till shore platform, and the build-up of the upper beach at the northern and southern ends often makes the limits of the ord difficult to define. After a period of easterly, onshore winds the lower beach tends to be flattened and the fine sand commonly covers much of the previously exposed till shore platform. The upper beach is generally reduced and lowered. Northerly, onshore winds of light and moderate strength produce little alteration within the ords, but strong winds over 20 knots from this direction, which have been shown to be capable of producing the most powerful waves along the Holderness coast produce the greatest changes within the ords and will be considered in detail below as they form an important part of longer term changes.

(b) Long Term Changes

The short term processes of beach build up and destruction mainly effect changes in the cross-profile. Larger scale, long term changes result from important longshore movements of beach sediment. The greatest and most rapid short term changes associated with the destruction of features characteristic of build up, and the resultant lowering and smoothing of the profile have been shown to result from northerly storms. These same storms result in the maximum possible rate of longshore drift along the Holderness coast through the formation of waves of maximum energy content (over 88,970 joules/m of wave crest/wave length) which break on this coast at the high average angle of 18.3^{o}, as measured at north-east Spurn Head. The rate of longshore drift is likely to be similar along the whole of the exposed north-east facing Holderness coast, with the exception of the Bridlington Bay section which is sheltered from the north by the 7 km long chalk promontory of Flamborough Head. It is only severe northerly storms which are capable of moving the ords along the coast, together with the intervening higher beaches, and the direction of movement is always southwards.

The most important morphological feature associated with rapid longshore drift has already been noted briefly in connection with periods of beach destruction, namely the construction of a tongue of coarse sand and shingle drawn out south-eastwards from the upper beach to lie at a low angle to the normal beach alignment. These features appear to form mainly along the southern half of the Holderness coast and along the east side of Spurn Head where the beaches in general are higher and broader, and therefore contain a greater volume of sediment. Whilst these tongues of material may be found at intervals along a high section of beach between two ords, it is the tongue which may develop from the upper beach near the northern boundary of an ord which is most important in inducing the ord's southward movement (see Fig. 4). During a period of strong northerly winds, with the associated high energy waves and storm surge, the tongue is built up approximately parallel to the crests of the breaking waves which approach the coast at a marked angle. After the storm when the angle of approach of the waves is reduced, and they may even approach parallel to the coast, the tongue gradually swings round to lie parallel to the main beach and is steadily moved landwards by the transfer of material from the seaward to the landward side by the swash of the breaking waves. Unless this movement is interrupted by a further northerly storm, the tongue will become incorporated in the upper beach within a few weeks.

At the same time as the tongue develops at the northern end of an ord, the upper beach at the southern end becomes rapidly diminished during northerly storm conditions. The result is that as the northern end is infilled, the southern end is eroded and the centre is therefore moved southwards without the overall form of the ord being destroyed. Thus the ords and the intervening higher sections of beach are moved southwards as a continuous system.

Rhythmic features have been described from a wide variety of coasts throughout the world during the past 40 years but the majority of the features are smaller in size than ords and show well developed transverse bars below low water mark, which are not a feature of ords. Investigations into the mechanisms controlling them have generally focused attention on relatively short term changes in wind and wave conditions, affecting rates and directions of longshore drift, the development of edge waves, and the establishment of circulation cells in the nearshore zone, associated with rip currents (Dolan, 1971; Sonu, 1973; Komar, 1976; Wright and others, 1979; and Chappell and Eliot, 1979). Whilst short term changes have been shown to occur along the Holderness coast, they are superimposed on the long term changes associated with ords.

In a recent paper which examines the development and migration of beach pads on St. Joseph Peninsula, Florida, U.S.A., Entsminger (1977), was concerned with features which appear to be related to the built-up sections of beach between the ords along the Holderness coast. Beach pads are rhythmic forms which he believes are uniquely definable and should not be confused with sand waves, transverse bars, giant cusps or beach cusps. The beach pad is described as an asymmetrical, roughly triangular shoreline feature with its base parallel to the coast, the next longest side being updrift and the shortest side downdrift. A bar is attached at the seaward point of the pad with its axis approximately parallel to the updrift side of the pad. The embayment between beach pads is arcuate and may extend back to the dune ridge which is commonly eroded by waves especially during storms. The St. Joseph Peninsula beach pads had a spacing of several hundred metres and their width was 10 to 15% of their spacing. Their internal structure indicated that they grew from an updrift sand source by swash action. The beach pads migrated at a rate inversely proportional to their size, from 0.61 to 1.23 m per day. Entsminger believes that beach pad development occurs when a coast undergoing erosion experiences unidirectional wave attack with a high breaker angle and that the beach pads are formed in an attempt to realign the shoreline in such a manner that the breaker angle will be reduced.

The Holderness ord commonly has a triangular form which would fit between the triangular shape of two adjacent beach pads, that is its base is parallel to low water mark, its next longest side is downdrift and its shortest side is updrift. The tongue of upper beach material or bar which forms at a low angle seawards from the northern upper beach during northerly storm conditions does indeed reduce the angle of the waves breaking upon it. The size and spacing of the ords and intervening built-up beach are clearly of a larger order than those of the beach pads but Entsminger vizualises these being found at different scales on different coasts. Individual ords do however appear to be a much more persistent feature than beach pads and their intervening hollows. The absence of bars below low water mark along the Holderness coast and the general absence of rip currents do not suggest the presence of cellular flows off this coast. Nor does it seem likely that ords are related to edge wave formation because of their wide spacing.

As ords are not found along the extreme northern section of the Holderness coast, north of between Barmston and Skipsea, it seems probable that the shelter provided there by Flamborough Head against northerly storms may be of fundamental importance. Whilst further field investigations are required, it does seem probable that ords develop at the point where the high energy northerly waves refracted around Flamb-

---

Fig. 4. Form of Ord movement along the southern part of the Holderness coast.

a. Characteristic ord morphology after absence of northerly storm conditions.

b. Ord morphology during northerly storm with wind velocity over 15 knots.
N.B. Northern upper beach tongue.

c. Ord morphology immediately following such a northerly storm as northern upper beach tongue is pushed shorewards.

d. Ord morphology a few weeks later after upper beach tongue has been pushed to cliff foot to form new upper beach.

orough Head first meet the coast south of Bridlington Bay (Phillips, 1964). There, because of the high angle of approach they will set in motion a rapid rate of longshore drift of beach material southwards. Longshore drift in Bridlington Bay will be much slower, (assuming it does exist under these conditions) and thus a gap will

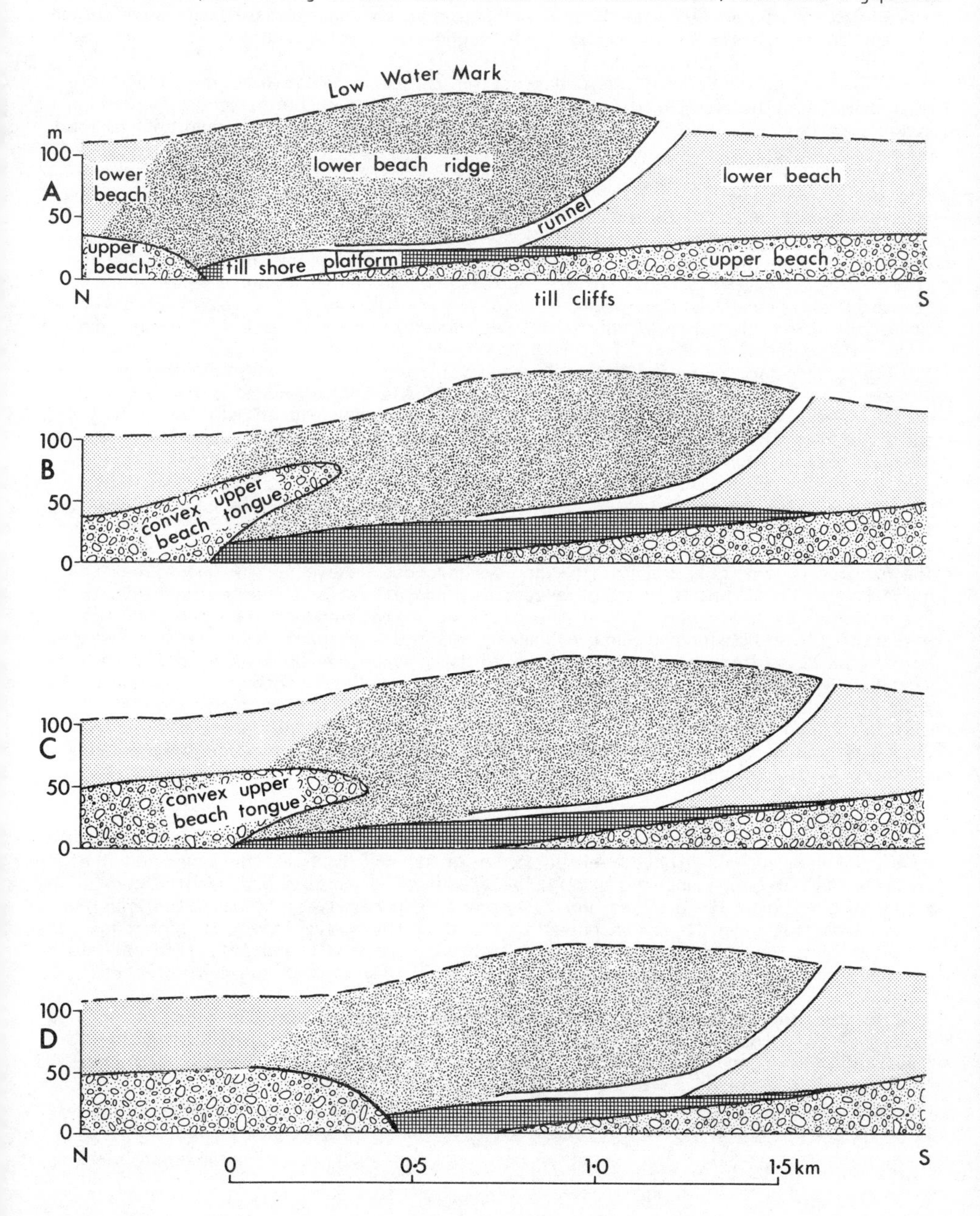

form in the beach where the shelter from Flamborough Head ceases. This exact position will vary according to the precise angle of approach of the refracted waves. An ord which has formed relatively far south is likely to be moved southwards by a subsequent storm which may be able to move the northern higher beach southwards. Once an ord has developed its infilling will be almost impossible if the assumption is correct that the rate of longshore drift along the entire length of this only slightly curving coast is approximately the same. In a series of wave refraction diagrams, Scott (1976) has confirmed that under northerly storm conditions an energy differential does exist between Bridlington Bay and the more exposed coast southwards. The long waves which commonly have a 10 to 12 second period are refracted round Flamborough Head to reach the coast first between Barmston and Skipsea.

THE ROLE OF THE BEACH IN COASTAL EROSION

The average rate of erosion along the Holderness coast, calculated as 1.20 m per year by Valentin, masks considerable variation from year to year depending on the frequency and severity of storms, especially from the north. It also masks marked variation along the coast at any one time, depending on the location of the ords.

The high beach, with its upper and lower divisions, which is the most common form between Barmston and the tip of Spurn Head, affords considerable protection to the cliffs from wave attack. As already noted it is only during high water spring tides and sometimes with a storm surge that waves reach the cliff foot with this form of beach. However when an ord with its low beach level is present there is little protection for the cliffs, which suffer wave attack at high water on both spring and neap tides. The maximum rate of cliff erosion is therefore achieved where an ord is located.

The erosion of the till cliffs is a complex process. Wave attack undermines the cliff at its foot, which creates instability above. This may sometimes lead to massive collapse of the upper part of the cliff or it may lead to rotational slumping of masses of the till along arcuate planes of weakness. Especially during the wetter conditions of winter large mud flows commonly develop on the sloping cliff face carrying fine material onto the beach to be removed rapidly in suspension by the waves. With till cliffs there is therefore usually a time lag between the erosion of the cliff foot and the retreat of the cliff top. This is not the case where dunes back the beach, as at Spurn Head. Wave attack of the dune foot causes almost immediate collapse of the higher part of the dune, to produce the most rapid rate of coastal retreat.

Along limited sections of the coast, most notably at Hornsea, Withernsea and Spurn Head, engineering works have been built in an attempt to halt the erosion. The presence of an ord along such an artificially protected section may result however in considerable damage to groynes and revetments, especially if their piles and foundations have not been firmly anchored in the till shore platform. In some cases revetments have been undermined by the unimpeded wave attack and infilling material has been removed to be used by storm waves to further batter the sea defences.

A very close inter-relationship has been demonstrated between the Pleistocene till cliffs of Holderness which supply the sediment for beach development, and the form taken by those beaches which itself controls the rate of cliff erosion and therefore the rate of release of sediment to the beaches. An appreciation of the changes in beach morphology resulting from varying wind, wave and tidal conditions is therefore fundamental in understanding the very rapid erosion of this coast.

REFERENCES

Catt, J.A., and L.F. Penny (1966). The Pleistocene deposits of Holderness, East Yorkshire. *Proc. Yorks. geol. Soc.* 35, 375-420.

Chappell, J., and I.G. Eliot (1979). Surf-beach dynamics in time and space - An Australian case study, and elements of a predictive model. *Mar. Geol.*, 32, 231-250.

Dolan, R. (1971). Coastal landforms: crescentic and rhythmic. *Bull. geol. Soc. Am.*, 82, 177-180.

Entsminger, L.D. (1977). Migration of beach pads on St. Joseph Peninsula. In W.F. Tanner (Ed.), *Coastal Sedimentology*. Geology Dept., Florida State Univ., Tallahassee.

Komar, P.D. (1976). *Beach Processes and Sedimentation*. Prentice-Hall, New York. 429 pp.

Madgett, P.A., and J.A. Catt (1978). Petrography, stratigraphy and weathering of Late Pleistocene tills in East Yorkshire, Lincolnshire and North Norfolk. *Proc. Yorks. geol. Soc.*, 42, 55-108.

Penny, L.F., G.R. Coope and J.A. Catt (1969). Age and insect fauna of the Dimlington Silts, East Yorkshire. *Nature, Lond.*, 224, 65-67.

Phillips, A.W. (1962). *Some aspects of the coastal geomorphology of Spurn Head, Yorkshire*. Unpublished Ph.D. thesis, University of Hull.

Phillips, A.W. (1964). Some observations of coast erosion: studies at South Holderness and Spurn Head. *Dock Harb. Auth.*, 45, 64-66.

Scott, P.A. (1976). *Beach development along the Holderness coast, North Humberside, with special reference to ords*. Unpublished Ph.D. thesis, University of Lancaster.

Sonu, C.J. (1973). Three-dimensional beach changes. *J. Geol.*, 81, 42-64.

Valentin, H. (1954). Der Landverlust in Holderness, Ostengland, von 1852 bis 1952. *Erde, Berl.*, 6, (3-4), 296-315.

Wright, L.D., J. Chappell, B.G. Thom, M.P. Bradshaw and P. Cowell (1979). Morphodynamics of reflective and dissipative beach and inshore systems: Southeastern Australia. *Mar. Geol.*, 32, 105-140.

Dr. Ada W. Pringle,
Department of Geography,
University of Lancaster,
Bailrigg,
Lancaster. LA1 4YR.

19. Spurn Point: Erosion and Protection after 1849

G. de Boer

To be asked to contribute to this collection of essays is an honour that I much appreciate, for I have been a colleague of Lewis Penny's during the whole of his time with the Department of Geology at the University of Hull. His interests in the Quaternary deposits of Holderness inevitably drew him frequently to Dimlington and so very close to Spurn, a feature in which we have long shared an interest both physiographical and as members of the Yorkshire Naturalists' Trust. Most of what I have written about Spurn hitherto has been concerned with its earlier history, and so it seemed appropriate here to trace the more immediate antecedents of the Spurn that he and I have come to know well, and have seen change significantly within our own experience of it.

The breaching of Spurn Point by the sea on 28 December, 1849 marks a turning point in the long history of this sand and shingle spit at the mouth of the Humber; what followed was to be radically different from what had happened before. It was a major breach and was torn across the peninsula during a north-westerly gale and an exceptionally high tide (Vetch, 1850). The same storm seriously damaged the low lighthouse which, as a result of earlier erosion had stood since about 1838 on what was at high tide an artificial island defended by a rampart. The storm swept the rampart away completely and attacked the foundations of the lighthouse, leaving it dangerously unsupported (de Boer, 1968). It was evidently a storm surge that caused all this damage, and it is such surges, usually at their strongest development when gales from between the north and northwest coincide with a high spring tide, that generally accompany the most serious occurrences of erosion and flooding along the coast of Holderness and indeed along the east coast from the Humber to the Thames.

This breaching was the climactic event of a protracted period of severe weather. The wind had reached gale force a day or two before the breach, and strong winds continued to blow for five weeks afterwards. Once opened, the breach, which lay between half and three quarters of a mile north of the circular compound surrounding Smeaton's high lighthouse and was where the chalk bank now is, continued to develop in a way that seemed clearly to presage the complete washing away of Spurn Point. The lifeboat first went through the breach instead of going round the tip on 29 February, 1850, and on 29 June six or seven craft going to Spurn for cobbles or gravel, and drawing at least six feet of water, passed through from the Humber to the sea on a single tide. A survey made on 20 July showed that the breach was 320 yards wide and 12 feet deep at ordinary high water. It was thought that the depth was no more than this because a layer of firm clay in the bed of the breach was resisting erosion so far, but would soon be cut through. Witnesses estimated the strength of the tide through the gap at six or seven knots. Long stretches of sand dune had been swept away on each side of the gap so that nearly three quarters of a mile of the spit were submerged at high water spring tides. A month later, it was reported on 20 August that the breach was still growing wider and deeper.

The situation became still worse the following year. The lightkeeper reported on 28 February, 1851, that "the breach is very much wider and deeper than last August, but not much deeper in the centre as it is on clay. The beach is gone away very much on the sea side to the north of the breach and there is another place where the sea crosses the bank five hundred yards north of the breach, but that place has been open some four or five months, but as the beach has come along it has partly stopped it again, but now I see it is getting deeper. I saw the place this morning and I should think that at high water spring tides there would be four and a half feet of water in the centre. I think that a strong north west wind on the top of a spring tide would take away a great deal of this five hundred yards that is between the two places. The low light is well banked both on the north and south sides and on the east side the piles are only ten feet out of the sand to the top; as the tide in shore goes through the breach, we do not have such a strong tide on the beach by the low light.".

Within a week of this report, a violent storm on 5 and 6 March added further severe damage to that of December 1849 to the low lighthouse and flooded the cottages of the lifeboat crew nearby. These cottages, built in 1819 some way back from the Humber shore, and still a little way from it in 1825, had suffered flooding in 1836 and 1845 as erosion on this side as well as the sea side of the peninsula had become more severe. Loads of chalk from Hessle had been laid down by the edge of the river after the 1836 flooding but had checked neither erosion nor flooding (Storey, 1967). The second breach referred to by the lightkeeper was found to be 400 feet wide after the storm. Deterioration continued throughout the year. By September the main breach was 500 yards wide and 16 feet deep at highwater spring tides and in the following December there were gales in which the low lighthouse was washed down completely.

The end of 1851 indeed probably saw the worst conditions at Spurn before remedial action had become effective, and the situation is recorded in detail on the first edition of the Ordnance Survey Six Inch Map and on Calver's chart of the Humber; both were published in 1852 but are based on surveys made in 1851. They depict Spurn as a string of islets at high water. A wide gap extends over the site of the breaches and separates the head of the peninsula, Spurn Island as it is called on the Ordnance Survey map, from its proximal end. Spurn Island, as a result of erosion on both sides, is very narrow. The lifeboatmen's cottages, originally built well away from the Humber shore now project on to it and are protected by a chalk embankment. The bulbous tip, usually a feature of the spit, has gone completely. Increasingly material carried to it along the beach seems to have been washed over the narrowing neck or through the breach or swatchway once formed. Widening of the breaches and wasting of the tip seems to have gone so far indeed to make it seem in the highest degree improbable that the breaches would have closed naturally and a resumption of littoral drift to the tip would have restored its bulbous character. It seems much more probable that in a few more years Spurn Island would have disappeared altogether and a new spit begun to grow inside the line of its predecessor.

Such an event seems to have happened several times earlier in the history of Spurn at intervals of about 250 years (de Boer 1964). The peninsula breached in 1849 began to grow after the breaching of its predecessor about 1608, is first called Spurn Point about 1675, and although retreating westwards all the time, grew and developed during most of the first two centuries of its existence. Signs of its impending destruction began to show themselves more and more evidently from about 1790 onwards. Whereas before then there were dunes and marram grass all the way from the tip to Kilnsea, from about 1800 especially high seas had washed over and stripped the neck of the spit ever more frequently. Hewett's chart of 1828 marks it as regularly cov-

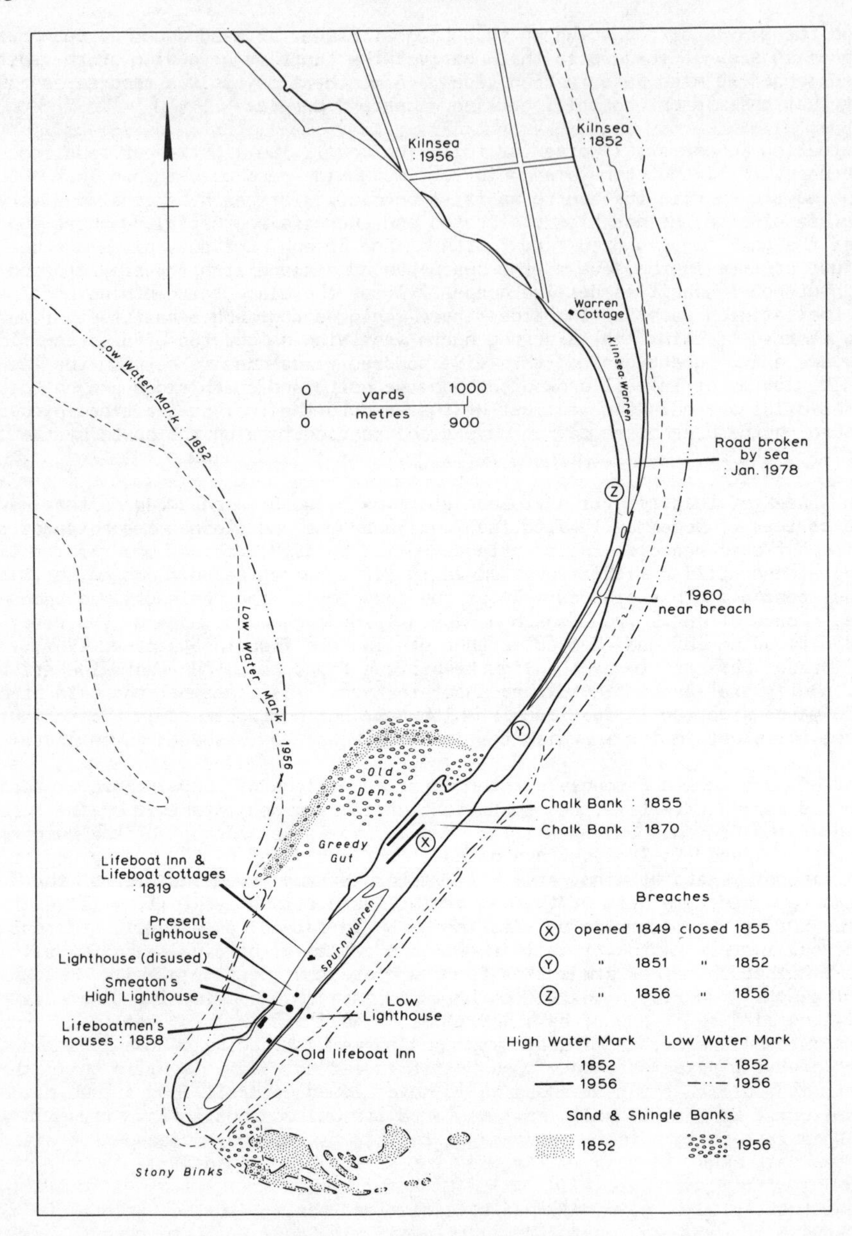

Fig. 1. Spurn after 1849

Outlines taken from Ordnance Survey six inch maps of 1852 and 1956 have been superimposed, and principal features mentioned in the text indicated. Based upon the 1852 and 1956 Ordnance Survey Six Inch maps with the permission of the Controller of Her Majesty's Stationery Office. Crown Copyright Reserved.

ered at high springs, and a map of 1830 already calls the distal end an island. The breach of 1849 was therefore the beginning of the final stage of a natural development that had taken place more than once before. What makes it a turning point is that this time this natural development was not allowed to proceed. The breach was closed artificially and the subsequent history of Spurn rendered different from that before.

It was not that there had been no expression of concern for the future of Spurn until the development of the breach. In 1845 the Commission on Tidal Harbours had pointed out that Spurn had lost half its breadth in the last 20 years and that perhaps 50,000 tons of cobbles and gravel were being shipped from the beaches each year. Nothing was done, however, until the opening of the breach produced a situation that could not be ignored. Captain Vetch R.E., who had made the original survey and reports on the breach in July, 1850, concluded that the stability of the lighthouse and of the whole peninsula was threatened and if it were lost the Hawk roadstead, the only harbour of refuge in an easterly gale between Harwich and the Forth, would lose its shelter and the channel become liable to change. He recommended four lines of action: a prohibition on taking shingle, the closing of the breach, the erection of groynes, and reclamation between Spurn and Sunk Island. Implementation of three of these followed promptly; the fourth has been discussed at intervals but not so far undertaken, and Spurn survives, and survives as a peninsula. The Admiralty had indeed issued an order on 21 March, 1850, before Vetch's report, prohibiting the removal of ballast or shingle from Spurn from the low lighthouse to the extreme tip at low water and northwards for 2½ miles on each side of the spit. Because this bore severely on the lifeboat crew and their families who earned their livings by loading with cobbles and gravel ships which beached on Spurn for the purpose, the order was relaxed in July to allow the crew to load gravel from the extreme tip. Iveson, the agent of Sir Clifford Constable who owned Spurn, challenged the Admiralty's right to forbid him to take material from his own foreshore, and it was not until September that Constable ordered abstraction from the prohibited area to stop. Vetch proposed that the breach or swatchway be stopped by a double line of stakes and wattling to promote the accumulation of sand, a suggestion that Iveson rejected as quite useless. "Nothing but large cliff stone (i.e. chalk from Hessle Cliff) to the extent of 1500 to 2000 tons will do any good there against the weight of water held up in the Humber." Groynes he conceded would be of service in accumulating a fuller beach.

An engineer, James Walker, was made responsible for closing the gaps, and seems to have been on the scene very soon after the submission of Vetch's report for it was he who reported in August on the widening and deepening of the breach. Little beyond investigation could have been done in the last months of 1850, and the storms of 5 and 6 March, 1851, already described, induced him on 7 March to ask the Harbour Department of the Admiralty for £10,000. A vote granting this sum in order to resist the inroads of the sea and protect the channels of the Humber was passed in the House of Commons on 19 July, and steps to remedy the situation soon followed for a letter of 27 November, 1851 tells how "from the experiments made by Mr. Walker, he has found at the little breach that the placing of stakes and wattling backed in by chalk has been successful and accumulated the matter passing along or thrown up by the sea to stop the breach. The stakes and wattling are not of themselves proposed to constitute the defence but only to be the means of accumulating material. The breaches stopped, the next operation will be to throw out groins from the shore to intercept travelling matter.". By May 1852 it was reported to the House of Commons that the breach had been closed; it was the smaller breach however and not the main swatchway. Walker had also been planting marram grass along the neck of the peninsula since April.

Spurn Island, however, was still losing ground. A temporary lighthouse had been put up to replace that washed down in December 1851, but it was only 90 yards from the high light, too close to give a satisfactory indication of direction to ships, and so a new permanent low lighthouse was built on the Humber foreshore, and first showed its light on 24 June. It was reached by a bridge from the shore. In February 1853 groynes were put up on the sea side of Spurn in front of the high lighthouse. Although the little breach had been closed, Walker's stakes and wattling were less successful at the main breach. Here a screen of stakes and wattling extended across most of the breach and acted like a weir at high water when the tides flowed over the top of it. A narrow section between the completed part of the barrier and the northern end of Spurn Island still defied all attempts at closure. It seems that beach material was still being carried through the breach so that the head continued to dwindle. Despite the earlier building of chalk barriers around the lifeboatmen's cottages, they still suffered from flooding, and were in such a poor state that in 1853, the Hull Trinity House decided to pull down the three cottages nearest, indeed by now almost projecting into the Humber, and build three new ones at the other end of the row. Nevertheless they were flooded again in 1855.

The main breach was still unclosed in 1854 when all the £10,000 had been spent, and in August of that year £6,000 were voted, to be expended under the supervision of London Trinity House, whose lighthouses were threatened. Finally, in 1855, after great trouble, it was reported, the main breach was closed. Even so a third breach opened the following year to a width of 80 yards and depth of 13 feet at high water; it was closed in the same year. In this year, 1856, a further sum of £6,000 was voted for the works at Spurn, but the size of the vote of 1857, £500, reflects the view that by then no further substantial works were necessary. The sea, it was believed, would wash up sand and shingle and so strengthen the barrier between it and the Humber. The accumulation of dunes was to be encouraged by erecting wattle screens and sowing marram grass seeds.

The closing of the breaches did not at once check the diminution of Spurn Island, and the lifeboat station suffered from its continuance. The sea side of the head of the peninsula between the high lighthouse and the point was pushed back so far that the house of the master of the lifeboat, and the public house that he ran, in a building originally erected as a barracks during the Napoleonic wars, were now on the high water mark, and their rear parts collapsed. The crew's cottages were again in poor shape, and so the building of a new row of cottages for them on a different site was begun in the same year. Because of the extent of erosion on the sea side, the defences of the high lighthouse were strengthened by a timber revetment.

Walter White, whose *A Month in Yorkshire* was published in 1858 provides an excellent description of Spurn at this time (pp. 29-36): "[Spurn] narrows and sinks as it projects from the main shore for about two miles, and this part, being the weakest and most easily shifted by the rapid currents, is strengthened every few yards by rows of stakes driven deeply in and hurdle work. You see the effect in the smooth drifts accumulated in the spaces between the barriers which only require to be planted with grass. As it is, the walking is very laborious: you sink ankle-deep and slide back at every step... A little farther on and we are on a rugged embankment of chalk: the ground is low on each side, and a large pond rests in the hollow between us and the sea on the left, marking the spot where a few years ago the sea broke through and made a clean sweep across all the bank. Every tide washed it wider and deeper, until at last the fishing vessels used it as a short cut in entering or departing from the river. The effect of the breach would, in time, had a low water channel been established, have seriously endangered the shore of the estuary besides threatening destruction to the site of the lighthouse. As speedily therefore as wind and weather would permit, piles and stakes were driven in and the gap was filled up with big lumps of chalk brought from the quarry at Barton, forming an embankment sloped on both sides to render the shock of the waves as harmless as possible. The trucks, rails, and sleepers with which the work had been accomplished were still lying on the sand awaiting removal...

By and by we come to firm ground mostly covered with thickly matted grass; a great irregular, oval mound... Near its centre is a fenced garden and a row of cottages

- the residences of the lifeboat crew. A little further, on the summit of the ridge stands the lighthouse, built by Smeaton in 1776, and at the water's edge on the inner side, the lower light...

[The mound] slopes gently to the sea and is somewhat altered in outline by every gale. At the time of my visit, rows of piles were being driven in, and barriers of chalk erected to secure the outer side between the tower and the sea; and a new row of cottages for the lifeboat crew, built nearer the side where most wrecks occur than the old row, was nearly finished. Beyond, towards the point, stands a public house in what seems a dangerous situation, close to the water. There was once a garden between it and the sea; now the spray dashes into the rear of the house; for the wall and one half of the hindermost room have disappeared along with the garden, and the hostess contents herself with the rooms in front, fondly hoping that they will last her time. She has but few guests now, and talks with regret of the change since the digging of ballast was forbidden on the Spurn. The trade was good for the diggers were numerous and thirsty. That ballast digging should ever have been permitted in so unstable a spot argues a great want of forethought somewhere."

The new cottages for the lifeboat crew were occupied in 1858 and are those which remained in use until 1975. Soon after they moved in, the public house was transferred to one of the cottages they had vacated, the building, in fact, referred to as the Lifeboat Inn until its demolition in 1978.

All these developments of 1858 indicate perhaps a renewal of confidence in the continued existence of Spurn after success was achieved in the closing of the breaches, yet doubts remained, as Walter White's description of the view of Spurn from the top of the lighthouse implies:- "Most remarkable is the tongue of sand... now visible in its whole extent and outline. It is lowest where the breach was made and, now that the tide has risen higher, the chalk embankment seems scarcely above the level of the water".

The doubts received more explicit expression after 1862. In that year the Harbour Transfer Act handed over to the Board of Trade the powers given to the Admiralty by the Harbours Act of 1814 to prohibit the removal of sand or shingle from the shore and powers also relating to the construction of harbour works. "About £22,000 had been expended from votes without any satisfactory result when the superintendence ofthe works was undertaken by the Board of Trade", T.H.W. Pelham, an Assistant Secretary to the Board told the Royal Commission on Coast Erosion in 1906. "At this time the design was changed. The work was undertaken by Messrs. Coode and a successful result obtained".

We have an account from Coode himself, (Shelford, 1869. pp. 501-4) of his proceedings at Spurn after he took charge on Walker's death in 1863. Walker, he pointed out, had not erected a single groyne, and had stopped the breach by means of a chalk bank, which because it projected into the Humber, checked the movement of beach material northwards along the inner side of Spurn, so that the foreshore still further north was weakened on the Humber side. The bulging edge of Spurn inside the Old Den caused by the bank remains a feature of the peninsula to the present day. Coode himself put initially, in 1864, five groynes along the sea side of Spurn, three covering the site of the main breach, two defending the narrow neck or High Bents. By 1870 another six had been added, and a straight line of chalk embankment, that called the Chalk Bank at the present time, was built across the site of the major breach inside the curving line of Walker's bank. Pickwell (1878) commented on how successful these groynes had been, more indeed than their designer had hoped, for whereas he had intended to conserve what was left, in fact the edge of marram covered sand dunes had advanced seawards in places, up to 80 yards near the point, and up to 40 yards along the entire neck. These dunes had been encouraged to grow by the building of small embankments of sand 6 to 8 feet high and planting their tops and seawards faces with marram. Pickwell noted also with approval the appearance and spread during the last four or five years of another dune coloniser, the sea buckthorn "of rapid growth and greatly to be recommended for similar situations". Altogether he estimated that there had been an accumulation of vegetated dune 60 yards wide for two miles along the neck of Spurn, and credited the groynes with causing the accumulation outside the dunes of a growth of beach 100 yards wide, 7 feet deep and 2½ miles long.

Nevertheless further groynes were added in the years up to the outbreak of war in 1914, extending the systems southwards to the lighthouse and northwards to the beach off Warren Cottage. A considerable length of timber revetment was added to the sea side of High Bents in about 1883-4. By this time the older groynes had been in position long enough for there to be a quite perceptible degree of aggradation north of each groyne and a recession of the upper edge of the dunes to the south. This has had the result that, as air photographs of Spurn as it is now show clearly, the revetment steps en echelon from each groyne to the next. Smaller sections of revetment were added also to the groynes protecting the low flat expanse of Kilnsea Warren south of Warren Cottage. The closure of the breaches and the replenishment of the beaches restored littoral drifting of material to and round the point which seems to have recovered its bulbous shape by the end of the century and to be extending further into the Humber.

During the 1914-18 war Spurn was fortified and the Spurn railway built to connect the batteries placed at the Point and at Kilnsea. This seems to have involved modifying the narrow neck at High Bents almost to a railway embankment. It is difficult to ascertain precise dates for these wartime events, though 1915 seems the most probable (Hartley, 1976). The sea wall sheltering the Godwin Battery at Kilnsea, now in ruins and largely swept away, probably belongs to the same year. Two groynes were added to the sea side of Spurn in 1915 and 1917, one at the northern end, the other at the southern end of the narrow neck. In the early years of the century before the war, the low ground between Kilnsea and Easington was liable to be covered by high tides especially in surge conditions, and an earth bank seems to have been built along its seaward margin at some stage during the war to check this and perhaps to ensure that Spurn could always be reached from Holderness.

Between the wars the erosion defences were fully maintained and several groynes were renewed or extended. It appears that the Board of Trade retained responsibility until 1939 when it was assumed by the Royal Engineers. In particular five groynes were erected on the beach in an array from opposite the site of the 1858 lifeboat cottages towards the point.

The outbreak of war in 1939 saw further developments apart from military armament, in particular the construction of the concrete road completed in 1942 and some strengthening of the narrow neck (or High Bents) revetment. In 1942 a storm surge caused severe erosion here on the sea side and about 30 yards of railway were left hanging in the air across a gap. The damage was made good by a seawall of concrete filled sand bags backed by a filling of chalk rubble and sand. The road formed a loop at the end of Spurn, parts of which now lie deeply buried beneath the sand. At the time of its construction it was not far from the point, which now however lies about 80 yards further away. The railway remained in service until 1951 and was found useful for transporting materials for maintaining groynes and revetments. It was dismantled shortly after.

These erosion defences appear to have been in good enough condition to have withstood the great storm surge of 31 January - 1 February, 1953, without serious damage on the sea side of Spurn. As in so many other instances of breaching or severe erosion, the gale-force winds were north-westerly, and these, blowing down the lower Humber, generated waves which damaged the lifeboatmen's cottages and demolished the river side end wall of the Lifeboat Inn. It was subsequently rebuilt and a timber revetment provided as further protection. More serious were the effects at Kilnsea where the sea bank protecting the low ground between Kilnsea and Easington was overwhelmed

Fig. 2. Air view of Spurn from ENE, summer 1977.

Traces of the breaches 1849-56 and evidence of the present vulnerable condition of the peninsula are clearly shown. The bulge on the river side just beyond the curving neck marks the bank thrown up by means of which the 1849 breach was finally closed in 1855. The faint white line connecting its two curved ends is the chalk bank built in 1870. The muddy shingle bank of Old Den can be seen lying in the Humber opposite the site of the 1849 breach, and a line of breakers arcing seawards from the tip traces the line of the Stony Binks.

The road has needed to be protected from wave attack from the river where it lies by the shore half way round the curving neck and just beyond the tip of the fringe of *Spartina* covered mudflat which shows almost black in the picture. From this stretch the road can be seen slanting across the narrow neck towards the camera to the sea side. It is just to the camera side of this point that collapses of sections of the wall of concrete filled sand bags pose another threat. Still nearer to the camera can be seen groynes; these are now ruinous and the inner edge of the beach has been pushed back beyond the ends of the groynes. Nearer still, where the line of surf comes further inshore is where a future breach seems most likely to occur. The road which is seen running very close to the beach in the summer of 1977 was six months later undermined and broken in January 1978. Thence to the edge of the picture can be seen the rapidly retreating boulder clay cliffs and the ruinous concrete sea wall, the recession of which seems to be depriving Spurn of the shelter necessary for its existence. (Photo: M.J. Wilkinson)

---

and sea and Humber joined in one great sheet of water. The northern half of the sea wall protecting the Godwin Battery at Kilnsea may well have been severely mauled at this time for it had collapsed and been largely swept away by 1958.

When Spurn became the property of the Yorkshire Naturalists' Trust from 1960 onwards, the new owners had neither the resources nor any legal obligation to maintain the erosion defences, a charge which had become burdensome during the final stages of War Department ownership. In January, 1960, the road nearly suffered the same fate as the railway in 1942, for during a period of stormy weather including surge conditions, the waves burst through the angle where a length of already partly collapsed revetment met the inner end of a groyne at High Bents, and washed away the dune behind until the edge of the road lay only inches from the edge of a high steep cliff of loose sand. A change of weather and wave conditions fortunately checked erosion before the road was undermined and caused accretion to make up a little of the damage.

The threat, however, seemed merely to have transferred itself elsewhere. From 1962 to 1965, continued erosion of the beach opposite the Chalk Bank eventually removed completely the dunes sheltering the road here so that waves were at times splashing on to the road itself. Here the stabilizing of sand by laying hawthorn hedge prunings encouraged the regrowth of lyme and marram grass and a redevelopment of the dunes. After a brief respite, erosion has become an increasingly serious menace during the 1970's in several places. There has been erosion from 1976 onwards which has seemed likely to undermine the road on the river-side from High Bents southwards; rubble from demolished derelict buildings has been placed here to hold at least the situation here for the time being. Gaps have been made in the concrete-filled sandbag wall to the north and on the sea side of High Bents and large circular craters from which the sea has washed sand are extending back towards the road.

Even more premonitory of grave problems ahead is the continued erosion of the low boulder clay cliffs from Kilnsea southwards along the coastline of Kilnsea Warren. After the dilapidation of the sea walls at Kilnsea, the cliff line has been trimmed back rapidly depriving this northern end of Spurn of a degree of shelter. The groynes at Spurn have become ruinous and the shore has been driven back beyond their inner ends so that they no longer retain the beach. The boulder clay cliffs seem to have been driven back some tens of yards even in the last decade so that they now lie west of the former line of railway. In the six weeks after Christmas 1976 the retreat was as much as 20 yards.

The most dramatic impact has been on the low flat stretch of Kilnsea Warren between Warren Cottage and High Bents. For some years the edge of these very low dunes has been driven back by the sea towards the road. In January, 1976, a high tide when fortunately there was little wind, covered the road and buried it under sand. On 11 January, 1978, a storm surge accompanied by a northeasterly gale forced the Nature Reserve Warden to leave his house, flooded the fields and road between Spurn and Kilnsea, and undermined the road a short distance south of Warren Cottage, cutting off water supplies and telephone communications to the point. Sections of the road collapsed, and a track to bypass this now useless length of road was made. This new bypass loop was itself drowned by a high tide on 31 December, 1978, and the undermined section of the old road suffered further damage and is now lost to sight beneath the beach.

This accelerated tempo of events is ominous indeed. The thread of low dune and beach linking the spit to Holderness is now extremely tenuous. A strong storm surge could cut it at any time and initiate a tidal swatchway. Spurn has been held artificially very nearly in one position while the coast to the north has continued to retreat so that the peninsula has become more and more exposed to destructive sea conditions. The natural cyclic destruction and reconstruction of Spurn having been deferred artificially for about 130 years, a much more radical destruction and rebuilding is becoming progressively more difficult to resist.

REFERENCES

De Boer, G. (1964). Spurn Head: its history and evolution. *Trans. Inst. Br. Geogr.*, 34, 71-89.
De Boer, G. (1968). A history of the Spurn lighthouses. *East Yorkshire Local History Series* No. 24, East Yorkshire Local History Society, 72 pp.
Hartley, K.E. (1976). The Spurn Head Railway. *Industrial Railway Record* No. 67, 249-292.
Pickwell, R. (1878). The encroachments of the sea from Spurn Point to Flamboro' Head, and the works executed to prevent loss of land. *Proc. Instn. civ. Engrs.*, 51, 191-212.
Shelford, W. (1869). On the outfall of the river Humber. *Proc. Instn., civ. Engrs.*, 28, 481-515.
Storey, A. (1967). *Trinity House of Kingston upon Hull.* Privately printed, 146 pp.
Vetch, J. (1850). *The report of Captain James Vetch, R.E., to the Lords Commissioners of the Admiralty on the subject of the breach made by the sea across Spurn Point at the entrance of the river Humber, 29 July, 1850.*
White, W. (1858). *A month in Yorkshire.* London. xv + 272 pp.

Other sources

Humberside County Record Office, *DDCC* 89/95-99/120/137
Hull Trinity House, *Spurn Lifeboat File.*
Yorkshire Naturalists' Trust, *Annual Reports* 1959-79

G. de Boer, Esq.,
Department of Geography,
University of Hull,
Cottingham Road,
Hull HU6 7RX

20. Episodes of Local Extinction of Insect Species During the Quaternary as Indicators of Climatic Changes

G. R. Coope

In recent years there has been a tendency to interpret palaeoclimatic evidence as indicative of sudden events rather than the traditional view of gradational change. The evidence has come from ocean bottom sediments (Ruddiman, Sancetta and McIntyre, 1977) from isotopic changes in the ice cores (Dansgaard and others, 1971) from palynology (Woillard 1979) and from palaeoentomology (Coope 1970, 1977). In many cases the effect of these sudden changes of climate was to throw the whole flora and fauna out of balance with the physical environment since different species were able to take advantage of the new conditions at very different rates. The resulting disharmony between the biota and the physical conditions means that it is not permissible to make direct comparisons between the fossil assemblages and the present day flora and fauna since the latter have had a long time of relatively stable climatic conditions during which an equilibrium state can have been at least approached. After a period of sudden climatic change, the large scale physical conditions may indeed be precisely comparable with those of the present day but the habitats on the ground, being made up of both physical and biological components may have been very different from any available at the present day because the biota takes time to adjust to the new conditions. What is becoming increasing clear is that, at least for the terrestrial ecosystems, this difference in timing is significant because of the suddeness and magnitude of Quaternary climatic changes.

Since it is evident that different species both of plant and animal colonise newly available areas at different speeds the incoming of a species in a stratigraphical sequence is influenced not only by the arrival of suitable physical conditions but also by the rate of spread of the species and the distance that it has had to traverse. The two latter variables are difficult to assess and so, though the presence of a species indicates the presence of suitable environmental conditions, these conditions may have prevailed for a considerable time prior to the arrival of the species in the area. Thus the use of the time of colonization of species is an unreliable index of the timing of climatic changes.

The local extermination of species by unacceptable conditions is not subject to the two problems of time and distance mentioned above. Extinction is sudden and can be precisely timed with the arrival of intolerable physical conditions, be they too cold or too warm. Of course there are problems of interpreting negative evidence. The absence of a species from the fossil record does not necessarily mean that the species is really extinct; the living coelacanth and monoplacophorans are a timely warning against excessive confidence. But when the absentees are both numerous and fit a consistent picture of ecological change, it is difficult to dismiss their disappearance as the result of sedimentological caprice. If we grant that the cessation of the fossil record of a species represents the real absence of that species locally it is still not permissible to use this local extermination of individual species as evidence of climatic change. The reasons for local extinction are legion and many involve factors quite unrelated to climate. When however, the extinctions involve a large number of species with widely differing ecological requirements but with broadly similar biogeographical ranges - say, the elimination of all exclusively arctic/alpine species - then the conclusion seems inescapable that some large scale change in the physical environment has taken place and the most likely of these is a change in the climate.

I have decided to concentrate attention here on changes in the insect fauna during the period of time that followed immediately on the retreat of the ice sheets of the last major glacial period, namely on the "Late Glacial" and early Flandrian. A great deal of interest has recently been shown in this period and it involves, certainly in Western Europe, several episodes of marked climatic change.

The insect fauna that followed up the retreating ice front in Britain was dominated by species that are today either restricted to arctic or alpine areas or else cosmopolitan species that are able to accept a wide range of climates. It is the former group that is of interest here. Two fossil insect assemblages can, with certainty, be attributed to this period since they were obtained from sediments that immediately underlie well authenticated Late Glacial sequences. The first of these was from Glanllynnau, North Wales (Coope and Brophy 1972) and the second was from Glen Ballyre, Isle of Man, (Coope 1971, Joachim 1978). Both sites were characterized by an ecologically varied assemblage of arctic/alpine species and since the sites are in the Irish Sea basin and relatively close geographically, the fossils will be treated as members of a single population.

Only those species actually involved in the episodic extinction will be mentioned. The Carabidae are a family of active ground beetles that are either carnivores or general scavangers. They were represented by *Bembidion fellmanni*, *B. hasti*, *B. lapponicum*, *Amara alpina* and *A. quenseli*. The Dytiscidae, carnivorous water beetles, include *Agabus arcticus*, *Potamonectes griseostriatus*, *Coelambus mongolicus*. The Hydrophilidae have carnivorous aquatic larvae but adults that feed on a variety of rotting plants, they include *Helophorus sibiricus*, *H. splendidus* and *H. obscurellus*. The Staphylinidae are chiefly predators of small soil arthropods and were represented by *Boreaphilus henningianus*. All these species are carnivorous at least at some stage in their life history and their food webs may be traced back to the algae and lichens as important primary producers of the arctic/alpine ecosystem. The numbers of phytophagous insect species at this time were conspicuously low. Most abundant were the Byrrhidae, whose larvae and adults are exclusively moss feeders, they included *Simplocaria metallica* and *Syncalypta cyclolepidia*. The weevils were rare with the exception of *Hypera obovatus* a species which is often associated with *Astragalus alpinus*. All the above species disappear suddenly from our fossil assemblages. They are ecologically diverse and include terrestrial and aquatic species, carnivores and plant feeders, so their almost simultaneous extermination can not be attributed to some local ecological change. The unifying feature that they all have in common is that they are arctic/alpine in their geographical ranges at the present day suggesting that a climatic change was responsible that was large enough to cross the tolerance limits of a number of species at the same time. The fact that this climatic change was able to eliminate such moderately northern species as *Agabus arcticus* suggests that the change was not only sudden but intense also.

To put an absolute date on this sudden cessation of arctic climatic conditions in central Britain, is difficult because the sediments that accumulated at this time are poor in plant remains. Some horizons, however, do yield sufficient organic matter for radiocarbon dates. We may therefore be certain that this sudden amelioration of climate was later than 14,468 ± 300 B.P. (Birm-212), a date obtained from a moss and terrestrial seed layer near to the base of the Glanllynnau sequence. It was certainly earlier than 12,556 ± 230 B.P. (Birm-176), a date obtained from terrestrial seeds 20 cm above the last horizon containing the arctic/alpine insect fauna at Glanllynnau. A radiocarbon date of 12,645 ± 280 B.P. (Birm-412) was obtained from coarse plant debris from a layer 5 cm above the last of the northern insect fauna at Glen Ballyre. The episode of sudden climatic amelioration seems best dated, in round numbers, to about 13,000 B.P.

The insect species that colonized Britain after the climatic amelioration just discussed, do not show any progressive tendency for the less temperate species to be followed later by the more thermophilous forms; rather the arrival of the first comers seems to depend on their being good fliers and able to inhabit open habitats with but a sparse or patchy vegetation cover. The present day geographical ranges of some of these species contrasts with those of their immediate predecessors. Thus two of the earliest species to reach Britain at this time were *Asaphidion cyanicorne* and *Metabletus parallelus* of mediterranean and south east European distribution respectively. From these beginnings the fauna became greatly enriched by the continual influx of temperate forms which, nevertheless, represent occupants of a wide variety of ecological niches.

The extermination of this temperate assemblage was as sudden and all embracing as that of the earlier arctic/alpine fauna. Because of the large number and diversity of species involved only a selection will be considered here. Insect faunas from four sites will be included: Glanllynnau; Glen Ballyre; Church Stretton, Shropshire (Osborne 1972), and St. Bees, Cumbria (Coope and Joachim 1980). Amongst the Carabidae *Bembidion octomaculatum, B. fumigatum, B. quadripustutatum, Microlestes minutulus, Cymindes angularis, C. macularis* and *C. humeralis* are all temperate species whose ranges extended further north during this part of the Late Glacial than they do at the present time. Amongst the water beetles were *Hydroporus granularis, Berosus signaticollis* and *Ochthebius foveolatus*. The phytophagous species were abundant at this time; *Donacia bicolor* and *D. marginata* both live on *Sparganium ramosum* but do not range today north of lat 62°N, well within the distribution of their host plant. Similarly the weevil *Sibinia sodalis* is a central and southern European weevil that lives on *Armeria* and its absence from northern Europe can not be accounted for by lack of its food plant. *Bruchidius debilis* is a beetle that lives on the seeds of various legumes but again its modern range does not include northern Britain, Denmark or Fennoscandia. The species named here are intended merely as representatives of an ecologically varied and numerically rich suite of thoroughly temperate insects that was largely exterminated with great rapidity in all of the four sites mentioned above and in other yet unpublished localities. The survivors of this catastrophe were almost all the widespread species whose climatic tolerance is seemingly broad based. Here again, therefore, it is the mass extinction of this temperate fauna, at least on a local scale, that indicates a rapid cooling of the climate.

It is easier to date this episode of climatic cooling because the sediments are richer in plant debris. At Glen Ballyre the faunal extinction can be dated to about 12,150 B.P. (GRO 1616). At Church Stretton the same episode can be dated at 12,135 ± 200 B.P. (Birm-158). At St. Bees the episode of extinction can be dated to about 12,000 B.P. after an allowance was made for hard water error. A satisfactory date could not be obtained for Glanllynnau because of rootlet penetration from above. From these dates (and others; see Coope 1977) it is clear that the sudden climatic cooling that brought about the widespread extermination of temperate insect species is best dated to a time rather before 12,000 B.P., that is at about the same time as coarse rock fragments entered the Late Glacial sediments at Low Wray Bay, Windermere suggesting climatic cooling and at about the commencement of the Older Dryas chronozone of continental authors (Mangerud and others, 1974).

After this sudden climatic deterioration there is no evidence from the insect fauna of any return of the temperate species mentioned earlier. For the next two thousand years there is a gradual loss of insect species, especially cool-temperate species, but these losses are made good by the gradual return of the arctic/alpine element. At no time, however, was there a period of mass extermination to compare with the earlier events, and the transition from "Allerod" time to "Younger Dryas" time is marked chiefly by a sudden invasion of truly arctic species and the increase in individual numbers of the northern species already present in the fauna.

The insect fauna of the Loch Lomond Stadial (= younger Dryas, = pollen zone III) is characterized by the presence of numerous arctic/alpine species many of which were present here in the insect fauna that followed the retreating ice front three thousand years earlier. Many sites have now yielded insect faunas of this period and they are very similar to one another throughout central Britain. Over one hundred and forty species have been recognised of which thirty four (25%) are now absent from the British Isles; all but two of which are, however, still living in Fennoscandia. The two exceptions are Siberian species. The richest insect assemblage from this period was from Drumurcher, Co. Monaghan (Coope and others, 1979) and similar faunas have been described from Church Stretton and St. Bees. At Glanllynnau the deposits of this age were poorly fossiliferous. This widespread, varied and exotic insect fauna was drastically cut back at the opening of the Flandrian period. All the "non-British" species were suddenly eliminated from lowland Britain and many other northern British species were exterminated at the same time.

Radiocarbon dates suggest that the arctic/alpine fauna thrived throughout Britain until about 10,000 B.P., an unfortunate figure because its roundness hints at a crude approximation, whilst in fact there are many radiocarbon dates available to pinpoint this period. At Croydon this fauna was dated to 10,130 ± 120 B.P. (Birm-101), (Peake and Osborne 1971). In the Gipping valley, Suffolk, this fauna was dated to 9,880 ± 120 B.P. (Har-259), (Rose and others, 1980). At Newtown Lane, West Bromwich, this fauna had a date of 10,025 ± 100 B.P. (St-3686) (Osborne 1980). However, at Newtown Lane all the arctic/alpine species had apparently gone by 9,970 ± 110 B.P. (St-3060). It is curious that the complimentary influx of temperate species shows little delay after the episode of mass extermination of the northern fauna, so that by about 9,500 B.P. (average of six 14C dates) (Osborne 1974) and by 9640 ± 180 B.P. (Q-398) (Bishop and Coope,1977) a thoroughly thermophilous assemblage of insect species was widespread, even as far north as south west Scotland.

In summary, there were three episodes of local extinction of insect species during the Late Glacial period, two of which correspond to sudden climatic ameliorations (at about 13,000 B.P. and at 10,000 B.P.) and one to a sudden cooling (at just before 12,000 B.P.). Although it is difficult to interpret the magnitude of the climatic changes involved because of the negative character of the evidence used, it seems likely that the climatic warmings at 13,000 B.P. and 10,000 B.P. were on the same scale as one another, as they wiped out, to a similar degree each time, a characteristic arctic/alpine fauna and also eliminated species that find conditions acceptable in the north of Britain today. The scale of this extermination suggests that an increase of at least 5°C occurred in the average July temperature at these times. That these ameliorations no doubt involved a greater change than this is indicated by the incoming southern British species at this time extending their ranges further north than they seem able to do at the present day.

The single episode of mass extermination due to a cooling of the climate at just over 12,000 B.P. is interesting because it involves such a large number of species. The survivors however, suggest that it did not involve as great an amplitude of change as the episodes of amelioration. A drop of about 3°C in average July temperatures would have been adequate to eliminate many of the southern species that disappeared at this time. This is an inadequate drop in summer warmth to have exterminated much of the flora of these times and not enough to have inhibited the spread of birch woodland. There is thus no reason why vegetational development should not have continued in spite of the cooling climate.

An interesting outcome of this study is that there is little evidence for any mass extinction of insect species at the start of pollen zone III when the birch woodland began to be eliminated from England. This period is marked chiefly by the further incursion into Britain of arctic/alpine insect species even as far as the extreme south of England. After their mass extermination 10,000 years ago they never again returned to lowland regions.

REFERENCES

Bishop, W.W. and G.R. Coope (1977). Stratigraphical and faunal evidence for late glacial and early Flandrian environments in South West Scotland. In J.M. Gray and J.J. Lowe (Eds), *Studies in Scottish Late Glacial Environments*, Pergamon Press, Oxford. pp. 61-88.

Coope, G.R. (1970). Climatic interpretation of Late Weichselian Coleoptera from the British Isles. *Revue Géogr. phys. Géol. Dyn.*, 12, 149-155.

Coope, G.R. (1971). The fossil Coleoptera from Glen Ballyre and their bearing upon the interpretation of Late Glacial environments. In G.P.S. Thomas (Ed.), *Q.R.A. Field Guide to the Isle of Man*, Univ. Liverpool. pp. 13-15.

Coope, G.R. and J.A. Brophy (1972). Late glacial environmental changes indicated by coleopteran succession from North Wales. *Boreas*, 1, 97-142.

Coope, G.R. (1977). Fossil coleopteran assemblages as sensitive indicators of climatic changes during the Devensian (last) Cold Stage. *Phil. Trans. R. Soc.*, B B 280, 313-340.

Coope, G.R., J.H. Dickson, J.A. McCutcheon and G.F. Mitchell (1979). The late glacial and early post glacial deposit at Drumurcher, Co. Monaghan. *Proc. R. Ir. Acad.*, B 79, 63-85.

Coope, G.R. and M.J. Joachim (1980). Late glacial environmental changes interpreted from fossil Coleoptera from St. Bees, Cumbria, N.W. England. In J.J. Lowe, J.M. Gray and J.E. Robinson (Eds.), *Studies in the Late glacial of North West Europe*, Pergamon Press, Oxford.

Dansgaard, W., S.J. Johnson, H.B. Clausen and C.C. Langway Jr. (1971). Climatic record revealed by the Camp Century ice core. In K.K. Turekian (Ed.), *Late Cenozoic glacial ages*, Yale University Press, New Haven and London. pp. 37-56.

Joachim, M.J. (1978). *Late glacial coleopteran assemblages from the west coast of the Isle of Man*. Ph.D. thesis University of Birmingham.

Mangerud, J., B.E. Andersen, B.E. Berglund and J.J. Donner (1974). Quaternary stratigraphy of Norden, a proposal for terminology and classification. *Boreas*, 3, 109-127.

Osborne, P.J. (1972). Insect faunas of late Devensian and Flandrian age from Church Stretton, Shropshire. *Phil. Trans. R. Soc.*, B 263, 327-367.

Osborne, P.J. (1974). An insect assemblage of early Flandrian Age from Lea Marston, Warwickshire, and its bearing on the contemporary climate and ecology. *Quat. Res.*, 4, 471-486.

Osborne, P.J. 1980. The late Devensian-Flandrian transition depleted by serial insect faunas from West Bromwich, Staffordshire, England. *Boreas*, 9, 134-147.

Peake, D.S. and P.J. Osborne (1971). The age of the Wandle Gravels in the vicinity of Croydon. *Proc. Trans. Croydon nat. Hist. scient. Soc.*, 14, 145-175.

Rose, J., C. Turner, G.R. Coope and M.D. Bryan (1980). Channel changes in a lowland river catchment over the last 13,000 years. In R.A. Cullingford, D.A. Davidson and J. Lewin (Eds), *Time Scales in Geomorphology*, Wiley and Sons, Chichester, New York, Brisbane and Toronto. pp. 159-175.

Ruddiman, W.F., C.D. Sancetta and A. McIntyre, Glacial/Interglacial response rate of subpolar North Atlantic waters to climatic change: the record of oceanic sec-iments. *Phil. Trans. R. Soc.*, B 280, 119-142.

Woillard, G. (1979). Abrupt end of the last interglacial *s.s.*, in north-east France. *Nature Lond.*, 281, 558-562.

Dr. G.R. Coope,
Department of Geological Sciences,
University of Birmingham, P.O. Box 363,
Birmingham B15 2TT

21. Scottish Late-Glacial Marine Deposits and their Environmental Significance

J. D. Peacock

INTRODUCTION

In his great monograph on the fauna and flora of the British Isles, Edward Forbes (1846, p. 365) acknowledged his debt to the Glasgow amateur geologist, James Smith of Jordanhill, who was the first worker to realise the significance of the fossil remains of cold water mollusca in Scottish raised marine deposits and thus to lay a foundation for the study of Pleistocene biogenic deposits in Britain (Smith, 1838). This work soon bore fruit in the researches of T.F. Jamieson (1865) and particularly those of Crosskey and Robertson (1867-74) in the latter part of the 19th Century. These workers, together with an enthusiastic band of amateurs centred chiefly on Glasgow and Paisley applied the rapidly growing body of knowledge concerning the distribution of modern plants and animals, especially the marine mollusca, to the study of the raised marine and littoral deposits found on both the east and west coasts of Scotland. It was soon realised that these beds contained three distinct faunas. Strata with truly arctic faunas were found to be widespread around the Firth of Forth and Firth of Tay, whereas 'post-Tertiary' beds about the Clyde estuary carried fossils of a more boreal character, though still with arctic elements. Both of these were distinct from the post-glacial beaches and estuarine deposits which carried fossils of species to be found around the British Isles today. These researches were confirmed and extended during the mapping of coastal districts of Scotland by the Geological Survey, but it is probably fair to say that, with a few exceptions, little was added to the picture supplied by the 19th Century workers until the advent of radiocarbon dating and the exploration of the continental shelf. In the following account most attention is given to the raised marine deposits found on shore or readily accessible from the shore.

The raised shorelines around Scotland have been the subject of intensive study in recent years, chiefly by J.B. Sissons and his co-workers (summaries in Sissons 1976, and Jardine 1977). In brief, such shorelines comprise a *late-glacial series* formed as isostatic recovery following the decline of the Devensian ice-sheet exceeded the eustatic sea-level rise, and a *post-glacial series*, formed after the eustatic rise in sea-level temporarily overtook isostatic recovery. Both series decline outwards from a centre or centres of uplift over the West Highlands and are not found in the Outer Hebrides, Orkney and Shetland. To date, the ages of the late-glacial beaches are speculative, and the post-glacial beaches have been satisfactorily dated only in the Forth and Tay estuaries. Perhaps the most important of these features are the Main Post-glacial Shoreline, with an age in the Forth of about 7,000 B.P. and the late-glacial Main Perth Shoreline in eastern Scotland undated, but with an inferred age in the period 13,000 - 13,500 B.P. Another important feature, widespread in both east and west Scotland, is the so-called Main Late-glacial Shoreline comprising a rock platform and cliff with an associated gravel layer. Its formation has been ascribed to intense marine erosion aided by the periglacial conditions accompanying the Loch Lomond Readvance (Sissons, 1974; Gray, 1978). Implicit in accounts of the late-glacial shorelines is the assumption that some of these can be related to ice-front positions, but there are few localities where this can be satisfactorily demonstrated in detail (see Armstrong, Paterson and Browne, 1975).

Raised marine, estuarine and littoral deposits of late- and post-glacial age are widely distributed on land near the coast of Scotland (Fig. 1) and have been recorded offshore in a number of bores. Because the Scottish shoreline has migrated considerably with the interplay of isostasy and eustasy following the disappearance of the late-Devensian ice, the deposits display lateral and vertical variation greater than those to be expected even in a complex nearshore environment. Consequently it has not yet proved practicable to set up formal lithostratigraphic units tied to type sections. Sections in these deposits are also transitory by their very nature, the beds usually being exposed in temporary excavations. In this account only informal names are used, but, as discussed below, there is evidence to suggest that climatostratigraphical subdivisions occur within parts of the succession.

The late-glacial marine strata are subdivided into the *Errol beds*, found on the east coast of Scotland in the Forth and Tay estuaries, and the *Clyde beds*, which occur chiefly in the Clyde estuary and various localities northwards to Loch Linnhe. The former are thought to have been laid down between the Devensian maximum at 18,000 B.P. B.P. and roughly 13,500 B.P. and the latter between 13,500 B.P. and the late-glacial/post-glacial boundary at 10,000 B.P. (see below). Correlatives of these beds occur offshore and at other localities on land. The distribution of post-glacial estuarine deposits (carse clays), found in all east coast estuaries, but only sparsely north of the Solway coast in western Scotland, seems to reflect a fundamentally different hydrographic regime on west and east coasts which is not evident in the distribution of late-glacial deposits.

ERROL BEDS

The term 'Errol beds' was introduced by Peacock (1975) to replace terms such as 'Arctic Clay of Errol' (Davidson, 1932) and is applied informally to raised marine clays, silts and sands which contain a distinctive arctic fauna and which are widely distributed around the east Scottish coast (Fig. 1). The sediments are commonly well to poorly laminated and contain scattered pebbles and boulders which are probably ice-rafted. In colour they are dominantly reddish brown, but grey and 'blue' clays have also been described as well as black, organic, probably sulphide-rich intercalations of sand (Davidson, 1932). At many localities the deposits are several metres thick, but in channels and basins much greater thicknesses are known. Sections currently available for inspection are weathered to the full depth and in spite of intensive searching in recent years, no material suitable for radiocarbon analysis has been found. The deposits often extend well above the highest raised beaches (Armstrong, Paterson and Browne, 1975; Tait, 1934) to a marine limit as high as 40 m O.D. Away from the Forth and Tay estuaries the upper limit of these beds probably descends below sea-level, e.g. they are not known on shore north of Montrose (Jamieson, 1865).

The fauna of the Errol beds, which are probably partly glaciomarine, is of low diversity and of mid- to high arctic aspect (*sensu* Feyling-Hanssen, 1955). It is characterised by molluscs and ostracods such as *Portlandia arctica* (Gray)*, *Palliolum groenlandicum* (Sowerby), *Rabilimis mirabilis* (Brady), *Cytheropteron montrosiense* Brady, Crosskey and Robertson and *Krithe glacialis* Brady, Crosskey and Robertson.

*All species names are in a modern form in this account.

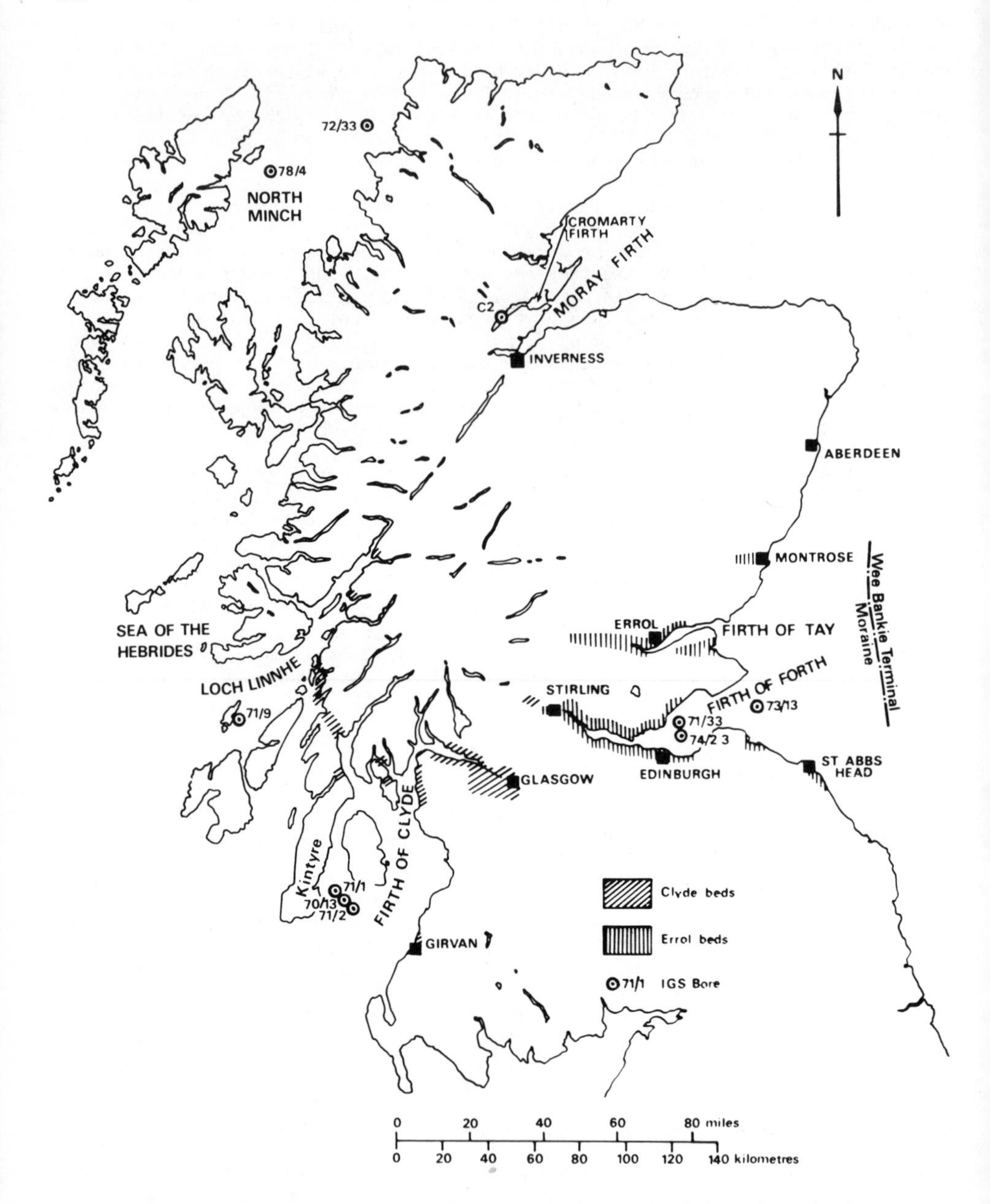

Fig. 1. Distribution of Scottish raised late-glacial marine strata.

Of these, *C. montrosiense* is thought to be extinct (Neale and Howe, 1975) and *R. mirabilis* does not occur in the Clyde beds (see below). The brittle star *Ophiolepis gracilis* Allman has been reported from several localities.

Offshore, the St. Abbs Beds up to 20 m thick have been correlated with the Errol beds on the basis of their similar fauna and lithology (Thomson and Eden, 1977). They occur chiefly within the limit of the Wee Bankie terminal moraine, which may mark the limit of late-Devensian ice near this part of the Scottish coast. Correlatives of the Errol and St. Abbs beds are probably present off northern Scotland, where *R. mirabilis* is known from offshore bores (see below), but only doubtfully present in SW Scotland, where *Portlandia arctica* has been reported in possible marine clays at two localities (Brady, Crosskey and Robertson, 1874).

14C age B.P.		Raised marine strata		Offshore Bores	
		W. Scotland	E. Scotland	W. Scotland	E. Scotland
10,000–11,000	Loch Lomond stade	Clyde beds	Ardullie beds† ----- Findon beds† (upper)	Formation 3*	Forth Beds (lower)
11,000–13,500	Windermere Interstadial		Findon beds† (lower)		
13,500–18,000		(None known)	Errol beds (Forth and Tay)	Glacio-marine* beds (Formation 2)	St. Abbs Beds

* Binns, Harland and Hughes 1974b for Sea of the Hebrides
† Peacock, Graham and Gregory 1980 for Cromarty

Table 1. Tentative Correlation Table for Scottish Late-Glacial Marine Strata.

CLYDE BEDS

Nearshore Evidence

The term Clyde beds is applied informally to late-glacial marine strata with a generally high-boreal fauna (*sensu* Feyling-Hanssen, 1955) which are found around the coast of the Firth of Clyde and northwards along the west coast of Scotland (Fig. 1). Most of the deposits are within a few metres of Ordnance Datum, but in the inner Clyde estuary there are extensive thick deposits above 10 m O.D., rising locally to 35 m O.D. No undisturbed marine deposits above O.D. have yet been reported north of Loch Linnhe. Southwards, Clyde beds are known from Girvan, (Crosskey and Robertson 1874), where they pass upwards into fresh water deposits.

At a number of localities, the succession begins with pink or brown laminated silts, sands and clays which are commonly less than a metre thick in the Clyde upstream of Greenock (e.g. Rose in Jardine, 1980), but between 2 and 3 m thick at some sites adjacent to the Firth of Clyde, for instance at Ardyne (Peacock, Graham and Wilkinson, 1978) and Portavadie (Figs. 1 and 3). The laminae in the lower beds are, at least locally, coarser-grained than in the upper (Peacock, Graham and Wilkinson, 1978) and Rose (*in* Jardine, 1980) reports a proximal basal lamina at one site, with an upward thinning of the laminations. Though at first sight very well sorted, the washed residues contain small angular rock fragments and there is locally evidence of current action. The sediments yield a sparse marine fauna of low diversity, consisting chiefly of Foraminifera and Ostracoda (Graham and Wilkinson, 1978), but with juveniles of infaunal molluscs (*Portlandia* and *Thyasira*) locally and fragments of these and other organisms. These facts suggest very rapid deposition by bottom traction currents and by turbid surface meltwater plumes in the neighbourhood of a retreating glacier front, probably adjacent to land where the superficial deposits were yet to be stabilised by a vegetation cover (*cf*. Peacock, Graham and Wilkinson, 1978).

The bulk of the Clyde beds consist of grey and greyish brown clayey silts and sands with black sulphide-rich patches and a variable proportion of gravel and cobbles. The sorting of the silts and clays is usually less good than in post-glacial estuary-fill deposits, and angular rock fragments and fractured quartz grains are commonly present. These, together with scattered cobbles and boulders at some locations suggest ice rafting of debris from the adjacent shore (*cf*. Jamieson, 1865) and thus the presence of winter sea ice. Though the sections currently available for study usually show less than 5 m of deposit, thicknesses of more than 20 m occur in the Glasgow area and more locally in buried valleys, for instance at Lochgilphead (Fig. 3) and adjacent to Loch Linnhe. The thick late-glacial marine sediments in the Glasgow district, which are often reported as laminated (presumably where little bioturbated), have probably been reworked from the extensive glacio-lacustrine clays and silts associated with, or lying on, the till in the middle Clyde valley (*cf*. Browne, *in* Jardine, 1980, p. 13) (see below).

From the marine shell radiocarbon dates obtained so far, it is considered that deposition of the Clyde beds began before 12,700 B.P.* and probably prior to 13,100 B.P. (Browne and others, 1977). However, the range of radiocarbon ages relating to the beginning of deposition of the shelly beds, whether or not the subjacent laminated beds are present, is disconcertingly large (13,100 to 11,550 B.P. in rounded-off units) and at first sight raises questions both as to the validity of the radiocarbon dates and interpretations of the stratigraphy. Peacock, Graham and Wilkinson, (1978) concluded that major errors in the radiocarbon dates are unlikely in view of the good fit of such dates with climatic events in the later part of the late-glacial sequence. The range of dates and the location of the dated samples precludes a relationship to any systematic scheme of deglaciation over an extended period and is more likely to be a function of sedimentation rate at the base of the shelly beds, as discussed next.

It is suggested that the sequence of events in near-shore marine situations was, firstly, rapid deglaciation with diachronous deposition of the lower, coloured laminated beds. Such deposition may have taken place within a few years at any one site. This was followed, secondly, by very slow (or no) deposition when the sediment supply was sharply reduced or cut off as the ice withdrew into the hills and a closed pattern of vegetation became established on low ground. The reduction in sediment supply is testified by the sudden appearance of a diverse marine fauna with accompanying bioturbation and by the incoming of species such as *Arctica islandica* (Linné) which flourish only in clear water (Schäfer, 1972, p. 376). This model, which differs slightly from that proposed by Peacock, Graham and Wilkinson, (1978), is further

---

*All marine shell ages are quoted as adjusted ages, roughly comparable to ages on terrestrial organic matter.

supported by the wide range of radiocarbon ages from the base of the shelly beds at two localities where multiple dating has been carried out. At Inchinnan west of Glasgow, Browne and others reported dates of 13,100 and 12,000 B.P. from the same horizon, and three samples from the base of the shelly beds at Geilston, west of Dumbarton yielded ages between 12,300 and 11,800 B.P. (Harkness and Wilson, 1979).

The range of radiocarbon dates referred to above suggests that marine sedimentation in nearshore areas was slow or absent for several hundred years following deglaciation. In the Glasgow and Paisley areas, however, much of the thick drape of Clyde beds, which extends up to 15 to 20 m O.D., was probably laid down before 11,800 B.P., as sea-level seems to have been below such levels after this date (Mitchell, 1952; Peacock, Graham and Wilkinson, 1978). Elsewhere in western Scotland, the thin, high -level deposits laid down during the fall of sea-level prior to this date may not have been preserved.

After roughly 12,000 B.P. there is evidence that the rate of deposition of marine sediments increased. In the Leven valley, between the south end of Loch Lomond and the Clyde (Fig. 3), a thick late-glacial sequence was laid down after 11,800 B.P. (Browne *in* Jardine, 1980, p. 11) and at Ardyne, Portavadie, Geilston and Inverkip (west of Greenock) most of the relatively thin deposits accumulated after 12,000 B.P. (Harkness and Wilson, 1979; Peacock and others, 1977). Moreover, the marine shell radiocarbon ages obtained so far from reworked marine sediments in the moraines of the Loch Lomond Readvance all fall in the period between 12,000 and 11,000 B.P. (Peacock, 1971; Rose *in* Jardine, 1980, p. 33). As the period between 12,000 and 11,000 B.P. was one when temperatures on land were low, if not arctic (Coope, 1977), it is possible to speculate that frost action, solifluction and perhaps rejuvenating glaciation in the Highlands at this time increased the load carried by the rivers and streams and thus the rate of sedimentation in the sea-lochs.

The fauna of the Clyde beds, which at Ardyne contains some 60 to 70 species of Mollusca, about half that number of Ostracoda and 50 to 60 species of Foraminifera, is considerably more diverse than that of the Errol beds, but less so than the post-glacial fauna. In addition to many eurythermal species, or species which reach their southern limits in the British Isles today, it includes cold water forms which are often dominant. Thus foı example *Elphidium clavatum* Cushman, commonly dominates the population of Foraminifera, and the cool eurythermal species *Elofsonella concinna* (Jones), *Eucytheridea bradii* (Norman) and *E. punctillata* (Brady) the Ostracoda. *Macoma calcarea* (Chemnitz) is a common, often dominant mid-boreal to arctic species among the Mollusca, and is joined by cold stenothermal species of *Portlandia* not now present around the British Isles, such as *P. lenticula* (Möller) and *P. fraterna* (Verrill and Bush) in faunas indicating deeper water.

In his last paper on the Garvel Park Dock sections, near Greenock, Robertson (1883) noted firstly that in some sections of Clyde beds, strata with offshore faunas were succeeded by those with shallow water or even littoral species, and secondly that at Garvel Park itself there was an incoming of warmer water molluscs near the top. In the same volume of the Transactions of the Geological Society of Glasgow, Scott and Steel (1883) described another section at Garvel Park in which beds characterised by arctic stenothermal molluscs overlay the beds with a boreal fauna. The relationships at Garvel Park are summarised in Fig. 2 and compared with the New Graving Dock section described by Bishop and Dickson (1970) and Bishop and Coope (1977).

Though no faunal determinations have been published, it is likely that the arctic bed of Scott and Steel is represented in the New Graving Dock section by thick deposits radiocarbon dated to the Loch Lomond (Younger Dryas) stade. Both beds are characterised by horizons with much seaweed and land plant debris, and the dated molluscs reported as *Astarte sulcata* (da Costa) are probably the similar *A. elliptica* Brown, said to be frequent in the bed with arctic molluscs. Excepting the basal laminated silts, strata older than the Loch Lomond stade are thin or absent in the New

Graving Dock section (Fig. 2).

The suggestion that the incoming of warmer water Mollusca at Garvel Park occurred immediately prior to the very cold period of the Loch Lomond stade has been confirmed in radiocarbon dated sections at Ardyne (Peacock, Graham and Wilkinson, 1978).

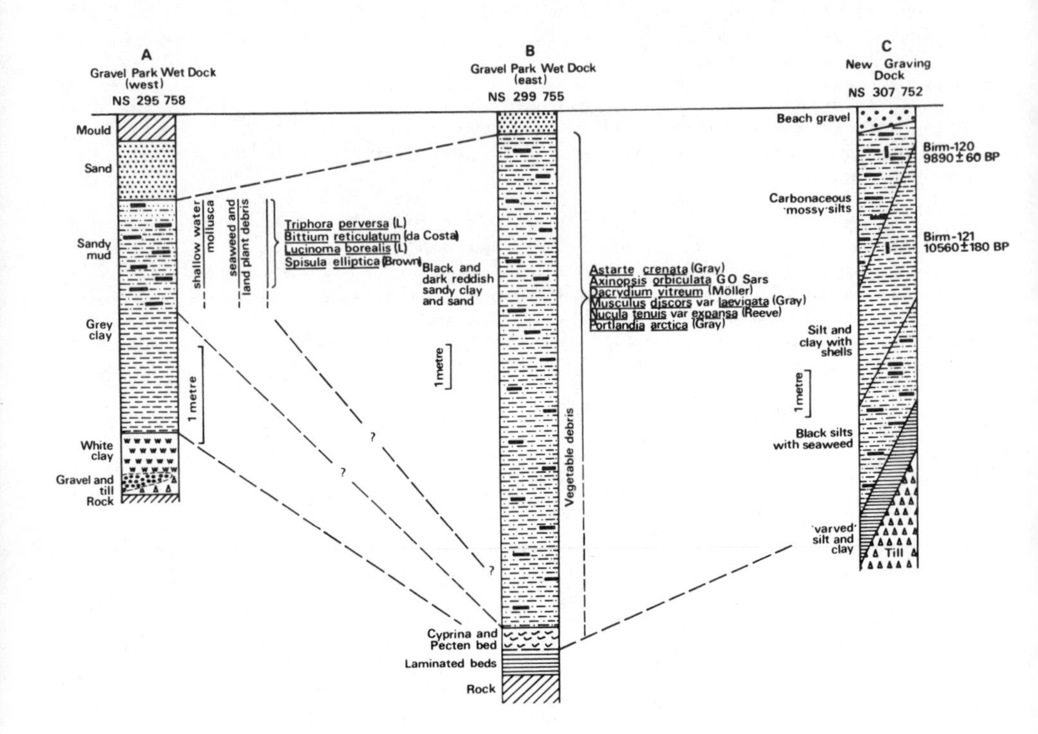

Fig. 2. Generalised sections of late-glacial strata at Garvel Park, near Greenock. Section A based on Robertson (1883), and Section B on Scott and Steel (1883). Section C based on Bishop and Coope (1977) and Bishop and Dickson (1970). Tops of Sections A and B probably slightly above O.D.; top of Section C - 0.5 m below O.D.

The warmer interval, during which a number of lusitanian to mid-boreal Mollusca and Foraminifera (in particular *Ammonia batavus* Hofker) immigrated into the area, is dated here and at Lochgilphead to about 11,000 B.P. However, though summer temperatures were *relatively* high at this time, the winters were probably very cold (Peacock and others, 1977).

In the strata referred to the Loch Lomond stade at Ardyne, low water temperatures were inferred from the fauna of boreo-arctic species, an interpretation reinforced by the occurrence of the arctic stenothermal bivalve *Portlandia arctica*. *P. arctica* is characteristic of glaciomarine conditions in East Greenland today (Ockelmann, 1958). However, its occurrence merely as juveniles at Ardyne may have been related to the presence of glaciers of the Loch Lomond Readvance some distance away rather than to glaciers nearby. It is noteworthy that at Garvel Park, situated in the drainage area of the Gareloch glacier and the great Loch Lomond glacier itself, the fauna included a much greater proportion of arctic species than that at Ardyne, though it was also said to contain *Arctica islandica* (Linné), a species which does not extend into mid- or high-arctic seas at present. The marine fauna of the Loch Lomond stade thus seems likely to have been boreo-arctic in character with only a few arctic stenothermal species. This implies that either the time interval was too short for the immigration of many arctic stenotherms, or, more likely, the marine environment was one with a considerable range of temperature, being the meeting place of polar and Atlantic water as well as glacier meltwater. Recognition of the Loch Lomond stade in marine strata, particularly in offshore bores, may prove difficult in the absence of radiocarbon ages.

Robertson's (1883) very generalised interpretation of shallowing with time in Clyde bed successions has been borne out by recent investigations, and sea-level curves have been constructed for two sites in the Firth of Clyde (Peacock and others, 1977 and 1978). These must be regarded as extremely tentative. They do indicate in a general way that sea-level fell rapidly before 12,000 B.P. From a little after this date to 11,000 B.P. the level seems to have been static at a few metres above Ordnance Datum, and may have remained so until 10,000 B.P. At Ardyne, there is evidence for an erosive episode post-dating 11,000 B.P. during which sediments were locally removed to a depth of about -6 m O.D. On the basis of statistical errors in the radiocarbon dates this interval could have been as much as 500 years (Peacock, Graham and Wilkinson, 1978). In the absence of beach deposits it was considered unlikely that this coincided with a period of reduced sea-level. At Garvel Park (Fig. 2) a possibly contemporaneous episode of non-deposition and/or erosion can be inferred at the base of strata referred to the Loch Lomond stade, but evidence for a change in the type of sedimentation and thus for major sea-level changes is lacking. Marine sedimentation at another site (Portavadie, also in the Firth of Clyde) apparently continued until 10,700 B.P. or later at an altitude only a little below Ordnance Datum (Harkness and Wilson, 1979, p. 255).

Other difficulties arise with the interpretation of depth in late-glacial marine strata. For example, in the Cromarty Firth, the microfauna found at depth in boreholes is of a type associated with shallow water, whereas the raised beaches suggest contemporaneous high sea-levels. At Cromarty, this has been explained by invoking sediment transport at a delta front (Peacock, Graham and Gregory, 1980), but could more generally reflect an unstable rapidly-changing late-glacial environment without a known modern analogue. The common occurrence in the Clyde beds at Paisley of the usually intertidal mussel *Mytilus edulis* with species indicative of sublittoral or offshore conditions (Robertson, 1883) might be explained on the same basis. Apparent anomalies such as these may disappear as the presently sketchy knowledge of modern environments and their faunal associations improves.

Inspection of available records shows that the Clyde bed fauna in the interval predating the Loch Lomond stade varies between the outer coasts of the Firth of Clyde and the inner estuary. There is a marked reduction in, for instance, molluscan species (60 - 70 at Ardyne, 35 at Inchinnan, near Renfrew and nil south-east of Glasgow). *Littorina littoralis* Linné is common in very shallow water deposits adjacent to the Firth (Fig. 3) but its place is taken in the estuary by *L. palliata* (Say). The latter today replaces *L. littoralis* northwards on the coast of north-west Norway (Sars, 1878) but in the Clyde bed context the critical factor would appear to have been the upriver decrease in salinity.

Little use has yet been made of shell form and clines in palaeo-environmental studies, both of which would probably repay examination, though attempts have been made to use the size of adult shells, e.g. *Hiatella* (Strauch, 1968) and *Portlandia lenticula* (Aarseth and others, 1975). As an example of the former, the shallow-water, weed-loving prosobranch *Rissoa parva* (da Costa) is represented in the Clyde beds almost entirely by the *interrupta* form of Adams (1798), thought to survive a more extreme environment than the 'normal' ribbed variety (Wigham, 1975). The great predominance

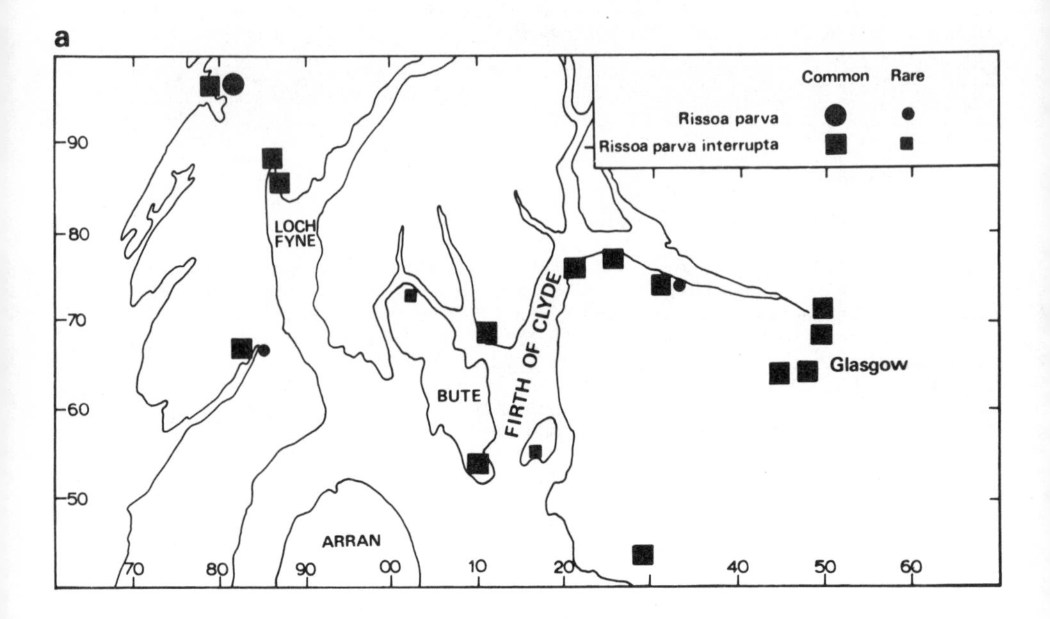

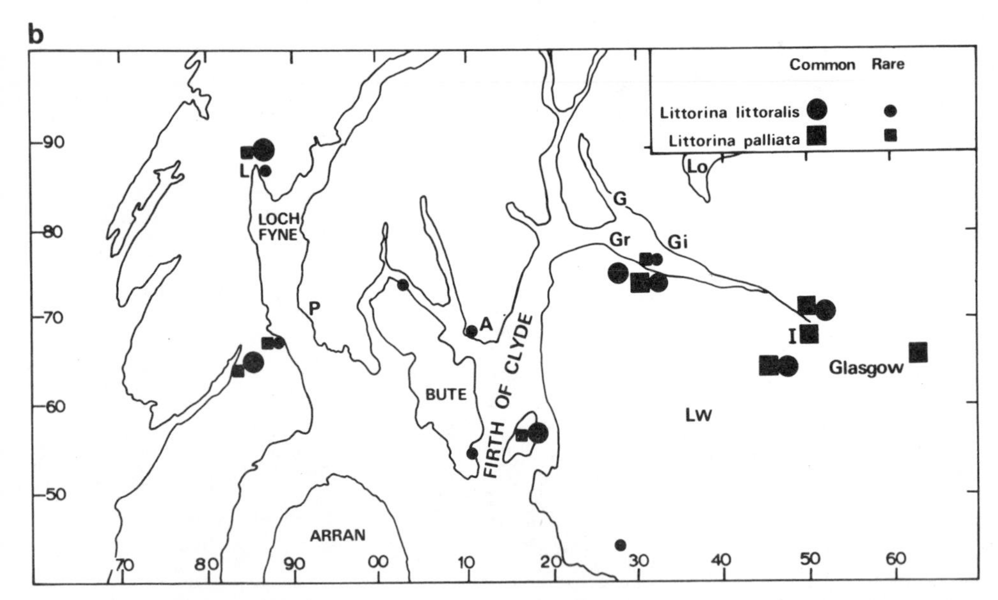

Fig. 3. Distribution of *a. Rissoa parva* (da Costa) and *R. parva interrupta* (Adams), and *b. Littorina littoralis* (Linné) and *L. palliata* Say in the late-glacial raised marine strata of the Firth of Clyde.

A Ardyne, G Gareloch, Gi Geilston, Gr Greenock, I Inchinnan, L Lochgilphead, LW Lochwinnoch gap, Lo Loch Lomond, P Portavadie.

of R. *parva interrupta* probably signifies cold conditions irrespective of salinity, and the isolated station where the ribbed form is common (Fig. 3) may relate to the 'warm' interval predating the Loch Lomond stade. The *Astarte montagui* cline (Jensen, 1912) might be a sensitive indicator of palaeotemperature in late-glacial strata, particularly as it is susceptible to statistical analysis. Investigations so far by the writer suggest that only forma *typica* (with a present-day boreal distribution) occurs in the Clyde beds, at least in strata predating the Loch Lomond stade. This reinforces the overall view that the marine climate during the bulk of the Windermere Interstadial was similar to that in western Norway between Lofoten and the North Cape today.

Offshore Evidence

Sediments of probable late-glacial age with faunas indicating a marine climate similar to that of the Clyde beds have been reported from a number of offshore boreholes as well as from boreholes in the estuary fills of the Cromarty Firth and the Firth of Forth (Fig. 1). In the Sea of the Hebrides, late-glacial and post-glacial muds disposed in basins locally reach thicknesses of over 150 m (Binns and others, 1974 a, 1974 b) including beds interpreted as glaciomarine (see below, p.232) and a comparison has been made with the drape of glaciomarine beds now accumulating in front of glaciers in Spitsbergen (Boulton, Chroston and Jarvis, 1980). Evidence for a 'temperate' interval with high percentages of *Ammonia batavus* within the upper part of the late-glacial succession have been found at two borehole sites (71/9, Binns, Harland and Hughes, 1974 b and 78/4, Gregory *in* Peacock, Graham and Gregory, 1980 (Fig. 1)). In 71/9 Robinson (1980) records cold-water ostracods, including the arctic stenothermal species *Krithe glacialis* (Brady, Crosskey and Robertson) in the beds overlying those with *A. batavus* and reinforces the suggestion that the former were deposited during the Loch Lomond stade. Similar variable, but locally very thick late- and post-glacial deposits in which a 'temperate' late-glacial interval has been identified on the basis of dinoflagellate cyst assemblages occur between Kintyre and the mainland (Harland *in* Deegan and others, 1973).

On the east coast of Scotland, boreholes in the Cromarty Firth have passed through strata which have been correlated with the Windermere Interstadial and the Loch Lomond stadial on the basis of their lithology and fauna (Peacock, 1974; Peacock, Graham and Gregory, 1980). A horizon rich in *Ammonia batavus* occurs in one of these bores immediately below strata referred to the Loch Lomond stade, which is here represented in part by strata in which the fauna is characterised by low productivity and diversity. Late-glacial sediments with a high-boreal marine fauna also occur at Inverness below bouldery deltaic deposits referred to the Loch Lomond stade (Peacock, 1977), and it is likely that the shells of cold-water molluscs dredged east of Aberdeen (Jamieson, 1865) came from similar strata exposed on the sea floor.

In the Firth of Forth, strata probably equivalent to the Clyde beds are known in boreholes at the head of the estuary (Read *in* Francis and others, 1970; Peacock,1974) and have been found in other boreholes in the central and eastern parts of the Firth (Thomson and Eden, 1977, Thomson, 1978). In these latter, the St. Abbs Beds are overlain by the Forth Beds, the lower part of which contains a more diverse cool water marine fauna (Gregory and others, *in* Thomson, 1978) similar to that of the Clyde beds. Four boreholes (Fig. 1) have yielded evidence of a 'temperate' interval within the late-glacial succession, apparently at or near the Forth Beds/St. Abbs Beds boundary.

SOME GENERAL CONSIDERATIONS

The replacement of the Errol faunal assemblage by that characteristic of the Clyde beds marks an important change from arctic to boreal conditions in nearshore marine sediments comparable to that recorded about 13,000 - 13,500 B.P. by the planktonic Foraminifera in deep sea cores west of the British Isles and the Coleoptera on land (Ruddiman and McIntyre, 1973; Peacock, 1974; Coope, 1977). No radiocarbon dates are yet available to date this event in eastern Scotland, but on the other side of the North Sea, the transition from the Danish Younger Yoldia Clay to the overlying Upper Saxicava Sand, which seems to mark the equivalent change in marine climate, may be as young as 12,800 B.P. (Knudsen, 1978). On the coast of western Norway, the oldest boreal faunas are dated about 12,700 B.P. (Mangerud, 1977), which is probably a minimum date. As the oldest Clyde beds faunas have yielded ages as high as 13,100 B.P. (Browne and others, 1977), the possibility must be considered that the change from arctic to boreal conditions took place in the North Sea several hundred years later than in the open Atlantic. However, it seems unlikely that an abrupt climatic change taking place on the west Scottish coast would be unrecorded in the east, at a time when the ice there had already shrunk back into the Highland valleys (Armstrong, Paterson and Browne, 1975), with a reduction in the very cold water available to support cold stenothermal species. The apparent diachroneity is likely to be at least partly a function of uncertainties in the few radiocarbon assays dating the climatic transition. Moreover, as parts of the central North Sea may have been dry land (Sissons, 1980) or perhaps very shallow water at this time, the Danish deposits referred to above were possibly laid down in a separate basin, where conditions regarding, for instance, the proximity of a large continental ice-sheet, may have been different from those on the east coast of Scotland. It follows from the above that the arctic to boreal change in coastal waters is likely to have been more or less simultaneous on both east and west coasts and can be used as a time-marker for deglaciation (Peacock, 1975).

On the west coast of Scotland, there is no evidence of arctic faunas in raised marine strata, other than those of the Loch Lomond stade, but offshore (Fig. 1), arctic faunas are known in two boreholes, one (72/33) south-west of Cape Wrath (Robinson *in* Peacock, 1975) and the other (78/4) off Lewis (Gregory, 1980). In the former, the arctic stenothermal ostracods *Rabilimis mirabilis* and *Cytheropteron arcuatum* typical of the Errol beds are accompanied *inter alia* by a low diversity cold water molluscan fauna, and in the latter the fauna includes abundant *Cytheropteron montrosiense*. The last named is probably also a typical 'Errol' ostracod, but has also been reported as 'rare' at more than one Clyde bed locality, for instance at the Misk Pit (Robertson, 1877). Such records suggest that probable glaciomarine beds which underlie strata with a Clyde beds type of fauna in the Sea of the Hebrides and the Minch are the west coast equivalents of the St. Abbs Beds and Errol beds of the east coast. Moreover, it can be inferred from this that the ice front about 13,500 B.P. lay behind the coastline in north and northwest as well as east Scotland excepting for the sea lochs between Loch Linnhe and the Clyde, a conclusion which agrees with geomorphological evidence (Dawson, 1980). The apparent absence of Errol beds in the inner Cromarty Firth may mean that this area, like the sea-lochs of western Scotland, was under glacier ice until 13,500 B.P. (Peacock, 1974) but this interpretation requires further support.

The presence of 'temperate' episodes is one of considerable interest. As noted above, one of these is dated about 11,000 B.P. and has been recorded or inferred at three sites. Offshore, strata with a 'temperate' fauna seem to overlie deposits correlated with the Errol beds on both east and west coasts. In the absence of radiocarbon dates any correlations must be tentative, but it is reasonable to suggest that the 'temperate' interval recorded offshore (with the possible exception of that in Borehole 71/9 referred to above) is not that recorded at the inshore sites by virtue of its stratigraphical position at the base of and not within, strata correlated with the Clyde beds. It would seem to relate instead to the period of greatest warmth at the beginning of the Windermere Interstadial and thus to be the marine equivalent of the period of highest interstadial temperature on land (Coope, 1977). The apparent absence of this 'temperate' interval in land-based records of marine strata may be because of lack of suitable sites available for detailed study, the presence of glacier ice at some localities about 13,500 B.P. (see above), and the large volume of cold glacier meltwater pouring into the sea during the rapid retreat of the ice.

Data relating to the late-glacial to post-glacial transition are scarce in marine and estuarine sediments, where the succession sometimes includes non-marine beds. In the Cromarty Firth, the boundary could be taken at more than one undated horizon, depending on whether lithological or faunal evidence is followed. Here Peacock, Graham and Gregory, (1980) concluded that the later part of the Loch Lomond stade

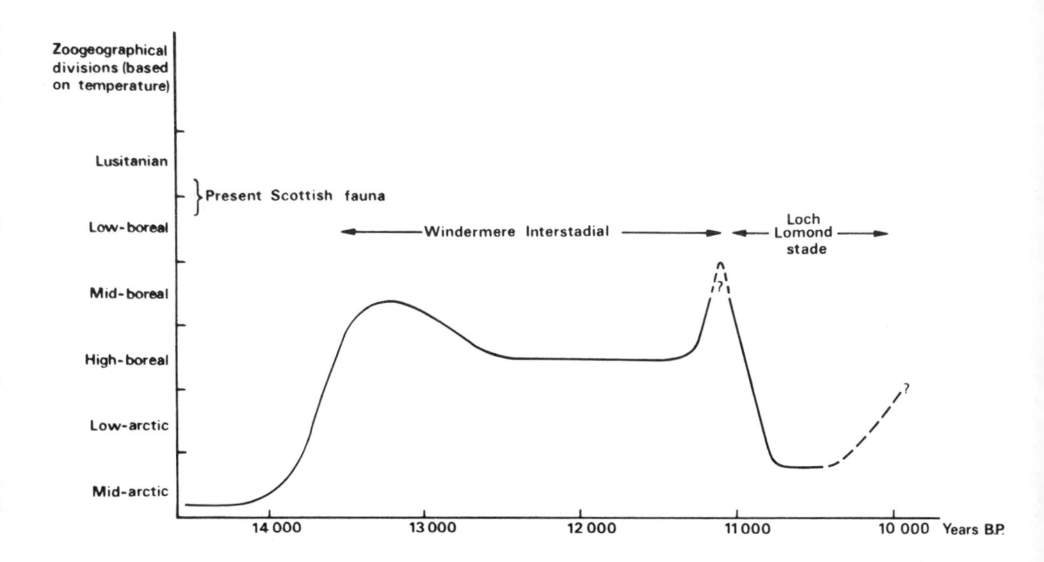

Fig. 4. Generalised curve showing changes in Scottish marine temperatures during late-glacial times.

had a less severe marine climate than the lower and that the post-glacial period began with the incoming of a warmer water fauna. Further south, in the Firth of Forth, it has been suggested that the post-glacial succession begins within the Forth Beds, probably the middle Forth Beds (Thomson and Eden, 1977), but the correlation of the seismic stratigraphy with the palaeontology here is not clear.

ACKNOWLEDGEMENTS

The writer is grateful to Miss D.M. Gregory, Dr M. Armstrong and Messrs. M.A.E. Browne, D.K. Graham, I.B. Paterson and J. Rose for critical discussion of the draft manuscript.

REFERENCES

Aarseth, I., K. Bjerkli, K.R. Björklund, D. Böe, J.P. Holm, T.J. Lorentzen-Styr, L.A. Myhre, E.S. Ugland and J. Thiede (1975). Late Quaternary sediments from Korsfjorden, western Norway. *Sarsia*, 58, 43-66.

Armstrong, M., I.B. Paterson, and M.A.E. Browne (1975). Late-glacial ice limits and raised shorelines in east central Scotland. pp. 39-44. In A.M.D. Gemmell (Ed.), *Quaternary Studies in North East Scotland*. Dept. of Geogr., Univ. Aberdeen

Binns, P.E., R. McQuillin, and N. Kenolty (1974a). *The geology of the Sea of the Hebrides*. Inst. geol. Sci. Rept, No. 73/14, 43 pp.

Binns, P.E., R. Harland and J.J. Hughes (1974b). Glacial and post-Glacial sedimentation in the Sea of the Hebrides. *Nature, Lond.*, 248, 751-754.

Bishop, W.W. and G.R. Coope (1977). Stratigraphical and faunal evidence for late glacial and early Flandrian environments in south-west Scotland. In J.M. Gray and J.J. Lowe (Eds), *Studies in the Scottish Lateglacial Environment*. Pergamon Press, Oxford. pp. 61-88.

Bishop, W.W., and J.H. Dickson (1970). Radiocarbon dates related to the Scottish Late-Glacial Sea in the Firth of Clyde. *Nature, Lond.*, 227, 480-482.

Boulton, G.S., P.N. Chroston and J. Jarvis (1980). A marine seismic study of late Quaternary sedimentation and glacier fluctuations along part of the west coast of Inverness-shire, Scotland. *Boreas* (in press).

Brady, G.S., H.W. Crosskey and D. Robertson (1874). A monograph of the Post-Tertiary Entomostraca of Scotland. *Palaeontogr. Soc. [Monogr.]*, 232 pp.

Browne, M.A.E., D.D. Harkness, J.D. Peacock and R. Ward (1977). Deglaciation chronology of the Paisley-Renfrew area, Scotland. *Scott. J. Geol.*, 13, 301-303.

Coope, G.R. (1977). Fossil coleopteran assemblages as sensitive indicators of climatic changes during the Devensian (Last) cold stage. pp. 313-337 in 'The changing environmental conditions in Great Britain and Ireland during the Devensian (Last) cold stage.' *Phil. Trans. R. Soc.*, B 280, 103-374.

Crosskey, H.W., and D. Robertson (1867-1874). The Post-Tertiary fossiliferous beds of Scotland. *Trans. geol. Soc. Glasg.*, 2, 267-281; 3, 113-129 and 321-341; 4, 32-45, 129-133, and 241-257; 5, 29-35.

Davidson, C.F. (1932). The arctic clay of Errol, Perthshire. *Trans. Proc. Perthsh. Soc. nat. Sci.*, 9, 55-68.

Dawson, A.G. (1980). Raised shorelines and Weichselian ice-sheet decay in the Southern Scottish Inner Hebrides. *Quaternary Newsl.*, No. 30.

Deegan, C.E., R. Kirkby, I. Rae and R. Floyd (1973). *The superficial deposits of the Firth of Clyde and its sea lochs*. Inst. geol. Sci. Rept, No. 73/9, 42 pp.

Feyling-Hanssen, R.W. (1955). Stratigraphy of the marine Late Pleistocene of Billefjorden, Vestspitsbergen. *Skr. norsk. Polarinst.*, Nr. 107, 1-186.

Forbes, E. (1846). On the connexion between the distribution of the existing fauna and flora of the British Isles, and the geological changes which have affected their area, especially during the epoch of the Northern Drift. *Mem. geol. Surv. U.K.*, 1, 336-432.

Francis, E.H., I.H. Forsyth, W.A. Read and M. Armstrong (1970). *The geology of the Stirling District*. Mem. geol. Surv. Gt Br., x + 357 pp.

Graham, D.K. and I.P. Wilkinson (1978). *A detailed investigation of a late-Glacial faunal succession at Ardyne, Argyll, Scotland*. Inst. geol. Sci. Rept, No. 78/5, 17 pp.

Gray, J.M. (1978). Low-level shore platforms in the south-west Scottish Highlands: altitude, age and correlation. *Trans. Inst. Br. Geogr., New Series*, 3, 151-164.

Gregory, D.M. (1980). Assemblages of the ostracod *Cytheropteron montrosiense* Brady, Crosskey and Robertson from offshore Devensian deposits in the Minch, west of Scotland. *Scott. J. Geol.* 16, 287-289.

Harkness, D.D. and H.W. Wilson (1979). Scottish Universities Research and Reactor Centre Radiocarbon Measurements III. *Radiocarbon*, 21, No. 2, 203-256.

Jamieson, T.F. (1865). On the history of the last geological changes in Scotland. *Q. Jl geol. Soc. Lond.*, 21, 161-203.

Jardine, W.G. (1977). The Quaternary marine record in south west Scotland, and in the Scottish Hebrides. In C. Kidson and M.J. Tooley (Eds), *The Quaternary History of the Irish Sea*. Geol. J. Spec. Issue, No. 7, pp. 99-118.

Jardine, W.G. (Ed.) (1980). *Q.R.A. Field Handbook to the Glasgow Region*, Univ. Glasgow. 70 pp.

Jensen, A.S., (1912). *The Danish Ingolf-Expedition*, 11.5. Lamellibranchiata (Part I), pp. 1-119.
Knudsen, K.L. (1978). Middle and Late Weichselian marine deposits at Nörre Lyngby, northern Jutland, Denmark, and their foraminiferal faunas. *Danm. geol. Unders.*, Series II, 112, 44 pp.
Mangerud, J. (1977). Late Weichselian marine sediments containing shells, foraminifera, and pollen, at Agotnes, Western Norway. *Norsk geol. Tidsskr.*, 57, 23-54.
Mitchell, G.F. (1952). Late-glacial deposits of Garscadden Mains, near Glasgow. *New Phytol.*, 50, 277-286.
Neale, J.W. and H.V. Howe (1975). The marine Ostracoda of Russian Harbour, Novaya Zemlya and other high latitude faunas. In F.M. Swain (Ed.), *Biology and Paleobiology of Ostracoda*. Bull. Am. Paleont., 65, Paleont. Res. Inst., Ithaca pp. 381-431.
Ockelmann, W.K., (1958). The Zoology of East Greenland: Marine Lamellibranchiata. *Meddr Grønland*, 122 no. 4, 256 pp.
Peacock, J.D. (1971). Marine shell radiocarbon dates and the chronology of deglaciation in western Scotland. *Nature Phys. Sci.*, 230, 43-45.
Peacock, J.D. (1974). Borehole evidence for late- and post-glacial events in the Cromarty Firth, Scotland. *Bull. geol. Surv. Gt Br.*, No. 48, 55-67.
Peacock, J.D. (1975). Scottish late- and post-glacial marine deposits. In A.M.D. Gemmell (Ed.), *Quaternary Studies in North East Scotland*. Dept. of Geogr., Univ. Aberdeen. pp. 45-48.
Peacock, J.D. (1977). Sub-surface deposits at Inverness and the inner Cromarty Firth - a summary. In G. Gill (Ed.), *The Moray Firth Area: Geological Studies*. Inverness Field Club, Inverness. pp. 103-105.
Peacock, J.D., D.K. Graham, J.E. Robinson and I. Wilkinson (1977). Evolution and Chronology of Lateglacial Marine Environments at Lochgilphead, Scotland. In J.M. Gray and J.J. Lowe (Eds), *Studies in the Scottish Late glacial Environment*. Pergamon Press, Oxford. pp. 89-100.
Peacock, J.D.,D.K. Graham and I.P. Wilkinson (1978). *Late-Glacial and post-Glacial marine environments at Ardyne, Scotland, and their significance in the interpretation of the history of the Clyde sea area*. Inst. geol. Sci. Rept, No. 78/17, 25 pp.
Peacock, J.D., D.K. Graham and D.M. Gregory (1980). Late- and post-Glacial marine *environments in part of the inner Cromarty Firth, Scotland*. Inst. geol. Sci. Rept, no. 80/7 (in press).
Robertson, D. (1883). On the Post-Tertiary Beds of Garvel Park, Greenock. *Trans. geol. Soc. Glasg.*, 7, 1-37.
Robertson, D. (1877). Notes on the Post-Tertiary Deposit of Misk Pit, near Kilwinning. *Trans. geol. Soc. Glasg.*, 5, 297-309.
Robinson, J.E. (1980). The marine ostracod record from the late-glacial period in Britain and NW Europe. In J.J. Lowe, J.M. Gray and J.E. Robinson (Eds), *Studies in the late glacial of North-West Europe*. Pergamon Press, Oxford. pp. 115-122.
Rose, J. (1975). Raised beach gravels and ice wedge casts at Old Kilpatrick, near Glasgow. *Scott. J. Geol.*, 11, 15-21.
Ruddiman, W.F. and A. McIntyre (1973). Time-transgressive deglacial retreat of polar waters from the North Atlantic. *Quat. Res.*, 3, 117-130.
Sars, G.O. (1878). *Bidrag til Kundshaven om Norges arktiske Fauna*. I. *Mollusca. regiones arcticae norvegiae*. Christiania. 466 pp.
Schäfer, W. (1972). *Ecology and Palaeoecology of Marine Environments*. Oliver and Boyd, Edinburgh. 568 pp.
Scott, T. and J. Steel (1883). Notes on the occurrence of *Leda arctica* (Gray); *Lyonsia arenosa* (Möller), and other organic remains, in the post-Pliocene clays of Garvel Park, Greenock. *Trans. geol. Soc. Glasg.*, 7, 274-283.
Sissons, J.B. (1974). Late-glacial marine erosion in Scotland. *Boreas*, 3, 41-48.
Sissons, J.B. (1976). *The Geomorphology of the British Isles: Scotland*. Methuen, London. 150 pp.
Sissons, J.B. (1980). The extent of the last Scottish ice-sheet. *Quaternary Newsl.*, No. 30.

Smith, J. (1838). On the last changes in the relative levels of the land and sea in the British Islands. *Mem. Wernerian nat. Hist. Soc.*, 8, 49-88.

Strauch, F. (1968). Determination of Cenozoic sea temperatures using *Hiatella arctica* (L). *Palaeogeogr., Palaeoclim., Palaeoecol.*, 5, 213-233.

Tait, D. (1934). Braid Burn, Duddingston and Portobello Excavations, 1929-31. *Trans. geol. Soc. Edinb.*, 13, 61-77.

Thomson, M.E. (1978). *IGS studies of the geology of the Firth of Forth and its approaches.* Inst. geol. Sci. Rept, No. 77/17, 56 pp.

Thomson, M.E., and R.A. Eden (1977). *The Quaternary deposits of the central North Sea, 3. Quaternary sequence in the west central North Sea.* Inst. geol. Sci. Rept, No. 77/12, 18 pp.

Wigham, G.D. (1975). Environmental influences upon the expression of shell form in *Rissoa parva* (da Costa). [Gastropoda: Prosobranchiata]. *J. mar. biol. Ass. U.K.*, 55, 425-438.

Dr. J.D. Peacock, F.R.S.E.,
Institute of Geological Sciences,
Murchison House,
West Mains Road,
Edinburgh EH9 3LA

22. On the Classification of Glacial Sediments

Edward A. Francis

INTRODUCTION

The complexity of Quaternary deposits has been realized for many years, as demonstrated by the statement: "Diluvium is chaos" (quoted by Charlesworth 1957, 360 from Lossen 1875). Some of the apparent complexity arises from inadequate data, but problems also result from the inadequate ordering of the available data, that is to say, insufficiently effective classification. By improving classification, we may hope to decrease the complexity which seems to face us.

The act of classification is intimately linked with the processing of information. As an oversimplification, the process of scientific study can be presented as beginning with perceptions or observations, followed by the formation of concepts, and the formulation of these in verbal terms. In fact, these are interrelated, and frequently the observer only perceives something because he already has a relevant concept in his mind. But the process is intimately bound up with classification. Thus, Radford and Burton (1974, 350) state: "one of the essential features of perception is that the perceiver should be able to categorize events consistently, to classify" and Bourne (1966) writes: "As a working definition, we may say that a concept exists whenever two or more distinguishable objects or events have been grouped or classified together and set apart from other objects on the basis of some common feature or property characteristic of each.". As Harvey (1973, 326) remarks: "Classification is, perhaps, *the* basic procedure by which we impose some sort of order and coherence upon the vast inflow of information from the real world.". Radford and Burton (1974, 18) also state: "Creative scientific work consists largely in *discovering* or *devising* new categories by which to order the phenomena; and routine work, on the other hand, involves a great deal of sorting into already known categories". Whether or not these views attract full agreement, it is clear that classification is an important activity and is not to be treated with disdain.

The classification or categorization of phenomena involves definition. Definitions should be framed with care, and it should be clear which characteristics are necessary for the definition to apply and, if excluded, would render the definition inapplicable - the defining characteristics -, and which others are additional factual statements - the accompanying characteristics. A parallel can be made with the essential and accessory minerals in a petrographic definition. It must, however, be recognized that these definitions relate to words and not to objects. This important point means that if, for example, a definition of *till* is framed, then this definition is simply a further set of words which could be used in place of the word *till* to mean the same thing. As Popper (1966, 18) writes: "... science does not use definitions in order to determine the meaning of its terms, but only in order to introduce handy shorthand labels". The material *till* is so variable that it is not possible to cover all the variations within a simple group of words, and consequently any definition which attempts to apply a descriptive formula must have such wide (and thus poorly defined!) limits that it will not be possible to use the term as an effective label. Till is so obviously variable that the difficulty of formulating a descriptive denotation is familiar to all those who have experience of this material, but perhaps it is less obvious that the difficulty exists for the definition of all materials. In consequence, definitive questions should not be posed in the form: "What is *till*?", but rather: "Under what conditions does a *till* form?" or "What is the mode of formation of the material to which the name *till* is applied?". This suggests that the basis for classification of glacial deposits should be essentially genetic, as appreciated by Chamberlin (1894) many years ago.

PROBLEMS OF CLASSIFICATION

The complexity of Quaternary deposits exists on at least three levels: the sedimentological, the stratigraphical, and the morphological, and is particularly apparent in in the case of glacial sediments and the landforms which they comprise. Classification in glacial geology is therefore of special importance.

In relation to the sedimentology of glacial deposits, problems arise from such difficulties as the application of fully descriptive terms to various sedimentary types, the study of present-day processes, ignorance of the manner of formation of various bodies of glacial sediment, and the possibility that bodies of similar external appearance may be formed in different sub-environments. Difficulties involved in the nomenclature and the positions of boundaries in particle grade scales are not peculiar to glacial sedimentology, but the character of glacial sediments commonly points up these difficulties.

The establishment of a particle grade scale acceptable to all and suitable for both field and laboratory use has still not been achieved. Even though there is general agreement upon the need for standardization, there is still considerable variety of usage upon the most important boundaries such as those separating clay and silt, silt and sand, and sand and pebbles or gravel. Since particle size analysis is a common and widely employed method of studying glacial sediments, this raises serious difficulties for efficient comparison and provides ample opportunity for misunderstanding. Raukas, Mickelson and Dreimanis (1978), report on the replies to a questionnaire concerning the methodology of investigations of tills in Europe and North America, and note that the preferred boundaries for North American investigators were: clay/silt: 0.004 mm (= 8ø) - 53.4%, 0.002 mm (= 9ø) - 43.1%; silt/sand: 0.062 mm (= 4ø) - 71.2%, 0.05 mm ($>$4ø) - 21.9%, 0.074 mm ($<$4ø) - 4.1%; sand/gravel: 2.0 mm (= -1ø) - 92.8%. A still greater variety of boundaries was found to exist in Europe.

Krumbein's (1934) phi scale introduced a measure of size ø equal to $-\log_2$ of the diameter in millimetres, and this provides a very useful and convenient reference system, which may also be valuable in avoiding irrational class limits and in simplifying statistical calculations. Udden's (1914) geometrical divisions work out on this scale as:- clay/silt: 7ø, silt/sand: 3ø, and sand/gravel: -1ø, only the latter being represented sixty years later in the North American replies noted above. The scheme presented by Wentworth (1922, 1935), though based largely in principle on that of Udden, put the clay/silt boundary on the phi scale at 8, silt/sand at 4, but the sand/gravel boundary also at -1. Much earlier, Atterberg (1904, 1905) had devised a system with these three boundaries at 0.002 mm (9ø), 0.2 mm (2ø), and 2 mm (-1ø) respectively, and some of these, particularly the former and the latter, have been used quite widely, especially in Europe. It will be seen that the sand/

gravel boundary most widely used is 2 mm (-1ø), but Cayeux (1929) placed this at 5 mm (<-2ø). Wentworth's scheme introduced the term *granules* for Udden's very fine gravel composed of particles between 4 mm (-2ø) and 2 mm (-1ø), but although Lane and others (1947) adopted the Wentworth scheme, they did not retain this term. Recently, in assessing sand and gravel resources, the sand/gravel boundary has been placed by the Industrial Minerals Assessment Unit of the Institute of Geological Sciences at 4 mm (-2ø), for example Nickless, Aitken and McMillan (1978). Thus, although there is at least fairly general agreement that the term for particles of coarser grain size than sand should be *gravel*, this is not matched by similar agreement on the position of the boundary nor on the nomenclature of the material a little coarser than this. The situation has arisen that particles between 2 and 4 mm may be named quite differently under various schemes and their modifications:- very fine gravel (Udden 1914, Lane and others, 1947), granules or granule gravel (Wentworth 1922, 1935), very fine pebbles (Dunbar and Rodgers 1957), or coarse sand (e.g. Nickless, Aitken and McMillan, 1978). As Wentworth himself aptly expressed the appropriate conclusion (in Twenhofel 1961, 200): "... general classifications of sedimentary products have usually been more or less unsatisfactory to all but the classifier".

Glacial sediments are commonly poorly sorted. In consequence, descriptive names applied on the basis of a simple grade scale are difficult to use, and generally require at least one or two qualifiers. In some sediments, the range of particle sizes is so great that a specific term, *till*, peculiar to glacial geology, came to be employed. However, as noted previously, the original descriptive designation has become inapplicable, and this term has now taken on a genetic meaning. Sediments containing a mixture of grain sizes have been notoriously difficult to classify, but workable schemes can be set out for specific purposes, though these are of course often subject to later modification as needs change. The problem was considered by Wentworth (1922), and his classes reflect the dominant proportion of the grade class, so that, for instance, *gravel* contains > 80 per cent gravel, *sandy gravel* contains gravel > sand > 10 per cent, others < 10, and *gravelly sand* contains sand > gravel > 10 per cent, others < 10. However, various authors have proposed modified schemes which in their opinion reflect more closely field usage rather than subdivision for laboratory or strictly sedimentological purposes. Thus, for Willman (1942), *gravel* contains 50 to 100 per cent pebbles, while a sandy mixture containing 25 to 50 per cent pebbles is referred to as *sandy gravel*, although this includes more sand than pebbles. A classification on these lines is now in common use and has been adopted for industrial purposes by the Institute of Geological Sciences (e.g. Nickless, Aitken and McMillan, 1978) with modifications to represent a content of 'fines' of grain size less than 4ø. Folk (1954), on the other hand, restricts *gravel* to mixed sediments with more than 30 per cent pebbles. As with particle grade scales, it would seem that the problem of arriving at an agreed classification of mixed sediments derives more from a resistance to general acceptance of a particular scheme than from the inherent complexity of the sediments.

Stratigraphical problems in glacial geology result from the rapid variation of glacial sediments both sequentially and areally, the impersistence of strata, repetition of sedimentary type, the short time-span in relation to rates of evolution, the general scarcity of fossils, and the inclusion of derived faunas and floras. The stratigraphy of glacial deposits is thus often very complicated, perhaps more so than in any other sedimentary environment, with the possible exception of the volcanic.

Problems associated with morphology relate to the difficulties of correlating form with mode of formation, the existence of homologous forms, the precise differentiation of land forms, the categorization of transitional forms, and the extent of modification of form with increasing age. Further problems arise in connection with the mapping of both sediments and landforms, and even if these are mapped satisfactorily, which is rare because field data are commonly insufficient for close control, the infrequent exposures and the relationships of information obtained from sections to the landform as a whole are often very difficult to interpret.

It is not surprising that, faced with these numerous and varied problems, even the experienced research worker may on occasion find enthusiasm somewhat dampened. As suggested above, however, some alleviation of the apparent complexity may be made by improving classification, and thus improving the conceptual basis of the organization of the data. In order to promote more complete understanding, introduction of new classifications must be purposive.

CLASSIFICATION AND FACIES

The ability to classify phenomena depends upon the means used to distinguish between the classes. The use of genetic considerations cannot be immediate in the study of ancient deposits, and consequently the classes must be identifiable on particular properties of those deposits, from which the origin may be inferred. Criteria for separation of classes in glacial geological deposits may be distinguished on purely sedimentary properties *i.e.* sedimentology, on the sedimentary sequence *i.e.* stratigraphy, and on the form of the accumulation *i.e.* morphology. Various modes of origin result in the formation of bodies of sediment with particular sets of properties which can be specified in terms of sedimentary structure and morphological expression. It is convenient and appropriate to refer to these as *facies*.

Dunbar and Rodgers (1957, 136) write: "... the term facies means the general aspect of the rocks, lithologic and biologic (and by extension, structural and tectonic and even metamorphic), as that aspect reflects the environmental conditions under which the rocks were formed". According to Reading (1978, 4): "A facies is a body of rock with specified characteristics. In the case of sedimentary rocks, it is defined on the basis of colour, bedding, composition, texture, fossils and sedimentary structures". However, as Reading points out the term *facies* is used in many different senses. In the context of glacial sediments, three may be given prominence here:- 1) in respect of the observed material, *e.g. gravel facies*, 2) in respect of the genetic process of supposed formation, *e.g. lodgement till facies*, 3) in an environmental sense, *e.g. proglacial facies*. Reading also cites a fourth sense: as a tectofacies related to phases of an orogenic cycle. There is no apparent example of this of course in the Quaternary glacial environment, though since glacial sediments are often disturbed as a result of the melting of ice with which they were deposited in contact, an analogous usage on a much smaller scale might be expressed by *stagnation facies*, or, in the case of glaciotectonic deformation as a result of lateral ice pressure, *ice push facies*.

If *facies* is used, it must be made clear in which sense the term is meant. Reading (1978, 4) states: "... each facies must be defined objectively in observable, and possibly measurable features", and a "facies should ideally be a distinctive rock that forms under certain conditions of sedimentation, reflecting a particular process or environment". The concept of facies may be applied in glacial geology, and it is suggested here that it should be used for classification, so that classes are specified in terms of sedimentary structure and morphological expression, reflecting the particular sub-environments in which the bodies were formed. Thus, the term *facies* in the present context is applied in a genetic and an environmental sense, and the classes are morphogenetic, distinguished as sediment-landform facies.

Facies associations are groups of facies that occur together and are considered to be genetically or environmentally related (Reading 1978, 5). Thus, Edwards (1978, 435) distinguishes terrestrial associations consisting entirely of facies formed in terrestrial environments and "subdivided into an inner association including basal till, a marginal association with mainly ice marginal facies, and an outer association excluding basal till, and including mainly marginal to non-glacial facies". In the latter, the defining characteristic is based upon exclusion (*i.e.* a negative property) and is consequently open to criticism. Edwards also distinguishes marine facies associations, but notes that "marine and terrestrial facies may occur in the same area and interfinger with one another".

Edwards subdivides terrestrial facies associations into three, but these do not correspond with the three principal sediment association and land systems proposed by Boulton and Paul (1976), which comprise the subglacial/proglacial, the supraglacial and the glaciated valley associations. Fookes, Gordon and Higginbottom (1975) had earlier also distinguished three glacial land systems: the till plain (lodgement till which may have a drumlinized surface), fluvioglacial and ice contact deposit (outwash), and glaciated valley (ablation till and moraine ridges) land systems, but Boulton and Paul (1976, 171) objected to this subdivision because they found that "lodgement tills often have push-moraines, drumlins and fluted moraines on their surface and may be intimately associated with proglacial outwash and eskers", and these should therefore all be grouped together in one association: the subglacial/proglacial.

Edwards (1978) refers to what is essentially the same group as Boulton and Paul's subglacial/proglacial as the marginal facies association, but presumably Boulton and Paul would object to his inner association for the same reason that they objected to the till plain system of Fookes, Gordon and Higginbottom (1975). It must be recognized that although a facies should be defined objectively, grouping into associations is often essentially subjective. Problems also may arise from the fact that facies associations founded upon environmental bases are not necessarily mutually exclusive because the boundary of each sub-environment is transitional. Thus, although Boulton and Paul (1976) object to the till plain land system of Fookes, Gordon and Higginbottom on the grounds that proglacial outwash is often intimately associated with lodgement tills and therefore should be grouped with them, their own grouping could also be objected to on the grounds that supraglacial deposits such as flow tills are often intimately associated with proglacial outwash. There is apparently no reason why proglacial should be preferentially grouped with subglacial rather than supraglacial, and indeed Boulton (1972) drew attention to the importance of the supraglacial/proglacial association. Similarly, Edwards (1978, 424) groups supraglacial and ice marginal proglacial in order to emphasize the genetic association of various facies, although his facies are primarily lithological.

It was pointed out previously that the introduction of new classifications must be purposive, and it may therefore be expected that different grouping of the same phenomena may be explainable in terms of their different purposes. Thus, it may be suggested that Fookes, Gordon and Higginbottom (1975) classified their land systems so that their engineering characteristics could be differentiated, while Boulton and Paul's (1976) grouping reflects their special interest in genetic processes. Edwards' (1978) classification is primarily based upon descriptive sedimentary considerations. Facies associations, unlike facies themselves, are not necessarily defined uniquely or objectively.

FACIES AND MODELS

Classifications are also linked in some cases with facies models. A facies model is a general summary of a specific sedimentary environment (Walker 1979, 3). Its usefulness is based upon the consideration that it integrates a wide range of observational data, but it is also valuable for its predictive power (Blatt, Middleton and Murray, 1972, 186), because a small amount of local data interpreted by means of a correctly applied and comprehended facies model "results in potentially important predictions about that local environment" (Walker 1979, 4). As Reading succinctly expresses it: "*Models* are idealized simplifications set up to aid our understanding of complex natural phenomena and processes".

In the glacial environment, Boulton (1972) has used current processes operating at the margins of Spitsbergen glaciers as a model for Pleistocene glacial processes, so that Pleistocene sequences and landforms can be understood by reference to the products of these processes. This model emphasizes the role of supraglacial processes resulting from a high level of transport of subglacially-derived debris and the transmission of a very large proportion of this to the glacier terminus, together with the absence of subglacial melting. If debris is distributed in distinct zones parallel to the margin, ('controlled') ice-cored ridges with intervening valleys are formed. With uneven distribution, hummocky topography is formed, resulting in a ('controlled' or 'uncontrolled') hummocky till surface, or, if tills are extremely fluid, a till plain. Tectonic structures of varying complexity and including folds and faults, some of which are in the form of thrusts, are explained as the result of settlement over or against melting ice. The 'controlled' ridges parallel to the ice margin were observed by Gripp (1929), who interpreted them by contrast as push-moraines, and Gripp (1978) has also noted the presence of recumbent folds and other structures in Spitsbergen and in north-western Germany, but explains them by lateral ice pressure. Both Boulton and Gripp in effect use Spitsbergen glaciers as models for Pleistocene processes, but these models incorporate contrasting interpretations. This illustrates the point that models are not simply descriptive summaries of observations of processes operating in a particular environment, but that they are also integrative conceptual frameworks.

Boulton and Eyles (1979) derive a genetic model of valley glaciers from observations of modern examples, and suggest a genetic classification of the sediments and landforms. They draw a primary distinction between tills formed predominantly from subglacially-derived debris which has been comminuted at the glacier sole and consequently has produced a relatively high proportion of silt and fine sand, and those formed predominantly from supraglacially-derived debris which has produced a 'supraglacial morainic till' relatively deficient in those fractions. The supraglacial morainic till is regarded as being deposited as two principal facies: the first, Facies A, deposited along the frontal margin during retreat, and the second, Facies B, deposited along the lateral and the latero-frontal margins during stationary or advance phases and typically forming dump moraines. The use of the term facies here is predominantly in an environmental sense, but also with a genetic overtone because the condition of the glacier is important.

Boulton and Eyles' Facies A is regarded as having three sub-facies:-
1) where till cover on the glacier is thin, direct dumping from the ice surface commonly produces a sporadic collection of boulders often concentrated in pushed ridges by winter readvances,
2) where till cover on the glacier is thick enough to delay melting, the till is let down, typically to form hummocky stagnation topography, either at the margin of an active glacier retreating upvalley by sequential stagnation of the margin, or at a margin downwasting *in situ* over the whole tongue, the final degree of relief being controlled by the thickness of till accumulating as mudflows,
3) till laid down in association with proximal ice-contact outwash as mudflows to deposit complexes (multiple sequences), and forming a pitted kame plain or outwash surface.

These so-called sub-facies are predominantly genetic rather than environmental, and thus the term facies is being used here in somewhat different senses in Facies A and in its component sub-facies. In so far as the "sub-facies" each comprise a sediment with specific properties reflecting a differentiated mode of deposition in associated sub-environments, they would perhaps be more appropriately referred to as facies and grouped together for purposes of classification.

The Spitsbergen model presented by Boulton (1972) has been adopted to explain various Quaternary glacial sequences, particularly those formed during the last glaciation. In consequence, it has been concluded that the last ice sheet in Britain was of sub-polar type and was characterized at its maximum extent by marginal cold ice and internal temperate ice. A similar regime has been incorporated in the model proposed by Clayton and Moran (1974). Other environmental conditions require different models, and Shaw (1977) has described a model of deposition under arid polar conditions, with a classification of tills based upon positions of transportation and deposition, process of deposition, and tectonic facies (*i.e.* degree of attenuation of foliation). Carey and Ahmad (1961) distinguish between dry-based and wet-based glaciers depending upon whether the basal ice is below or at the pressure melting point and thus whether freezing or melting takes place at the base. Edwards (1978) proposes a model of glaciomarine sedimentation adjacent to a wet-based tidewater glacier and bases this upon observations of tidewater glaciers in Spitsbergen by Boltunov (1970) and upon features of Pleistocene subaqueous outwash and current dynamics of modern marine deltas.

It is also important to consider the relationships between facies. The study of ancient and of modern sediments is usefully complementary because the relationship of facies in sequences is most clearly seen in sections in ancient successions, but the relationship between facies in modern sediments is most clearly seen in areal distributions. Nevertheless, in modern glacial environments limited sections do occur, particularly during relatively rapid deglaciation, and these provide important evidence on the relationships between various sedimentary types, while Quaternary glacial sediments, especially those of the last glaciation, can be mapped from a facies point of view so that the areal relationships between the various types can be worked out to a greater or lesser extent.

FACIES AND SEQUENCE

The importance of the study of the relationships between facies in sequence was appreciated by Walther (1894), and his so-called Law of Succession of Facies states (Blatt, Middleton and Murray, 1972, 187-188): "The various deposits of the same facies area and, similarly, the sum of the rocks of different facies areas were formed beside each other in space, but in a crustal profile we see them lying on top of each other... it is a basic statement of far-reaching significance that only those facies and facies areas can be superimposed, primarily, that can be observed beside each other at the present time". Important qualifications for this generalization are that it can only apply in the absence of major breaks in the sequence, and any particular section or sequence is unlikely to include all the facies that were developed beside each other at any one time. Consequently, it is necessary to decide what constitutes a major break in a sequence and to be able to identify one if it exists. It is also necessary to have access to a comprehensive series of sections in an area in order to ascertain the full range of facies.

It was noted previously that glacial sequences are commonly characterized by rapid variations of sediment type, that these take place both vertically and laterally, and that strata are often impersistent. The application of Walther's principle in glacial geology is therefore likely to be attended by difficulties, but nevertheless such application can be made. In Boulton's (1972) Spitsbergen model, a series of hypothetical sections (p. 367, Fig. 3) reflects elements actually observed in many morainic zones. Flow till and outwash are seen beside each other and are closely associated with melt-out till and subglacial till, although melt-out till tends to be low in the sections and subglacial till at the base. According to Walther's principle, these four facies can be superimposed, although it would be expected that subglacial till would lie at the base and melt-out till low in the deposited succession. The lateral contiguity of flow till and outwash taken in conjunction with their modes of deposition would suggest the probability of interbedding of these upper members. Thus, an expected full sequence might be as follows (reading from the base):- subglacial till, melt-out till, flow till, outwash, flow till, outwash.

In practice, the full facies sequence would rarely be achieved, and a partial sequence such as subglacial till overlain successively by outwash and flow till could be formed. Such a tripartite succession is commonly found in sedimentary sequences formed during the last glaciation, and was often misinterpreted as the result of two phases of ice advance, or even of glaciation, with recessional deposits sandwiched between. This misinterpretation stems from the flow till having been identified as a subglacial till, but it is also partly due to the inference that a change in sediment type implies a major break in the sequence. Close observation of the contacts between the units comprising such a sequence commonly demonstrates that interdigitation, interbedding, or transition by gradation takes place and therefore major breaks are absent. The component units are consequently successional facies of a single sequence reflecting sediments that were formerly laid down beside each other as implied by Walther's principle.

The presence of sharp contacts does not, however, necessarily imply that the succession is not composed of sedimentary units formerly laid down beside each other. Thus, some flow till units have sharply planar bases (*e.g.* Francis, 1978, 47, 78), but examination of subjacent sediments may reveal the presence of water escape structures or lateral transposition structures resulting from the increase in pore water pressures induced by the additional load when the flow till became superincumbent upon those sediments. Some flow tills with sharp contacts, when traced laterally, are seen to wedge out within waterlaid sediments, and others may pass laterally by gradation into waterlaid sediment. Such features imply contemporaneity and the sharp contacts are not indicative of major breaks in the sequences.

Interpretations of sequences must depend upon identification of facies and their associations, and also upon the structural relationships in addition to the purely sedimentary modes of deposition. Eyles and Slatt (1977) describe a Pleistocene sequence in Newfoundland which is interpreted by them to comprise lodgement till, melt-out till, flow till, supraglacial and proglacial outwash, and supraglacial rhythmites. The lodgement till, as suggested by the Spitsbergen model, lies close to the base of the glacial succession and contains both clasts from lower-lying head and lenses of overlying till. Overlying tills are thought to be melt-out tills which have sharp contacts interpreted as former ice-marginal shear planes, an inference supported by dips in the former up-glacier direction and convergence toward valley margins. Although previous interpretations visualized repeated advances to explain a stacked series of these melt-out tills, the whole succession is interpreted by Eyles and Slatt as having been deposited in a single episode of glaciation. The sharp boundaries between the melt-out tills, which could be seen as representing major breaks in the sequence, are identified simply as the result of structural disturbance. Representatives of the various facies now making up this Pleistocene succession in sequence can be observed forming side by side at the present-day margins of Arctic glaciers of subpolar type as in Spitsbergen. This example illustrates the need for clarity in assessment of the criteria for the identification of major breaks in a sequence, and that discontinuities, even if sharp, may not obviate the application of Walther's principle. It also demonstrates the utility of an appropriate model in interpretation.

OUTLINE OF A TERRESTRIAL GLACIAL SEDIMENT-LANDFORM FACIES CLASSIFICATION

The broad primary classification of terrestrial glacial facies reflects their association with four main sub-environments:- basal, marginal, proglacial and limnal. The boundaries of these sub-environments are transitional in many places and consequently some facies do not fit neatly into one particular category. Other glacial facies sometimes found on land, but not included in this outline, are glacio-aeolian and glaciomarine sediments.

I. Basal

The major proportion of debris in glacial transport is carried in the basal part of the ice and significant quantities are deposited directly at the base of the ice: such sediments are *subglacial*. The principal categories are: 1. lodgement till, 2. deformation till, 3. subglacial melt-out till.

II. Marginal

A. Some of the debris transported in the basal zone may leave this, become transferred to a supraglacial position near or at the margin and be deposited in the marginal sub-environment without the involvement of flowing meltwater. The sediments are generally referred to as *supraglacial*, and the principal categories are:-
1. Supraglacial melt-out till, 2. flow till, 3. supraglacial morainic till, 4. (some) eskers.
B. Meltwater rivers, streams and run-off deposit sediments in contact with the glacier ice. These are generally relatively coarse-grained, and are roughly equivalent to Flint's (1971, 184) 'ice-contact stratified drift', but stratification is commonly severely disturbed or totally lost as a result of the melting of the ice or by ice movement. These sediments are referred to here as *glaciotactual* (Francis, 1975, 47). The principal categories are:
1. glacio-aqueous moraines 2. glaciotactual fans 3. glaciotactual deltas 4. kames 5. moulin-kames 6. kame-terraces 7. eskers.

III. Proglacial

Meltwater rivers and streams flowing beyond the glacier margin deposit sediments as bodies of alluvium, which is generally relatively coarse-grained but becomes finer-grained with increasing distance of transport. The sediments are referred to here as *glaciofluvial*, and it is recommended that this term be restricted rather than extended to all kinds of meltwater deposits (Francis, 1975, 46). The general term for all sediments deposited by waters associated with the ice in the glacial environment is *glacio-aqueous* (Thwaites, 1957, 29; Francis, 1975, 44). The principal categories of glaciofluvial sediments are:
1. outwash plains, 2. valley trains.

IV. Limnal

Lakes provide the best traps for the deposition of relatively fine-grained glacial sediment, and though such bodies of meltwater are typically even more ephemeral than lakes in general, they may exist long enough for substantial thicknesses of deposits to be accumulated. In some cases, deposits of mixed particle sizes are also formed such as the lacustrotills of May (1977), but the typical sediments are deltas and rhythmites. The sediments are referred to generally as *glaciolacustrine*. Glacial lakes may be found in both the marginal and proglacial sub-environments, and probably also subglacially, but the processes of deposition are generally so distinctive and the sediments commonly so characteristic that they can be considered as a specific sub-environment in their own right. Glaciotactual deltas are also of course lacustrine, but it is considered here that these are more appropriately placed in the marginal group. Some other deltas formed in situations where the direct contact of ice and lake was very small in comparison with the length of shore bordered by land, and consequently deltas were constructed at the point of entry into the lake of proglacial streams, e.g. Glacial Lake Hitchcock (Ashley, 1975). The principal categories of glacio-lacustrine sediments are:
1. proglacial glaciolacustrine deltas, 2. rhythmites, 3. lacustrotills, 4. glaciolacustrine beaches.

ACKNOWLEDGEMENT

I should like to express my appreciation of the interest and encouragement received over nearly twenty years from the dedicatee of this volume, my friend Lewis Penny.

REFERENCES

Ashley, G.M. (1975). Rhythmic sedimentation in Glacial Lake Hitchcock, Massachusetts, Connecticut. In A.V. Jopling and B.C. McDonald (Eds.), *Glaciofluvial and Glaciolacustrine Sedimentation,* Tulsa, Oklahoma (Society of Economic Paleontologists and Mineralogists). pp. 304-320.

Atterberg, A. (1904). Sandslagens klassifikation och terminologi. *Geol. För. Stockh. Förh.* 25, 397-412.

Atterberg, A. (1905). Die rationelle Klassifikation der Sande und Kiese. *Chem. Zeitschr.*, 29, 195-198.

Blatt, H., G. Middleton and R. Murray (1972). *Origin of Sedimentary Rocks.* Englewood Cliffs, New Jersey (Prentice-Hall). 634 pp.

Boltunov, V.A. (1970). Certain earmarks distinguishing glacial and moraine-like glacialmarine sediments, as in Spitsbergen. *Int. Geol. Rev.*, 12, 204-211.

Boulton, G.S. (1972). Modern Arctic glaciers as depositional models for former ice sheets. *J. geol. Soc. Lond.*, 128, 361-393.

Boulton, G.S. and N. Eyles (1979). Sedimentation by valley glaciers: a model and genetic classification. In C. Schlüchter (Ed.), *Moraines and Varves,* Rotterdam (Balkema) pp. 11-23.

Boulton, G.S. and M.A. Paul (1976). The influence of genetic processes on some geotechnical properties of glacial tills. *Q. Jl eng. Geol. Lond.*, 9, 159-194.

Bourne, L.E. (1966). *Human Conceptual Behaviour.* Boston (Allyn and Bacon). 139 pp.

Carey, S.W. and N. Ahmad (1961). Glacial marine sedimentation. In G.O. Raasch (Ed.), *Geology of the Arctic,* Vol. II, Toronto (University of Toronto Press). pp. 865-894.

Cayeux, L. (1929). Les roches sédimentaires de France: Roches silicieuses. Paris (Imprimerie Nationale).

Chamberlin, T.C. (1894). Proposed genetic classification of Pleistocene glacial formations. *J. Geol.*, 2, 517-538.

Charlesworth, J.K. (1957). *The Quaternary Era.* 2 vols. London (Arnold). 1700 pp.

Clayton, L. and S.R. Moran (1974). A glacial process-form model. In D.R. Coates (Ed.), *Glacial Geomorphology.* Binghamton, New York (State Univ. of New York). pp. 89-119.

Dunbar, C.O. and J. Rodgers (1957). *Principles of Stratigraphy.* New York (Wiley). 356 pp.

Edwards, M.B. (1978). Glacial Environments. In H.G. Reading (Ed.), *Sedimentary Environments and Facies,* Oxford (Blackwell). pp. 416-438.

Eyles, N. and R.M. Slatt (1977). Ice-marginal sedimentary, glacitectonic, and morphologic features of Pleistocene drift: an example from Newfoundland. *Quat. Res.*, 8, 267-281.

Flint, R.F. (1971). *Glacial and Quaternary Geology.* New York (Wiley). 892 pp.

Folk, R.L. (1954). The distinction between grain size and mineral composition in sedimentary rock nomenclature. *J. Geol.*, 62, 344-359.

Fookes, P.G., D.L. Gordon and I.E. Higginbottom (1975). Glacial land forms, their deposits and engineering characteristics. In: *The Engineering Behaviour of Glacial Materials.* Birmingham (Midland Soil Mechanics and Foundations Engineering Society). pp. 18-51.

Francis, E.A. (1975). Glacial sediments: a selective review. In A.E. Wright and F. Moseley (Eds.), *Ice Ages: Ancient and Modern.* Liverpool (Seel House Press). pp. 43-68.

Francis, E.A. (1978). *Field Handbook, Annual Field Meeting 1978, Keele.* Keele (Quaternary Research Association). viii + 101 pp.

Gripp, A. (1929). Glaciologische und geologische Ergebnisse der Hamburgischen Spitz-

bergen - Expedition 1927. *Abh. Geb. Naturw., Hamburg*, 22, 146-249.
Gripp, A. (1978). Glazigene Press-Schuppen, frontal und lateral. In C. Schlüchter (Ed.), *Moraines and Varves*, Rotterdam (Balkema). pp. 157-166.
Harvey, D. (1973). *Explanation in Geography*. London (Arnold). 521 pp.
Krumbein, W.C. (1934). Size frequency distributions of sediments. *J. sedim. Petrol.*, 4, 65-77.
Lane, E.W. and others (1947). Report of the subcommittee on sediment terminology. *Trans. Am. geophys. Un.*, 28, 936-938.
May, R.W. (1977). Facies model for sedimentation in the glaciolacustrine environment. *Boreas*, 6, 175-180.
Nickless, E.F.P., A.M. Aitken and A.A. McMillan (1978). *The sand and gravel resources of the country around Darvel, Strathclyde*. Inst. Geol. Sci. Miner. Assess. Rept., No. 35. 153 pp.
Popper, K.R. (1966). *The Open Society and its Enemies*. Vol. II. Hegel and Marx. Fifth edition. London (Routledge and Kegan Paul). 420 pp.
Radford, J. and A. Burton (1974). *Thinking: Its Nature and Development*. London (Wiley). 440 pp.

Raukas, A., D.M. Mickelson and A. Dreimanis (1978). Methods of till investigation in Europe and North America. *J. sedim. Petrol.*, 48, 285-294.
Reading, H.G. (1978). Facies. In H.G. Reading (Ed.), *Sedimentary Environments and Facies*, Oxford (Blackwell). pp. 4-14.
Shaw, J. (1977). Tills deposited in arid polar environments. *Can. J. Earth Sci.*, 14, 1239-1245.
Thwaites, F.T. (1957). *Outline of Glacial Geology*. Madison, Wisconsin. 136 pp.
Twenhofel, W.H. (1961). *Treatise on Sedimentation*. 2 vols. New York (Dover). 926 pp.
Udden, J.A. (1914). The mechanical composition of clastic sediments. *Bull. geol. Soc. Am.*, 25, 655-744.
Walker, R.G. (1979). Facies and Facies Models 1. General Introduction. In R.G. Walker (Ed.), *Facies Models*, Geoscience Canada Reprint Series 1. Toronto (Geological Association of Canada). pp. 1-7.
Walther, J. (1894). *Einleitung in die Geologie als historische Wissenschaft*. Vol. III. Jena (Fischer Verlag).
Wentworth, C.K. (1922). A scale of grade and class terms for clastic sediments. *J. Geol.*, 30, 377-392.
Wentworth, C.K. (1935). The terminology of coarse sediments. *Rep. natn. Res. Coun. Comm. on Sedimentation*. 1932-34, 225-246.
Willman, H.B. (1942). Geology and mineral resources of the Marseilles, Ottawa and Streater Quadrangles. *Bull. Ill. St. geol. Surv.*, 66, 343-344.

Edward A. Francis, Esq.,
Department of Geology,
University of Keele,
Newcastle,
Staffs. ST5 5BG

Subject Index

Geographical Index

CANCELLED
CANCELLED
20 JUL 2000
-6. FEB. 2001
CANCELLED
CANCELLED
2 0 APR 2004